Springer Proceedings in Earth and Environmental Sciences

Series Editor

Natalia S. Bezaeva, The Moscow Area, Russia

The series Springer Proceedings in Earth and Environmental Sciences publishes proceedings from scholarly meetings and workshops on all topics related to Environmental and Earth Sciences and related sciences. This series constitutes a comprehensive up-to-date source of reference on a field or subfield of relevance in Earth and Environmental Sciences. In addition to an overall evaluation of the interest, scientific quality, and timeliness of each proposal at the hands of the publisher, individual contributions are all refereed to the high quality standards of leading journals in the field. Thus, this series provides the research community with well-edited, authoritative reports on developments in the most exciting areas of environmental sciences, earth sciences and related fields.

More information about this series at http://www.springer.com/series/16067

Sergei Votyakov · Daria Kiseleva ·
Viktor Grokhovsky · Yulia Shchapova
Editors

Minerals: Structure, Properties, Methods of Investigation

9th Geoscience Conference for Young Scientists, Ekaterinburg, Russia, February 5–8, 2018

Springer

Editors
Sergei Votyakov
Institute of Geology and Geochemistry
Ural Branch of Russian Academy
of Sciences
Ekaterinburg, Russia

Viktor Grokhovsky
Ural Federal University
Ekaterinburg, Russia

Daria Kiseleva
Institute of Geology and Geochemistry
Ural Branch of Russian Academy
of Sciences
Ekaterinburg, Russia

Yulia Shchapova
Institute of Geology and Geochemistry
Ural Branch of Russian Academy
of Sciences
Ekaterinburg, Russia

ISSN 2524-342X ISSN 2524-3438 (electronic)
Springer Proceedings in Earth and Environmental Sciences
ISBN 978-3-030-00924-3 ISBN 978-3-030-00925-0 (eBook)
https://doi.org/10.1007/978-3-030-00925-0

Library of Congress Control Number: 2018967750

This Springer imprint is published by the registered company Springer Nature Switzerland AG
The registered company address is: Gewerbestrasse 11, 6330 Cham, Switzerland

Preface

This volume of *Springer Proceedings in Earth and Environmental Sciences* contains selected papers presented at the 9th Geoscience Conference for Young Scientists "Minerals: Structure, Properties, Methods of Investigation", which took place during February 5–8, 2018, at the Institute of Geology and Geochemistry, Ural Branch of Russian Academy of Sciences, Ekaterinburg, Russia.

The conference, which organized by the Institute of Geology and Geochemistry, UB RAS in cooperation with the Ural Federal University named after B. N. Yeltsin and the Institute of Mineralogy, UB RAS, was held under the auspices of the commission for X-ray diffraction, crystal chemistry, and spectroscopy of the Russian Mineralogical Society.

The central idea behind the conference program is the application of contemporary physicochemical methods of analysis to the study of terrestrial and extraterrestrial minerals. The conference was dedicated to the 5th anniversary of the "Chelyabinsk" meteorite which fell to Earth on February 15, 2013.

More than 70 participants, including young scientists, postgraduates, and students, as well as members of the Russian Academy of Sciences, professors, doctors, and candidates of sciences, represented academic and industrial institutes in Moscow, St. Petersburg, Kazan, Ekaterinburg, Novosibirsk, Syktyvkar, Apatity, Petropavlovsk-Kamchatsky, Miass, Ulan-Ude and elsewhere.

Plenary sessions were held for four days over three sections "Crystal chemistry of minerals," "Typomorphism and methods for studying minerals," and "Meteorites, asteroids, comets". More than 40 oral reports were presented, including 15 plenary lectures of invited specialists—S. K. Filatov, Yu. L. Voytekhovsky, N. N. Eremin, N. R. Khisina, V. P. Lyutoev, V. I. Grokhovsky, Yu. V. Shchapova, N. V. Selezneva, S. L. Votyakov et al. More than 25 posters were presented.

Oral reports were presented by young scientists, postgraduates, and students at specially organized sessions. Participants discussed relevant issues of crystal chemistry and mineral typomorphism, structural studies, physical and chemical and technological properties of minerals, nanoscale formations and inclusions, techno- and biogenic objects, non-crystalline mineral matter, and the problems in applied mineralogy.

The importance of biomineralogical formations in the human body, the role of bacteria in geological processes, the search and study of new mineral formations and their synthetic analogs with specified physical properties were considered alongside the analysis of inclusions in minerals. Much attention was paid to the discussion of the foundations and features of the application of contemporary physicochemical methods in mineralogical research, and to the integration of solid-state spectroscopy methods (Mössbauer, Raman, and radio spectroscopy) with mass spectrometry and electron microscopy.

Along with the experimental works, reports were presented concerning the atomistic modeling of the composition, structure, and properties of minerals, including diffusion; these considerations are extremely relevant to an understanding of the post-crystallization evolution of a particular mineral.

In addition to the main theme of the conference, which related to the analysis of the Earth minerals, reports devoted to the study of meteorites (in particular, the "Chelyabinsk" meteorite) were received by the audience with great interest. The use of modern methods of analysis allowed the authors of reports to reveal ancient mineral formations in the Ural "Northern Kolchim" meteorite, as well as to discover a new mineral in the "Uakit" meteorite from Buryatia. Several reports presented by young scientists raised questions of studying and modeling cosmic dust, as well as posing an intriguing question: "Are there any particles of meteoritic nature in the mud sediment of the city of Ekaterinburg?" Alongside problems relating to the potential comet–asteroid threat, a report on the latest UrFU meteorite expedition to the Chilean Atacama Desert attracted much interest from listeners.

According to the results of the conference, the best poster prize of young scientists was awarded to Dr. D. A. Khanin (Moscow State University, Moscow); for the best oral report on geological topic the prize was awarded to A. V. Kutyrev (Institute of Volcanology and Seismology, FEB RAS, Petropavlovsk-Kamchatsky); and for the best applied spectroscopic research the prize was awarded to E. A. Pankrushina (IGG UB RAS, Ekaterinburg).

Great interest was caused by three English workshops, conducted by the staff of the Department of Foreign Languages of the Institute of Philosophy and Law, UB RAS, Ekaterinburg—A. N. Oveshkova, T. A. Beavitt, and N. N. Koptyaeva.

On the last day of the conference, excursions were organized to the "Geoanalyst" Center for Collective Use of the UB RAS, the museum of scientific collections of the IGG UB RAS as well as to the Yeltsin Center.

This proceedings volume contains the full texts of 10 invited plenary lectures and of 31 young scientist papers presented at the conference. All the papers have been subjected to peer review by at least two referees.

Ekaterinburg, Russia

Sergei Votyakov
Daria Kiseleva
Yulia Shchapova
Viktor Grokhovsky

Acknowledgements

On behalf of the Organizing Committee of the 9th Geoscience Conference for Young Scientists "Minerals: Structure, Properties, Methods of Investigation", we would like to thank all the invited speakers, session chairs, and participants for making the conference successful. We also would like to express our gratitude to Irina Vlokh for the linguistic assistance and Dr. Natalia Popova, the Head of the Department of Foreign Languages, Institute of Philosophy and Law, UB RAS, Ekaterinburg, for valuable advice and discussion during the preparation of this proceedings volume.

The conference was supported by the Russian Foundation for Basic Research (Grant No. 18-35-10003-mol g).

About This Book

All papers published in this volume of *Springer Proceedings in Earth and Environmental Sciences* "Minerals: Structure, Properties, Methods of Investigation—9th Geoscience Conference for Young Scientists, Ekaterinburg, Russia, February 5–8, 2018" have been peer-reviewed over the course of processes administered by the proceedings Editors. Reviews by expert referees conformed to the professional and scientific standards expected of a proceedings journal published by Springer.

Contents

Zoned Olivines of Bronze Age Metallurgical Slags of Southern Urals According to LA-ICP-MS Mapping

Maxim N. Ankushev, Dmitry A. Artemyev and Ivan A. Blinov

Abstract Metallurgical artifacts including Cu-containing slag fragments were discovered at the Bronze Age settlements of Southern Urals. We have analyzed the slags of Sintashta culture from Kamenny Ambar settlement. The slags are mainly composed of olivine, magnetite and ultrabasic glass with rare inclusions of metallic copper; quartz, chrome spinels and serpentinite clasts are relic. The main rock-forming mineral of slags is olivine, in which the compositions of macro components and trace elements have been examined. The olivine is zoned: its central part is characterized by elevated forsterite minal with Co and Ni prevalence, while fayalite predominates on the peripheral parts and contains elevated concentrations of the remaining elements. The glass is characterized by higher Li, Ca, Al, Na, K, Sc, Ti, Ga, Nb, P, Pb, Rb, Sr, Ba, Zr, U and REE contents and indicates specific crystallization features as well as protolith composition and the fluxes used. Based on the LA-ICP-MS mapping, major and trace element zoning in olivine crystals has been considered in detail, and a crystallization scheme of metallurgical slags has been identified.

Keywords Metallurgical slags · Bronze age · Sintashta culture · Ultrabasic rocks · Olivine · LA-ICP-MS · Trace elements

Introduction

Southern Urals, with its rich mineral resources, was the part of the Eurasian mining and metallurgical province in the Bronze Age (Chernykh 1970). Earlier, the fragments of metallurgical slags were found on many regional archeological monuments, the research of which can give an answer to the question of mineral raw

M. N. Ankushev (✉) · D. A. Artemyev · I. A. Blinov
Institute of Mineralogy, Ural Branch of RAS, Ilmeny Reserve, Miass, Russia
e-mail: ankushev_maksim@mail.ru

D. A. Artemyev
South Ural State University, 10, 8 July Str., Miass, Russia

S. Votyakov et al. (eds.), *Minerals: Structure, Properties, Methods of Investigation*, Springer Proceedings in Earth and Environmental Sciences,
https://doi.org/10.1007/978-3-030-00925-0_1

material sources and the features of metallurgical process in ancientry. The main rock-forming mineral of most ancient slags is olivine, which often forms large zoned crystals and is also found in contemporary metallurgical slags, as well as in magmatic and pyrometamorphic rocks. The research of the morphology and the geochemical features of the olivine crystallization can shed light on the conditions and features of its crystallization.

The aim of the work was to study the trace element distribution in olivine and glass-like phase using the interpretation of geochemical maps of trace elements derived by laser ablation—inductively coupled plasma mass spectrometry (LA-ICP-MS).

Materials and Methods

In order to study the olivine composition, the slags corresponding to the previously allocated chromite-containing mineralogical type of the Sintashta culture of the Bronze Age were used. This type of slag corresponds to the using of azurite-malachite ores from the oxidation zone of massive sulfide and copper skarn deposits located in the ultrabasites of the Urals (Grigoriev et al. 2005). The studied samples of slags were collected from the Bronze Age settlement of Kamenny Ambar (Kartaly district of Chelyabinsk region), where they are widely represented across different cultural and historical horizons.

The major element composition was determined using the Tescan Vega 3 sbu scanning electron microscope with Oxford Instrumets X-act EDS (IMIN UB RAS, analyst Blinov I.A.). The olivine formulas were calculated using an anionic method of 4 O atoms.

The geochemical maps of the trace element distribution in olivine crystals are compiled using the IOLITE software package based on the interpretation and calculation of LA-ICP-MS analyses obtained by successive linear burning of the investigated area with a 12 μm laser spot during its movement at a speed of 10 μm/s with the distance between the burned lines of 12 μm.

The trace elements were determined by laser ablation using the Agilent 7700x inductively coupled plasma mass spectrometer (IMIN UB RAS, analyst Artemyev D.A.) with the following parameters: RF Power −1550 W, Ar sampling gas with flow rate of 1.0 l/min, Ar plasma gas with flow rate of 15 l/min, Ar cooling gas with flow rate of 0.9 l/min. New Wave Research UP-213 laser ablation system parameters: Nd:YAG laser, wavelength 213 nm, beam energy 10–12 J/cm^2, pulse repetition frequency 10 Hz, He carrier gas with flow rate of 0.65 l/min. USGS BCR-2G and NIST SRM-612 international reference materials were used for concentration calculation and instrument calibration. ^{29}Si was used as the internal standard.

Results and Discussion

Mineralogical and geochemical characteristics of metallurgical slags. Slags are characterized by porphyry structure, the bulk of which consists of newly formed crystals of olivine, magnetite and glass-like phase, at times the inclusions of metallic copper occur; relic minerals are represented by serpentine and chrome spinels.

The olivine forms prismatic idiomorphic crystals, often with well-defined zoning; skeletal individuals also occur. A lighter rim is visible in reflected light around the periphery of olivine grains characterized by an increased Fe content in comparison with the central parts. The olivine is represented by fayalite (Table 1), however, the amount of forsterite minal increases in the central parts of the crystals. Glass-like phase, representing from 10 to 40% of slag, includes a large number of small chain and feather olivine crystals. Magnetite forms idiomorphic aggregates of 5–15 microns in size, skeletal crystals and small symplectites in olivine crystals. The amount of magnetite in the sample usually does not exceed 5%.

Relic mineral inclusions are represented by chrome spinel grains of 0.1–0.5 mm in size, which are found both in the form of idiomorphic crystals and xenomorphic individuals, with eroded borders. Often, porous or partially destroyed grains are observed. A thin (3–5 μm) continuous or discontinuous chrome-magnetite rim usually develops on the periphery. There are isometric and elongated xenoclasts of quartz and serpentinized ultrabasites in the slags from 0.1 mm to 2 mm in size, often with melted borders. Melt inclusions in slags are represented by single-phase copper inclusions of isometric, round, less often elongated forms, having the size from the first microns to several mm. Its composition is represented by pure copper with a small fraction of Fe, Sn, As, Ag, Zn, Ni, and Co according to the electron microprobe data.

Trace elements in olivines. Besides major (Fe, Mg, Si) components, Ni, Co, Cr and Mn macro elements have a significant proportion in the olivine composition. In this case, the distribution of some of the mentioned elements demonstrates pronounced zoning. The mapping of relatively large (0.4–0.6 mm) olivine crystals made it possible to identify the trace elements confined to its magnesian part (with an increased forsterite minal content), fayalite rim, and slag glass-like phase (Fig. 1).

Fe. Due to the high concentration of Fe in olivine, the LA-ICP-MS data could not reveal a pronounced zoning. Nevertheless, it is well detected with the help of scanning electron microscopy (see Table 1). In the crystal's central zone (sample 3260 g/718, analyzes b, c) FeO content is in the range of 56–57 wt%, while the inner zone (sample 3260 g/718, analysis d), bright in reflected electrons, corresponds to FeO content of 63 wt%, and the rim (sample 3260 g/718, analyzes a, d) has 65–70 wt% of FeO.

Mg. According to the scanning electron microscopy data for the crystal's central zone (sample 3260 g/718, analyzes b, c), MgO content corresponds to 12 wt%, in the inner zone (sample 3260 g/718, analysis d), bright in BSE, corresponds to 6 wt%,

Table 1 Zoned olivine composition of Bronze Age metallurgical slags

№	Sample	Analyses	MgO	SiO_2	CaO	FeO	MnO	Al_2O_3	Total	Crystal chemical formula	Minal, %
1.	3260 g/718	a	1.00	28.12	0.67	69.84	0.13	0.24	100	$(Fe_{1.99}Mg_{0.05}Ca_{0.02})_{2.07}Si_{0.96}Al_{0.01}O_4$	$Fa_{96.18}Fo_{2.45}La_{1.18}Tf_{0.18}$
2.		b	12.30	31.00	0.19	56.33	0.18	–	100	$(Fe_{1.47}Mg_{0.57}Ca_{0.01})_{2.06}Si_{0.97}O_4$	$Fa_{71.60}Fo_{27.86}La_{0.31}Tf_{0.23}$
3.		c	11.51	30.78	0.28	57.22	0.21	–	100	$(Fe_{1.51}Mg_{0.54}Ca_{0.01}Mn_{0.01})_{2.06}Si_{0.97}O_4$	$Fa_{73.08}Fo_{26.19}La_{0.46}Tf_{0.27}$
4.		d	6.03	30.27	0.44	63.26	–	–	100	$(Fe_{1.72}Mg_{0.29}Ca_{0.02})_{2.03}Si_{0.99}O_4$	$Fa_{84.83}Fo_{14.41}La_{0.76}$
5.		e	0.91	33.28	0.65	64.75	–	0.41	100	$(Fe_{1.75}Mg_{0.04}Ca_{0.02})_{1.82}Si_{1.08}Al_{0.02}O_4$	$Fa_{96.35}Fo_{2.41}La_{1.24}$
6.	3214/718	a	6.16	30.34	–	63.5	–	–	100	$(Fe_{1.73}Mg_{0.3})_{2.03}Si_{0.98}O_4$	$Fa_{85.26}Fo_{14.74}$
7.		b	8.30	30.38	–	61.32	–	–	100	$(Fe_{1.65}Mg_{0.4})_{2.05}Si_{0.98}O_4$	$Fa_{80.57}Fo_{19.43}$
8.		c	3.96	31.00	0.28	64.76	–	–	100	$(Fe_{1.77}Mg_{0.19}Ca_{0.01})_{1.97}Si_{1.01}O_4$	$Fa_{89.73}Fo_{9.78}La_{0.50}$
9.		d	3.90	31.31	0.29	64.51	–	–	100	$(Fe_{1.76}Mg_{0.19}Ca_{0.01})_{1.96}Si_{1.02}O_4$	$Fa_{89.81}Fo_{9.68}La_{0.52}$
10.		e	1.30	30.69	0.33	67.69	–	–	100	$(Fe_{1.88}Mg_{0.06}Ca_{0.01})_{1.96}Si_{1.02}O_4$	$Fa_{96.11}Fo_{3.29}La_{0.60}$

Note. Analyses were carried out using VEGA3 TESCAN scanning electron microscope (analyst I.A. Blinov) in the Institute of Mineralogy UB RAS; dash – element is not detected

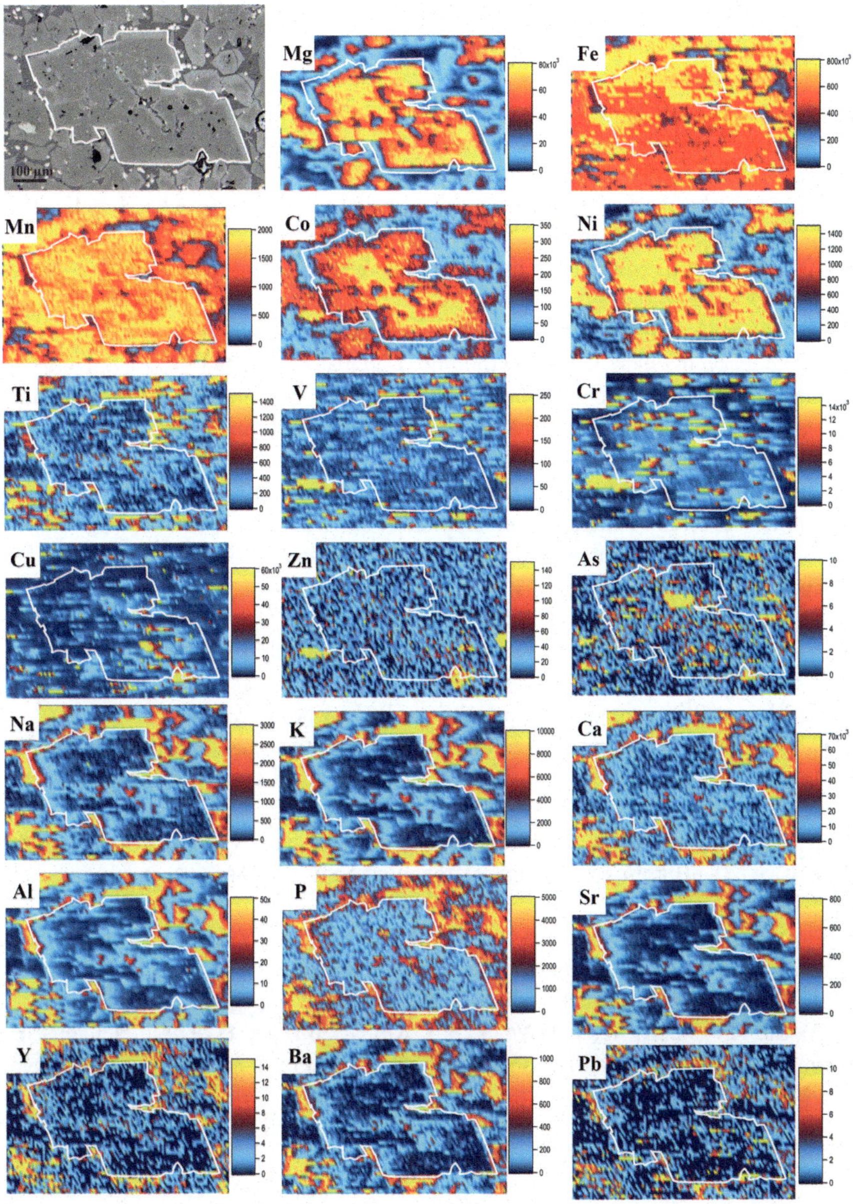

Fig. 1 Photo of olivine grains in metallurgical slag (reflected light) and geochemical distribution maps of some elements. The values are given in ppm

in the outer rim (sample 3260 g/718), analyzes a, d) to 0.9–1 wt%. Mg shows an inverse relationship with Fe, which is typical for olivine and indicated in many publications (Mineraly 1972).

Ni. In the crystal's central part Ni content reaches 1400 ppm and decreases to 800–1000 ppm in the inner zone and to 200–400 ppm in the rim.

Co. Cobalt exhibits the most pronounced zoning in olivine as compared to the rest of trace elements. In the crystal's central part, Co content is 300–350 ppm, it decreases to 150–250 ppm in the inner zone and to 50–100 ppm in the rim. Ni and Co are completely identical to Mg by the character of their distribution. Similar zoning for Ni is also evident in natural olivines of volcanic rocks (Landa et al. 1978) and kimberlites (Sobolev et al. 2015). NiO content varies within the range of 0.3–0.6% in olivine ultrabasitic massifs (Mineraly 1972), in magnesian olivines of kimberlites – up to 0.4%.

Cr. Zoning is extremely weak; Cr content reaches 3000 ppm in enriched zones. Bright spots on the map are due to the micro-inclusions of newly formed magnetite crystals. The capture of chromium by olivine at low pressure can be explained by the deviation of crystallization conditions from equilibrium, in particular by overcooling phenomena (Fodor and Keil 1976).

Mn. Mn concentration reaches 2000 ppm in the central magnesian part of the crystal, however, the manganese concentration decreases to 1200–1300 ppm at the fayalite rim – glass-like phase boundary. In the slag glass-like phase, the Mn concentration is 500–800 ppm.

Ti. A pronounced zoning is noted for Ti in olivines of kimberlite pipes (Sobolev et al. 2015), which is not observed in slags due to the element concentration in the glass-like phase; this may be due to the time of melt crystallization, which is negligibly small in comparison with the geological processes.

Cu is confined to copper metallic inclusions in the slag. It exhibits a weakly pronounced zoning in olivine crystals with the concentration reaching 1% in the enriched areas.

Zn, As are confined to metallic inclusions in slag and evenly distributed in the amount of 20–40 ppm for Zn and of the first ppms for As both in olivines and glass-like phase.

V, Sn are distributed evenly across the slag, both in glass-like phase and olivine; in the range of 50–70 ppm (V), and 1–5 ppm (Sn). The high values of V (200–250 ppm) are confined to the micro-inclusions of magnetite.

Li, Ca, Al, Na, K, Sc, Ga, Nb, P, Pb, Rb, Sr, Ba, Zr, U. These elements are confined to the slag glass-like phase with their content 4–6 times higher as compared to the olivine crystal.

Ta, Hf, Cd, Bi practically do not occur in slags; their content does not exceed several fractions of ppm.

REE. Olivines, crystallized from natural melts, concentrate an extremely small amount of rare-earth elements (Lesnov 2000), which also have been found in small amounts in the olivines of kimberlites (Mineraly 1972). REE content in metallurgical slags is very high according to local LA-ICP-MS analyzes; their uneven distribution is observed according to the data of geochemical mapping, which indicates that the olivine crystal is trapped in the melt inclusions of glass-like phase during crystallization. REE in slags are mainly confined to glass-like phase having their content 4–5 times higher in the glass than in the olivine, and indicate the use of carbonate (limestone) and phosphate (animal bones) fluxes, as is also pointed out by the increased concentrations of Ca and P.

Conclusions

As a result of the research of the olivine crystal morphology and zonation, the following crystallization scheme for slags has been established: firstly, a fayalite with a greater amount of forsterite minal is formed from the melt; then, with a decrease in temperature, a fayalite rim is formed around the grains, and then the xenomorphic and skeletal aggregates of magnetite fill the fissures and voids in the intergrowths; lastly, the hardening of the slag glass-like phase and the loss of metallic phases of copper occurs. Ni, Co, partly Cr and Mn are attracted by the magnesian part of the crystal. Fayalite rim is not fundamentally enriched with any admixtures as compared with other phases, but contains higher concentrations of elements characteristic for glass-like phase. Micro-inclusions of the newly formed magnetite concentrate Cr, V, and partly Ti. Lithophilic elements such as Li, Ca, Al, Na, K, Sc, Ti, Ga, Nb, P, Pb, Rb, Sr, Ba, Zr, U and REE are confined to the slag glass-like phase. Zn and As are also confined to the phase of metallic copper. High REE contents, uncharacteristic for olivines, and their uneven distribution in the crystal indicate the capture of glass-like phase inclusions by olivine during crystallization.

Acknowledgements The authors are grateful to L.N. Koryakova for providing the samples for research.

The research was conducted within the framework of the PFBR project № 18-00-00030 KOMFI.

References

Chernykh EN. Drevneishaya metallurgiya Urala i Povolzhya. M.: Nauka; 1970. [Chernykh EN. The historical metallurgy of the Urals and Volga region. M.: Nauka; 1970. (In Russ.)]

Fodor RV, Keil KA. Komatiite-like lithic fragment with spinifex texture in the Eva meteorite: origin from a Supercooled impact-melt chondritic parentage. Earth Planet Sci. Lett. 1976;29(1)

Grigoriev SA, Dunaev AYu, Zaikov VV. Chromites: an indicator of copper ore source for ancient metallurgy. Doklady Earth Sciences. 2005;400(1):95–98

Landa EA, Bagdasarov EA, Markovsky BA, Khotina MI. Ob osobennostyakh khimicheskogo sostava olivinov vulkanogennykh ul'tramafitov Meymecha-Kotuyskogo regiona i Kamchatki. Zapiski Vsesoyuznogo Mineralogicheskogo Obshchestva. 1978;CVII(3):280–290. [Landa EA, Bagdasarov EA, Markovsky BA, Khotina MI. On the chemical composition features of olivine volcanogenic ultramafites of the Meimecha-Kotui and Kamchatka regions. Proceedings of the USSR Mineralogical Society. 1978;CVII(3):280–290. (In Russ.)]

Lesnov FP. Zakonomernosti raspredeleniya redkozemel'nykh elementov v olivinakh. Zapiski Rossiiskogo Mineralogicheskogo Obshchestva. 2000;6:88–103. [Lesnov FP. Regularities in the distribution of rare-earth elements in olivine. Proceedings of the Russian Mineralogical Society. 2000;6:88-103. (In Russ.)]

Mineraly. III(I). Silikaty s odinochnymi i sdvoyennymi kremnekislorodnymi tetraerami. M.: Nauka; 1972. [Minerals. Silicates with single and double silicon-oxygen tetrahedra. M.: Nauka; 1972. (In Russ.)]

Sobolev NV, Sobolev AV, Tomilenko AA, Kovyazina SV, Batanova VG, Kuz'min DV. Paragenesis and complex zoning of olivine macrocrysts from unaltered kimberlite of the Udachnaya-East pipe, Yakutia: relationship with the kimberlite formation conditions and evolution. Russian Geology and Geophysics. 2015;56(1–2):260–279

Modern Urban Sediments: Identification of the Cosmic Spherules

Irina A. Danilenko, Elena M. Baglaeva, Evgeniya V. Petrova, Andrian A. Seleznev and Grigoriy A. Yakovlev

Abstract The cosmic spherules can be found in the modern urban sediments of the industrial city like Ekaterinburg. The experimental study procedures of spherule identification and extraction were developed. For 8 of the spherules studied their cosmic origin was confirmed using optical and electron microscopy observations as well as chemical analysis of their internal texture and inclusions.

Keywords Cosmic spherules · Ablation spherules · Micrometeorites · Urban sediments

Introduction

During the entering the Earth's atmosphere, meteoric bodies lose their mass significantly as a result of friction and ablation processes. Due to the action of the incoming air flow, the surface layer of a meteorite melts and is blown away from the surface of meteoroid body. Consequently, the particles of extraterrestrial origin, such as ablation spherules and dust particles, remain in the atmosphere. Gradually they precipitate on the surface of the planet (Tomkins et al. 2016; Genge et al. 2017; Cordier et al. 2011). Experiments on the collection of the cosmic spherules in a clean area such as Antarctica were carried out previously (Genge et al. 2017; Cordier et al. 2011). In this study, an attempt of the cosmic spherule identification in the modern urban sediments was made for the industrial center of Ekaterinburg (Russia).

I. A. Danilenko · E. M. Baglaeva · E. V. Petrova (✉) · G. A. Yakovlev
Ural Federal University, Mira str. 19/5, 620002 Ekaterinburg, Russian Federation
e-mail: evgeniya.petrova@urfu.ru

E. M. Baglaeva
e-mail: elenbaglaeva@gmail.com

E. M. Baglaeva · A. A. Seleznev
Institute of Industrial Ecology, Ural Branch of the Russian Academy of Science, S. Kovalevskaya str., 20, 620990 Ekaterinburg, Russia

S. Votyakov et al. (eds.), *Minerals: Structure, Properties, Methods of Investigation*, Springer Proceedings in Earth and Environmental Sciences,
https://doi.org/10.1007/978-3-030-00925-0_2

Ekaterinburg is the city in Russia with over a million population. It is located on the Middle Ural (N56.5°, E60.4°) in a humid continental climate (Dfb) in accordance with the climatic classification of Köpper-Geiger (Kottek et al. 2006). It is characterized by sharp variability in weather conditions, with well-marked seasons. The winter lasts from October until the middle of April and the temperature may fall to −45 °C, though rarely lower than −20 °C to −25 °C. Summer is short, only 65–70 days with an average temperature at about +18 °C. Ekaterinburg is an important transport and logistics hub on the Trans-Siberian Railway, the large industrial center of the machinery, metal processing, and ferrous and non-ferrous metallurgy.

Materials and Methods

The specimens of the surface mud sediments or puddle sediments of the contemporary anthropogenic sediments were collected at the territories of residential quarters of Ekaterinburg city in 2017. Sampling locations were chosen in the different parts of the city with a typical block of houses (multistory buildings), at the different lithogenic substrate areas. The selected blocks and their urban landscapes were constructed in the different periods of the second part of 20th century. Each location adjoined the street-road city network and had the inner courtyard and exterior contiguous territory. The urban landscapes were divided between the traffic area, parking, green (rest area or children's playgrounds), and pedestrian zones for the inner courtyard and the exterior contiguous territory.

Snow and snow-dust pulp specimens (a form of a liquid or semi-liquid mixture of water, loam, silt, and clay) were handpicked in March (I period). Soil, road dust, urban mud were collected in July (II period).

The sampling of the environmental components was conducted according to the following scheme. The five specimens were collected at the each selected residential block: two samples from the green zones (lawns), three samples from the parking, traffic area (road) and pedestrian zone (sidewalk). Each specimen of the environmental component was represented as the sample which was combined from 3–5 subsamples from different localizations at the given landscape zone.

The particle size composition analysis was performed for each specimen. Preparation of the urban specimens and subsequent granulometric (particle size distribution) analysis were conducted in a compliance with the standard requirements of the Russian Federal Certification System. Previously, the organic component (twigs, leaves, needles) was removed from the specimens.

The specimens of I period – snow and snow dust pulp – were melted at the ambient temperature. The solid mass of the samples of snow and snow-dust pulp was divided into the separate fractions depending on the particle size. The following order was applied: >1 mm, 0.25–1 mm, 0.1–0.25 mm, 0.05–0.1 mm, 0.01–0.05 mm, 0.002–0.01 mm. The particle size fractions 0.01–0.05 mm and 0.002–0.01 mm were separated by the decantation with vacuum filtering of the obtained solutions through the filters with the with pore sizes 0.008 mm and 0.002 mm, respectively. The wet sieving with an addition of the distilled water allowed the particle size fractions to be separated: 0.05–0.1 mm, 0.1–0.25 mm, 0.25–1 mm and >1 mm.

The same methods were applied for the separation of the specimens of II period – soil, road dust, and urban mud into particle size fractions. However, the sequence of fractioning was different: >1 mm, 0.25–1 mm, 0.002–0.01 mm, 0.01–0.05 mm, 0.05–0.1 mm, 0.1–0.25 mm. Some particles were selected from the fraction with the particle size of 0.25–1 mm by methods of visual diagnostics. Several of the following features were taken into account: shape, morphology, texture, shine, color, hardness, elasticity, and density of the particles (Seleznev 2017).

The experiment was initially aimed to study the composition of the industrial sediments. When spherical particles were revealed after the granulometric analysis of the modern sediments, the question about their origination arose.

The main object of the study is spherical particles. They were visually observed in the fractions of 0.1–0.25 mm and 0.25–1 mm. These particles could be formed during material solidification from the liquid state. Therefore, for some of the spherules, an extraterrestrial origin was suggested. Their formation proposed to be connected with the entry of meteoric bodies into the atmosphere of the Earth. Preliminary observations were presented in (Danilenko et al. 2018).

Exterior view of the chosen spherules was studied using JSM-6390/6390LV (Jeol) electron microscope. Then the spherules were glued on the glass plates with epoxy and sequentially polished on the sanding paper and diamond pastes. Spherules were polished up to the sections close to their diametric section. Because as section gets diametric, the spherule will escape from the epoxy glue and gets lost. The internal texture of the spherules was studied using optical microscope AxioVert 40 MAT (Carl Zeiss) and electron microscope ΣIGMA VP (Carl Zeiss) with the X-MAX EDS equipment (Oxford Instruments). Average linear sizes of chosen spherules were measured using AxioVision Ver 4.38 software on the optical images of the spherules.

Results and Discussion

Electron microscopy observations of the samples showed the dendritic texture on the surface for some of the small spherules, Fig. 1a. Following chemical analysis of the dendritic surfaces revealed the presence of Fe and O, mainly (with the traces of Si, Mg, Al, Ca and Mn). Similar looking spherules were observed in (Tomkins et al. 2016; Genge et al. 2017). Therefore, present study of the spherules interior was performed to determine whether the spherules have an extra-terrestrial origin or not.

Internal texture of 10 spherules was studied. Average linear sizes of spherules chosen for cutting was from 0.3 to 1.0 mm in diameter. Optical images of the typical spherules slices are shown in Fig. 2. Some of the studied spherules have glassy internal texture, which has been formed during rapid solidification of the melted material in the atmosphere. Most glassy spherules contain small

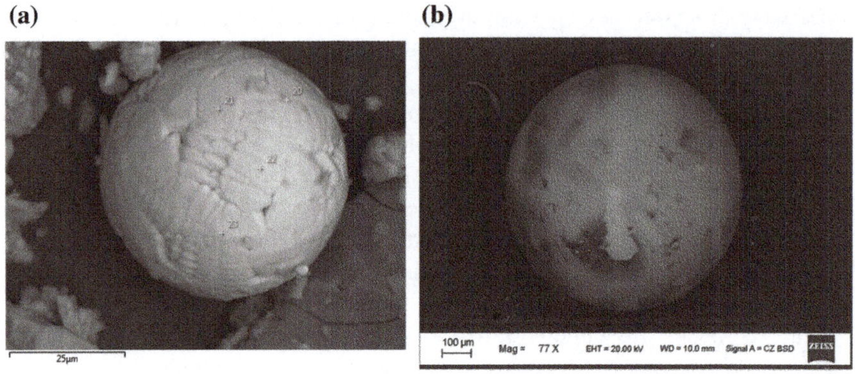

Fig. 1 Exterior view of the spherules: **a** – small spherule with a dendritic texture; **b** – bigger spherule with a smooth surface

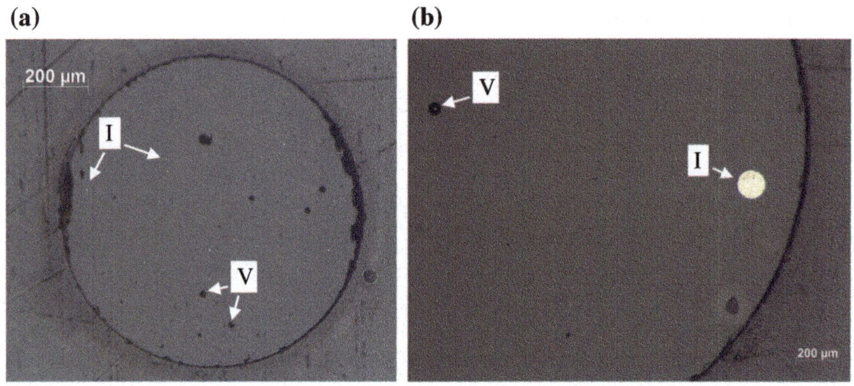

Fig. 2 Sections of the spherules: **a** – relatively large spherule with vesicles (V) and inclusions (I); **b** – spherical inclusion within the spherule

light-colored inclusions and vesicles. In some cases, their internal texture shows the presence of two different phases, which were marked as L2_1 and L2_2 in the table with chemical analysis data. Another type of the spherules appears as polycrystalline aggregates. Moreover, one two-phased spherule with fine-grained dendritic crystals of the second phase was observed.

Chemical analysis of the spherules was performed with EDS. Some results of the typical spherule matrix are shown in Table 1.

8 of the spherules contained small inclusions, which were few micrometers in size and consisted of Fe, Ni, Co, and S. They appeared as the small light-colored inclusions within the spherules sections under the optical microscope. Due to their small size, only the spatial distribution of chemical elements could be measured. A scanning electron image and a spatial distribution of chemical elements within

Table 1 The chemical composition of the spherules internal parts

Sample	O wt %	Na wt%	Mg wt%	Al wt%	Si wt %	P wt %	S wt %	K wt%	Ca wt%	Mn wt%	Fe wt %	Cr wt%	Ti wt%	Ni wt%	Total
L	59.04	-	9.96	3.43	17.49	-	0.25	-	6.28	-	3.55	-	-	-	100.00
L1	41.12	-	8.71	3.23	20.91	-	-	0.26	14.08	-	11.69	-	-	-	100.00
L2_1	41.44	-	9.78	3.12	22.48	-	-	-	-	-	10.27	-	-	-	100.00
L2_2	41.64	-	29.74	-	19.59	-	-	-	0.60	8.43	-	-	-	-	100.00

- amount is below detection limit

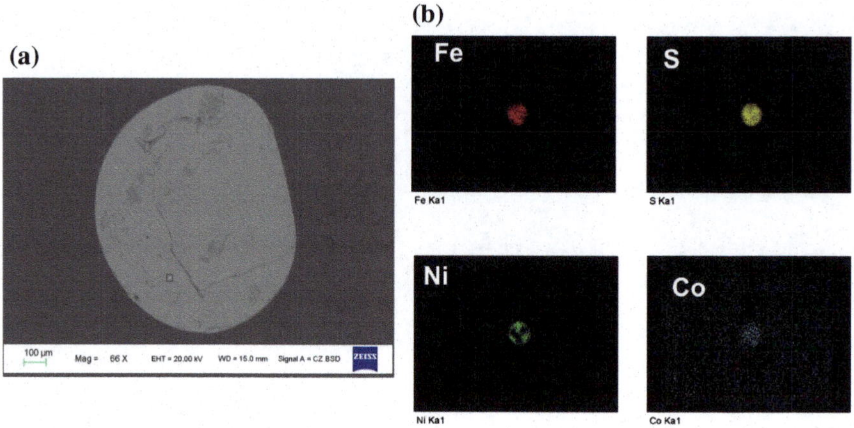

Fig. 3 Sections of the spherules: **a** – polycrystalline spherule with inclusions; **b** – spatial distribution of Fe, S, Ni, and Co within the inclusion in the rectangle in Fig. 3a

the inclusion are shown in Fig. 3. Chemical analysis was performed in the area marked in Fig. 3a.

Two of the spherules studied had relatively large inclusions, but one of them, unfortunately, was lost during the polishing of the sample. Another one, which was about 0.1 mm in diameter, was analyzed. Its composition was affordable for measurement: Fe – 41.40 wt.%, Ni – 32.0 wt.%, Co – 0.67 wt.%, and S – 25.92 wt. %. All the spherules with inclusions consist of the O ~ 40 wt.%, Si ~ 18%, Fe ~ from 4 to 16 wt.%, and small amount of K ~ 0.26-0.77 wt./%, Mn ~ 0.8–0.73 wt.%, Na ~ 0.65–2.47 wt.%, Al ~ 2.65–7.89 wt.%, Ca ~ 1.16–24.16 wt. %, Mg ~ 4.0–29.74 wt.%. The content is consistent with the content for cosmic spherules from the carboniferous deposits of the Usolka section (Sungatullin 2017).

For these spherules, the extraterrestrial origin was confirmed because their chemical composition was found to be in a good agreement with the cosmic spherule composition. Moreover, Fe, Ni, S, and Co presence in the composition of the inclusion had strongly supported this suggestion because these chemical elements are the usual component of the metal and sulfide phases of meteorites (Dodd 1981).

Three of the studied spherules contain Ti ~ 0.58–1.38 wt.%, which is supposed to be an indicator of the industrial origin of the spherule. The titanium is widely used in industrial alloys. Therefore, its products could appear in the atmosphere and precipitate in close regions where alloys have been produced. Nevertheless, Ti was revealed in inappreciable quantity in stony cosmic spherules (Cordier et al. 2011).

Conclusions

The spherules of the extraterrestrial origin are present in the sediments of the urban territories. However, their identification and extraction from the modern sediments are relatively complicated procedures. They could not be easily obtained by simple mud collection. The samples investigated were collected, filtered and divided into fractions according to the particle sizes. Then the spherules were found in the fractions of 0.1–0.25 mm and 0.25–1 mm using an optical microscope.

Investigation of the spherules using optical end electron microscopy with EDS confirmed cosmic origin for some of them.

Acknowledgements This work was partly supported by the Act 211 of the Government of the Russian Federation, agreement no. 02.A03.21.0006, and by the Russian Foundation for Basic Research, Project # 16-35-60044 mol_a_dc.

References

Cordier C., van Ginneken M., Folco L. Meteorit. Nickel abundance in stony cosmic spherules: Constraining precursor material and formation mechanisms Planet. Sci. 2011, 46 (8): 1110–1132.

Danilenko I.A., E.M. Baglaeva, E.V. Petrova et al. Identification of the cosmic spherules in the modern urban sediments. 81st Annual Meeting of The Meteoritical Society 2018 (LPI Contrib. No. 2067) Abstract# 6226.

Dodd R.T. Meteorites. A petrologic-chemical synthesis. Cambridge: Cambridge University Press; 1981.

Genge M.J. et al. The mineralogy and petrology of I-type cosmicspherules: Implications for their sources, origins and identification in sedimentary rocks. Geochimica et Cosmochimica Acta 2017, 218:167–200.

Kottek M., Grieser J., Beck C. et al. World map of the Köppen–Geiger climate classification updated. Meteorol Zeitschrift 2006, 15: 259–263.

R.Kh. Sungatullin i dr. Mikrosfery kosmicheskogo proiskhozhdeniya v kamennougol'nykh otlozheniyakh razreza Usolka, predural'skii progib. Geologiya i geofizika, 2017, 58(1): 74—85 [Sungatullin R.Kh. et al. Cosmic microspheres in the carboniferous deposits of the Usolka section (Urals foredeep). Journal of Geology and Geophysics. 2017, 58(1): 74—85. (in Russ)].

Seleznev A.A. i dr. Tekhnogennye obrazovaniya v poverkhnostnom gryazevom osadke v g. Ekaterinburg. Sbornik trudov XXIII Vserossiiskoi nauchnoi konferentsii "Ural'skaya Mineralogicheskaya shkola-2017" noyabr' 2-4, 2017, Ekaterinburg, Russiya, ss. 75-77 [Seleznev A.A. et al. Industrial creations in the fallout on the city of Ekaterinburg. Proceedings of the XXIIIth Young Researchers' Conference "Ural Mineralogical School-2017", November 2-4, 2017, Ekaterinburg, Russia, pp. 75-77. (in Russ)].

Tomkins A.G. et al. Ancient micrometeorites suggestive of an oxygen-rich Archaean upper atmosphere. Nature 2016, 533: 235–238.

Zircon Thermometry of the Yarot Granite Massif (The Subpolar Urals)

Yulia V. Denisova and Anna N. Vikhot

Abstract Zircon is one of common accessory minerals of igneous rocks. Geochemical and morphological features of this mineral reflect physicochemical properties of the mineral formation medium. It provides additional information about the nature and formation conditions of the rock. A considerable experience of using zircon as a geothermometer has been accumulated at present. It helps determine a more accurate rock-forming temperature regime. The results of the study of accessory zircon from the Yarot granite massif (the Subpolar Urals) are presented in this paper. The data obtained by different methods (the Watson saturation thermometry and the classical evolutionary morphological analysis according to J. Pupin and G. Tyurko) were compared. The formation of the Yarot massif granites took place at an average temperature of 841 °C (range: 806–926 °C) according to the Watson saturation thermometry. The study of the Yarot massif zircons gave a range from 750 to 900 °C using an evolutionary-morphological analysis. The average crystallization temperature of rocks was 843.75 °C according to Pupin. The Subpolar Urals granite formation temperatures were specified. The obtained temperatures were higher than the thermometric data of M. V. Fishman and his colleagues for this massif. Previously scientists assumed that the Yarot granite massif was formed at temperatures not higher than 720 °C.

Keywords Zircon · Granite · The Yarot massif · The Subpolar Urals

---

Y. V. Denisova (✉) · A. N. Vikhot
Institute of Geology of the Komi Science Centre, Ural Branch of the Russian Academy of Sciences, Pervomayskaya, 54, Syktyvkar, Komi Republic, Russian Federation
e-mail: yulden777@yandex.ru

A. N. Vikhot
e-mail: vikhot.anna@mail.ru

S. Votyakov et al. (eds.), *Minerals: Structure, Properties, Methods of Investigation*, Springer Proceedings in Earth and Environmental Sciences,
https://doi.org/10.1007/978-3-030-00925-0_3

Introduction

The Yarot massif is a narrow plate-like body that crosses the Malaya Yarota River valley. The Yarot rocks are predominantly strongly cataclased granites. Slightly cataclased granites occupy not more than 15% of the total rock volume and represent light gray massive medium-grained leucocratic differences. Enclosing rocks are the Upper Riphean deposits of the Moroin Formation for the Yarot massif (Denisova 2014, 2015, 2016). The granite massif testing was carried out using a point method with a selection of 5 samples. Unchanged rocks were taken with an average weight of 10 to 15 kg for each sampling material. Zircon content averaged 22–28 g/t in investigated rocks (Fig. 1).

1—mica-quartz schists, porphyries, porphyrites, marbles and quartzits interlayers; 2—terrigenous-carbonate deposits; 3—granites; 4—rhyolites, felsites; 5—contact lines: a—stratigraphic and magmatic, b—tectonic; 6—planar-structured bedding elements/planar structures of bedding elements. Massifs (numbers in circles): 1—the Badyayu massif; 2—the Yarot massif.

Materials and Methods

The methods of granite formation temperature evaluation on the basis of the zirconium saturation degree in the rock (Watson saturation thermometry) (Fishman et al. 1968) and the features of the zircon morphology (classical evolutionary

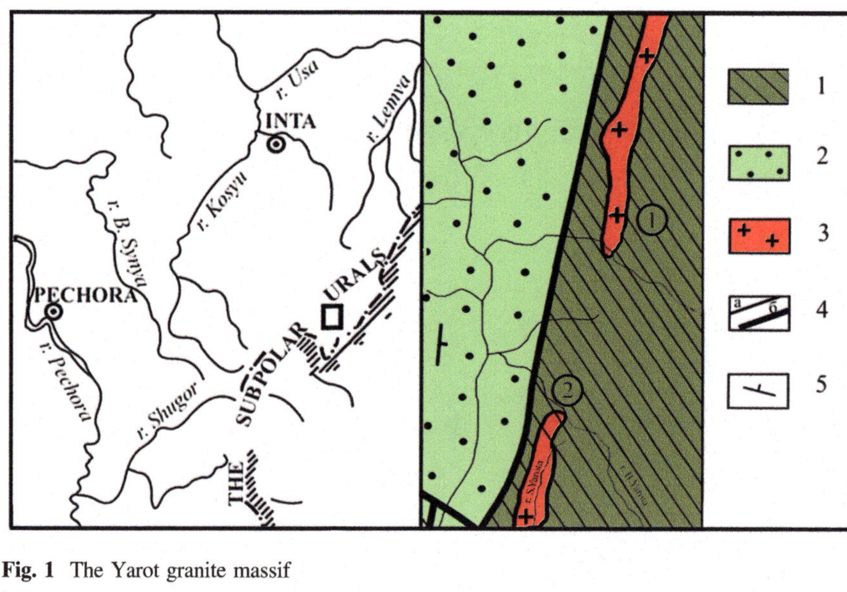

Fig. 1 The Yarot granite massif

morphological analysis by J. Pupin and G. Tyurko) (Watson and Harrison 1983) were used to determine the temperature regime of the Yarot massif zircon crystallization.

According to E. Watson and T. Harisson, the zirconium saturation level in the rock necessary for the mineral formation depends greatly on the molar ratio ($Na_2O + K_2O/Al_2O$) in the melt and weakly on the temperature and Si_2O and Na_2O/K_2O contents. It allowed the scientists to present the zirconium saturation degree in the rock as a function of the temperature and melt composition.

$$\mathrm{InD}_{\mathrm{Zr}}^{\mathrm{zircon/melt}} = (-3.80 - [0.85 * (\mathrm{M}-1)] + 12{,}900/\mathrm{T},$$

Where $\mathrm{InD}_{\mathrm{Zr}}^{\mathrm{zircon/melt}}$ is the ratio of Zr concentration in zircon and melt, M is the cation ratio (Na + K + 2 * Ca)/(Al * Si), and T is the temperature in Kelvin.

According to the morphological zirkon studies of J. Pupin and G. Tyurko (Pupin 1980), the prisms presence and the zircon crystal formation depends on the mineral crystallization temperature change. Change dipyramids is connected with features of formation of mineral medium chemicals. The identified correlation allowed researchers to construct a diagram of the basic typological zircon forms. Figure 2 illustrates the two-dimensional dependence of zircon crystal morphology on

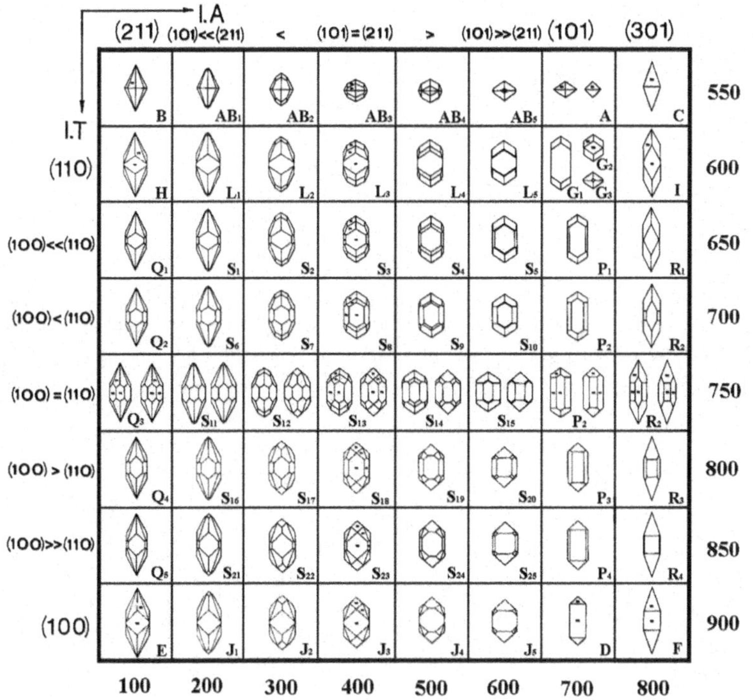

Fig. 2 Zircon classification by J. Pupin and G. Tyurko

temperature (conditional scale I. T.) and on mineral formation medium alkalinity (conditional scale I. A).

Results and Discussion

Saturation temperatures for zircon were obtained using the data on the chemical composition of the Subpolar Urals granites based on the GCDkit 2.3 program (The Comprehensive R Archive Network 2014) (Table 1).

The data obtained on the basis of Watson saturation thermometry can indicate that the Yarot massif granites were formed at temperatures in the range of 806–926 °C and the average rock formation temperature was 841 °C.

The temperature of granite formation was also obtained using the evolutionary morphological analysis. The study of accessory zircon monofractions showed that the high-temperature zircon types according to Pupin and Tyurko were typical for the investigated granites of the Subpolar Urals (Fig. 3).

The Yarot massif zircons correspond to such types as D, P3, P4, P5, F in their morphology. We can determine the formation temperature range of the massif rocks

Table 1 Chemical granite composition and saturation temperatures for zircon of the Yarot massif, mass. %

№	Major elements, mass. %						Zr, g/t	Saturation thermometry
	SiO_2	TiO_2	Al_2O_3	CaO	Na_2O	K_2O		
1	75.52	0.14	12.35	0.88	3.30	4.59	194.6	806
2	75.02	0.37	11.59	0.89	3.02	4.65	180.4	797
3	74.59	0.16	12.07	0.61	3.33	4.52	620.4	926
4	74.89	0.22	12.08	0.57	3.09	4.93	335.3	861
5	74.86	0.29	11.78	0.59	3.17	4.91	212.9	813

Note Major elements were obtained using the silicate method in CUC "Science" of the Institute of Geology of Komi SC UB RAS (Syktyvkar). Zirconium contents were obtained using an ICP-MS in the Institute of Geology and Geochemistry UB RAS (Ekaterinburg)

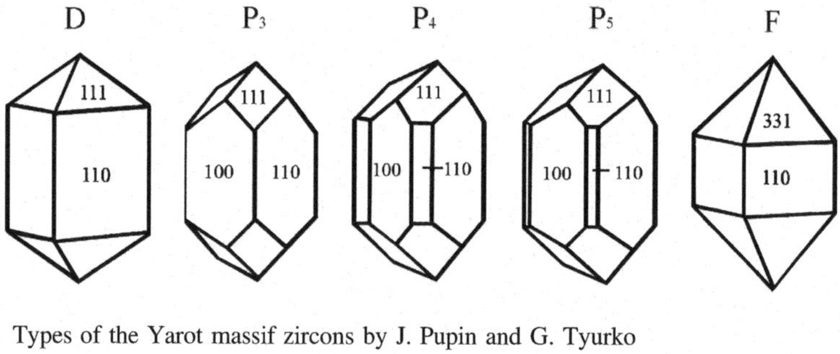

Fig. 3 Types of the Yarot massif zircons by J. Pupin and G. Tyurko

Table 2 The spreading degree of the Yarot massif zircon types (according to J. Pupin and G. Tyurko)

Zircon type	D	P_3	P_4	P_5	F	Average temperature
Temperature, °C	900	750	800	850	900	843.75
Quantity, pc	70	35	30	15	10	-

considering that each zircon type was crystallized at a certain temperature. Thus, morphotype D (a zircon with prism (110) and dipyramid (111)) is formed at high temperature (T = 900 °C). Morphotypes P5, P4, P3 zircons are characterized by faces (100), (110), (111). These minerals are formed at a temperature of 850, 800, 750 °C. The shape of F crystals is caused by the development of faces (100) and (331). It also gives evidence to a high crystallization temperature (T = 900 °C).

The studied granites were formed at a high temperature from 750 to 900 °C according to the evolutionary-morphological analysis of the Yarot massif zircons (Table 2). The average crystallization temperature of rocks 843.75 °C.

We have noted the same temperature ranges for the Yarot massif when comparing the temperatures obtained by different methods. The Watson method accuracy is 5% which implies that the discrepancies in the granites crystallization temperatures are insignificant.

Conclusions

According to the Watson saturation thermometry and the Pupin and Tyurko classical evolutionary morphological analysis, the formation of Yarot granite massif occurred at high temperature (from 750 to 926 °C). The obtained thermometric data did not confirm the temperature that was obtained by M. V. Fishman and his colleagues for this massif. Previously scientists assumed that the Yarot granite massif was formed at temperatures not higher than 720 °C (Fishman et al. 1968).

Acknowledgments The work was supported by the Basic Research Program of RAS № 18-5-5-19.

References

Denisova UV. Tipomorficheskie i tipohimicheskie osobennosti aktsessornyh tsirkonov granitoidov Pripolyarnogo Yrala. Vestnik IG Komi NC YrO RAN. 2014;5:9–16. [Denisova UV. Typomorphic and typochemical accessory zircons sings of the Subpolar Urals granitoids. Vestnik IG Komi SC UB RAS. 2014;5:9–16. (In Russ.)]

Denisova UV. Petrogeneticheskoe znachenie ZrO_2/HfO_2 otnosheniya v aktsessornom tsirkone granitov Pripolyarnogo Yrala. Vestnik IG Komi NC YrO RAN. 2015;2:23–31. [Denisova UV. The petrogenetic value of the ZrO_2/HfO_2 ratio in zircons (Subpolar Urals). Vestnik IG Komi SC UB RAS. 2015;2:23–31. (In Russ.)]

Denisova UV. Termometrya tsirkona iz granitoidov Pripolyarnogo Yrala. Vestnik IG Komi NC YrO RAN. 2016;11:11–22. [Denisova UV. Temperature survey of zircon from the granitoids of the Subpolar Urals. Vestnik IG Komi SC UB RAS. 2016;11:11–22. (In Russ.)]
Fishman MV, Yushkin NP, Goldin BA, Kalinin EP. Mineralogiya, tipomorfizm I genesis aktsessornyh mineralov izverzhennyh porod severa Yrala i Timana. M.-L.: Nayka; 1968. [Fishman MV, Yushkin NP, Goldin BA, Kalinin EP. Mineralogy, typomorphism and genesis of accessory minerals of igneous rocks in the north of the Urals and Timan. M.-L.: Science; 1968 (In Russ.)]
Pupin JP. Zircon and granite petrology. Contrib. Miner. Petrol. 1980;73:207–220
Watson EB, Harrison TM. Zircon saturation revisited: temperature and composition effects in a variety of crustal magma types. Earth Planet Sci Lett. 1983;64:295–304

Internet Source

The Comprehensive R Archive Network. 2014. https://cran.r-project.org. Accessed 15 Aug 2014

New Carbonaceous Chondrite from Northwest Africa

Kseniya A. Dugushkina and Stepan V. Berzin

Abstract The article presents the results of a study of a new meteorite from northwest Africa. The meteorite consists of ~20–30% of chondrules, 60–70% of fine-grained matrix, refractory inclusions (CAI, AOAs, forsterite rich objects) occupy no more than 3–5%. The meteorite is carbonaceous chondrite and belongs to the petrological type CM2. The meteorite is composed of olivine, orthopyroxene, clinopyroxene, chromite, troilite, kamacite and taenite. The average size of the chondrules is 0.3 mm. Porphyritic-type chondrules predominate. The meteorite matrix consists of small fragments of olivine, pyroxene, layered silicates and iron hydroxides. CAIs are small in size (0.05–0.3 mm). The main minerals of CAI are spinel, Ca-pyroxene, hibonite and perovskite. AOA mainly contains spinel-diopside inclusions. The forsterite-rich objects are bordered by low-Ca pyroxene, represented by enstatite. Meteorite is unshocked and has low grade of terrestrial weathering (metal partially oxidized). At present, the work is underway to study and register this new carbonaceous chondrite.

Keywords Carbonaceous chondrite · Refractory inclusion · CAIs · AOAs · Forsterite-rich objects

Introduction

Meteorite Northwest Africa 869 was found at the site of meteor rain fall in West Africa, Morocco in 2000–2001. The meteorite was classified as L4-6 ordinary chondrite (Connolly 2006). For our study, a fragment of the meteorite weighing 15.56 g, found in the place of precipitation of meteorite rain, was transferred to our institute by private collectors. During the research it has been established that this

K. A. Dugushkina (✉) · S. V. Berzin
A.N. Zavaritsky Institute of Geology and Geochemistry UB RAS,
15 Academic Vonsovsky st., Ekaterinburg 620016, Russia
e-mail: dugushkina.kseniya@mail.ru

S. Votyakov et al. (eds.), *Minerals: Structure, Properties, Methods of Investigation*,
Springer Proceedings in Earth and Environmental Sciences,
https://doi.org/10.1007/978-3-030-00925-0_4

fragment actually is not ordinary, but carbonaceous chondrite, and, accordingly, is a new previously unexplored meteorite.

Materials and Methods

The mineralogy and structural features of the meteorite were studied using the scanning electron microscope SM 6390LV (Jeol) with an energy dispersing attachment INCA Energy 450 X-Max 80 in the Common Use Center of the UB RAS "Geoanalyst" (Ekaterinburg, Russia). The chemical composition of minerals were studied using the electron microprobe analyser Cameca SX100 with five wavelength spectrometers and Bruker Xflash6 energy dispersive spectrometer in the Common Use Center of the UB RAS "Geoanalyst" (Ekaterinburg, Russia).

Results and Discussion

The meteorite has a dark brown color (Fig. 1). The fragment of the meteorite is covered with a melting crust of black color with a matte surface. The meteorite consists mostly of a fine-grained matrix, the chondrules comprise 20–30%, and refractory inclusions (calcium- and aluminum-rich inclusions (CAIs), forsterite-rich objects and amoeboid olivine aggregates (AOAs) occupy 3–5%. The mineral composition of the meteorite is represented by olivine, pyroxene, spinel, chromite, troilite, kamacite, melilite and hibonite.

The meteorite matrix consists of small fragments of olivine, pyroxene of layered silicates and iron hydroxides. The main amount of troilite, taenite, kamacite and iron hydroxides is confined to the meteorite matrix. The metallic iron is replaced by hydroxides. In addition to metal and sulfide, fragments of chromite are observed in the matrix.

Fig. 1 Meteorite Northwest Africa

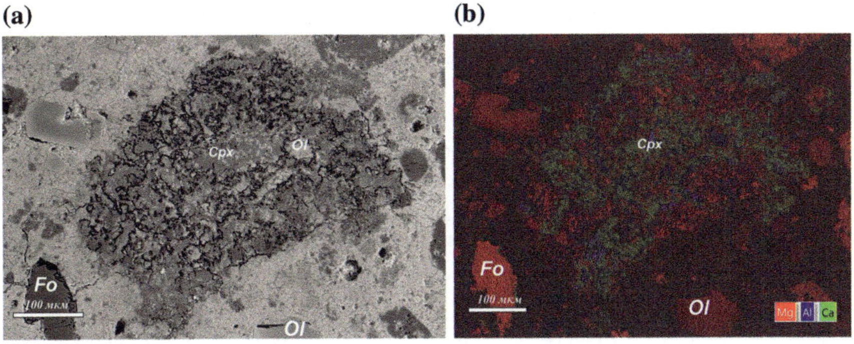

Fig. 2 The structure of calcium- and aluminum-rich inclusion (CAIs): **a** image in the back-scattered electron (BSE); **b** elemental distribution map Mg, Al, Ca

Chondrules on average have a size of up to 0.4 mm, sometimes reaching 1 mm. The main minerals of the chondrules are olivine and pyroxene. The accessories in chondrules are represented by chromite, spinel, taenite and kamasite. Porphyritic olivine (PO) and porphyritic olivine-pyroxene (POP) chondrules predominate. Chondrules mostly have uneven boundaries.

Calcium- and aluminum-rich inclusions (CAIs) vary in size from 0.05 to 0.3 mm (Fig. 2). CAIs have a highly porous. The mineral composition of CAIs is represented by pyroxene, spinel, hibonite and perovskite. Hibonite has the composition of SiO2 (1.7 wt%), MgO (0.4 wt%), Al2O3 (84.1 wt%), FeO (4.2 wt%), CaO (7.9 wt%), TiO2 (1.7 wt%). The chemical composition of perovskite includes SiO2 (2.92 wt%), MgO (5.69 wt%), Al2O3 (11.09 wt%), FeO (2.88 wt%), CaO (25.97 wt%), TiO2 (51.54 wt%).

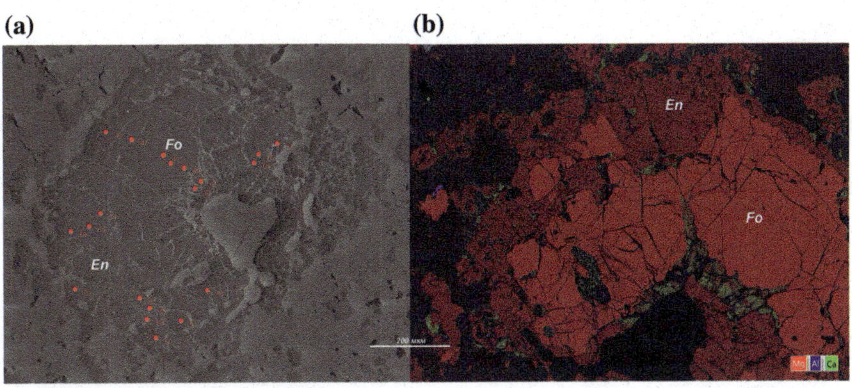

Fig. 3 The structure of forsterite-rich objects: **a** image in the back-scattered electron (BSE); **b** elemental distribution map Mg, Al, Ca

Table 1 Composition of forsterite, enstatite and low-Ca pyroxene in refractory forsterite-rich objects

No.	SiO_2	TiO_2	Al_2O_3	Cr_2O_3	FeO	MnO	MgO	CaO	Сумма	f
Forsterite										
46	42.3	0.08	0.17	0.39	0.86	0.14	55.11	0.35	99.41	0.005
47	42.5	0.06	0.31	0.11	0.39	–	55.34	0.72	99.5	0.002
48	41.64	0.09	0.27	0.09	0.32	–	56.02	0.67	99.13	0.002
49	41.57	0.09	0.37	0.11	0.39	–	55.58	0.78	98.93	0.002
50	41.64	0.08	0.22	0.09	0.49	–	55.94	0.69	99.2	0.003
51	42.64	0.04	0.08	0.38	0.81	0.08	55.28	0.39	99.72	0.005
52	42.37	0.06	0.12	0.25	0.68	–	56.06	–	99.56	0.004
53	42.13	0.09	0.17	0.16	0.68	–	55.08	0.55	98.91	0.004
64	42.67	0.04	0.07	0.47	0.91	0.12	54.82	0.27	99.4	0.005
Enstatite										
54	56.87	0.16	0.98	0.47	3.49	–	36.36	0.68	99.15	0.029
55	59.19	0.2	0.94	0.43	0.63	–	38.15	0.54	100.12	0.005
56	57.95	0.25	0.96	0.5	1.98	0.14	37.2	0.57	99.69	0.016
57	59.12	0.15	0.84	0.58	1.99	0.08	36.34	0.64	99.83	0.017
58	59.35	0.23	0.83	0.73	1.02	0.16	37.35	0.37	100.09	0.009
59	54.03	0.11	0.91	0.55	6.47	0.2	35.99	0.59	99.04	0.054
60	58.44	0.15	0.76	0.59	2.52	0.17	36.7	0.66	100.12	0.021
63	57.99	0.18	0.99	0.43	0.95	–	38.46	0.66	99.76	0.008
65	57.74	0.23	1.12	0.52	1.3	–	37.7	0.38	99.07	0.011
66	57.48	0.18	0.79	0.72	1.08	0.17	38.23	0.69	99.34	0.009
Clinopyroxene										
61	49.22	0.85	2.7	0.57	8.74	0.15	18.31	17.96	99.6	0.131
62	53.75	0.51	1.9	0.78	4	0.25	28.66	9.41	99.43	0.042

Forsterite-rich objects vary in size from 0.05 to 1.0 mm (Fig. 3). Unlike AOAs, they are composed mostly of high-magnesium forsterite and high-magnesian enstatite. The composition and iron content of forsterite and enstatite are presented in Table 1. Unlike chondrules, in these inclusions an insignificant amount of plagioclase glass is observed. In the inclusions there are grains of metal, which are replaced by iron hydroxides. The iron content of forsterite and enstatite rises from the center of the grain to the edge.

Acknowledgements The authors wish to thank A. V. Mikheeva and Dr L. V. Leonova (Common Use Center "Geoanalyst", IGG UB RAS) for the analytical support in this research. The study was funded by the Russian Foundation for Basic Research, project no. 17-05-00297.

Reference

Connolly HC., Zipfel J., Grossman JN., Folco L., Smith C., Jones RH., Righter K., Zolensky M., Russell SS., Benedix GK., Yamaguchi A., Cohen BA. The Meteoritical Bulletin, No. 90 // Meteoritics & Planetary Science. 2006. V.41. N 9. pp. 1383–1418

Scanning Electron Microscopy of Pyrite from Brown Coal (Mugun Coal Deposit, Irkutsk Basin)

Narine R. Dzhumayan and Aleksey V. Nastavkin

Abstract The pyrite specimens obtained from brown coal of the Mugun deposit have been studied by scanning electron microscopy (SEM) and X-ray fluorescence. Several pyrite generations have been identified. Pyrite of epigenetic origin has been found to prevail over other generations. Mineral inclusions of quartz and ilmenite and increased concentrations of Co, Ti and Sr have been found in the pyrite specimens studied. The presence of minerals and elements typical for igneous rocks may be indicative of the "Siberian traps" effect common in the deposit area on the conditions of peat accumulation.

Keywords Irkutsk coal basin · Mugun brown coal deposit · Pyrite · Ilmenite

Introduction

The Mugun brown coal depository is located in the North-Western part of the Irkutsk coal basin.

The Irkutsk coal basin is of platform type and genetically connected with tectonic basins. It contains coals of low coalification level, which, however, differ by their qualitative characteristics. The level of sulfur significantly affects variability of technological and chemical properties of coal.

The Mugun brown coal depository is confined to a syncline and characterized by monoclinal and slightly wavy sloping stratum bedding. Carboniferous deposits are imbedded in the Ordovician rocks in the erosive cavity in the North-Western area and characterized by irregular outline. The deposit area is rich with igneous rocks represented by traps deposited as sills. The igneous rocks of basic composition, which are known as "Siberian traps", are represented by fine-grained olivine

N. R. Dzhumayan (✉) · A. V. Nastavkin
Southern Federal University, Bolshaya Sadovaya, 105, Rostov-on-Don, Russia
e-mail: narinedzhumayan@mail.ru

A. V. Nastavkin
e-mail: nastavkin@sfedu.ru

S. Votyakov et al. (eds.), *Minerals: Structure, Properties, Methods of Investigation*, Springer Proceedings in Earth and Environmental Sciences,
https://doi.org/10.1007/978-3-030-00925-0_5

dolerites, which consist of plagioclase (45–60%), monoclinic pyroxene (20–50%) and olivine (up to 10%). The trap deposits were formed in the Pre-Jurassic period, and their thickness may reach 250 m.

The Cheremkhovskaya suite is considered to be industrially coal-bearing. It is represented by sandstones and, to a smaller degree, by siltstones, argillites and strata, and interlayers of industrially valuable coal. The petrographic analysis of sandstones forming the Cheremkhovskaya suite has shown that they consist of quartz, quartz-arkose, silty micaceous sandstone and grey wacke SSSR (1964). The coal from the Mugun depository contains medium levels of sulfur.

Materials and Methods

The objects of our study were polycrystalline pyrite aggregates incorporated into the coal mass. The experimental procedure consisted of macro- and microscopic analysis of pyrite crystals and was carried out in the Collective Center of Scientific Equipment "The Research Center for Mineral Raw and Environmental Condition Studies" of the Southern Federal University (Rostov-on-Don, Russia). The experiments were carried out with VEGA II LMU scanning electron microscope (Tescan, Czech Republic), INCA ENERGY 450/XT energy dispersion microanalysis system (OXFORD Instruments Analytical, United Kingdom) and INCA WAVE 700 wave dispersion microanalysis system (OXFORD Instruments Analytical, United Kingdom). The gross elementary composition of sulfides was estimated by X-ray fluorescence analysis with the Spectroscan Max GV X-ray crystal-diffraction vacuum spectrometer (Spectron, Russia). Concentrations of the following oxides and elements were analyzed: Al_2O_3, Fe_2O_3, SiO_2, P_2O_5, CaO, K_2O, MgO, TiO_2, MnO, Cr, Co, Ni, Cu, Zn, and Sr.

Results and Discussion

The results of our study have shown that sulfides are represented by several generations of pyrite and marcasite. Pyrite occupies cavities, splits and hollows in the coal stratum. Its crystal shape is regular octahedron or pentagonal dodecahedron, and their size may vary from several millimeters to two centimeters.

The first generation pyrite is represented by small granules of several micrometers in size dispersed within the coal matter itself. However, pyrite of later generations is more widely spread. It may be found in the form of thick tightly linked to each other smallest crystals accumulated in splits (Fig. 1).

Pyrite was, apparently, formed as a result of penetration of iron containing solutions into the stratum. Hydrosulfuric water penetrated the coal-bearing stratum through tectonic splits and weakened cracked zones. It appears, that the genesis of the Mugun coal sulphur is associated with hydrosulfuric water that penetrated it

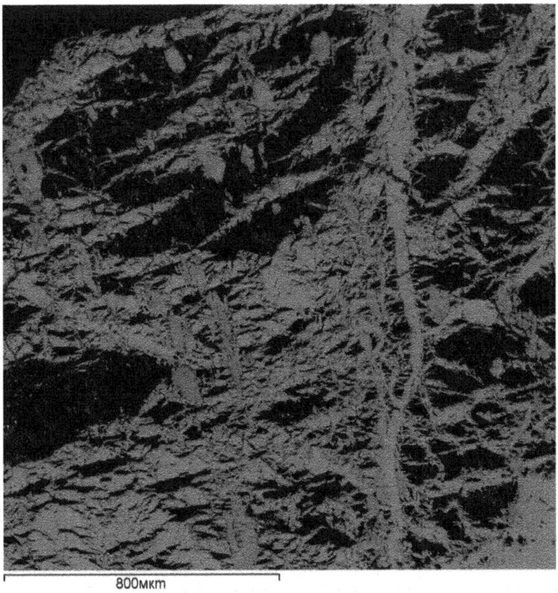

Fig. 1 Pyrite accumulates in splits and cavities in coal (SEM image)

from plastered and bituminous sedimentary strata of Lower Cambrian period with the help of sulfate reducing bacteria. One of the authors of this article has previously shown through the example of coal-bearing deposits of the Donetsk basin that iron sulfides were formed in situ singenetically in the bottom deposits of the sedimentation basin (Kizilshtein and Nastavkin 2003).

The chemical composition of pyrite is shown in Table 1. The absence of organogenic elements is quite possible, since it was collected from coals. Moreover, increased concentrations of Co, Ti, and Sr may be considered as the evidence of its interaction with magmatic rocks. Similar conclusions were made by Finkelman (1981), Ilenok and Arbuzov (2016). The latter study dealt with coals obtained from the Azeysk depository, which also belongs to the Irkutsk coal basin. The authors believed that enrichment with strontium was due to (1) destruction of pyroclastics with acid and (2) enrichment if peat with waters coming from the drift area and following authigenic mineralization. Several authors noted that the majority of microelements in coal (Cu, Zn, Ni, Mo, As, Se, Co, Cd, Sb, Hg, Pb) were associated with pyrite and other sulfidic minerals (Matjie et al. 2008, 2011; Hlatshwayo et al. 2009).

The microprobe analysis has shown that pyrite aggregates contains such mineral inclusions as quartz, a widespread mineral that forms country rocks, and ilmenite, which is rarely found in coal deposits. The presence of ilmenite may be due to the effect of widespread igneous rocks, which frame the coal deposit.

The size of these mineral inclusions may reach dozens of micrometers. Crystals of ilmenite are isometric and prolate in shape (Fig. 2); quartz grains are also isometric and elongated (Fig. 3), and granulated, which is indicative of deformations that took place in the coal strata.

Table 1 Pyrite gross composition (data of X-ray fluorescence analysis)

Oxide	Content, %
Al_2O_3	10.76
Fe_2O_3	81.26
SiO_2	5.69
P_2O_5	0.13
CaO	0.11
K_2O	0.20
MgO	1.48
TiO_2	0.35
MnO (г/т)	141.7
Element	**Content, g/t**
Cr	14.62
Co	1913.08
Ni	17.56
Cu	425.76
Zn	80.26
Sr	4320.88

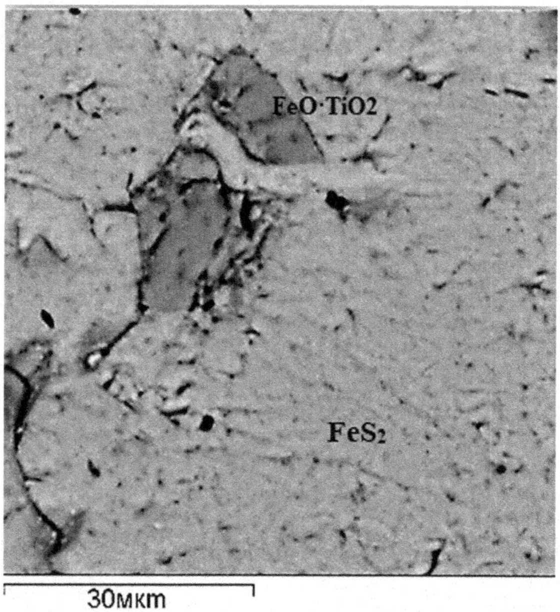

Fig. 2 The SEM image of ilmenite inclusions in pyrite

Many authors have reported that titanium, a non-carbonphilic element, is concentrated in non-carbon strata and its deposits are migratory in nature (Yudovich and Ketris 2006). Usually it is accumulated by a mineral carrier, which is represented by TiO_2. The sulfide form of natural titanium is presently unknown.

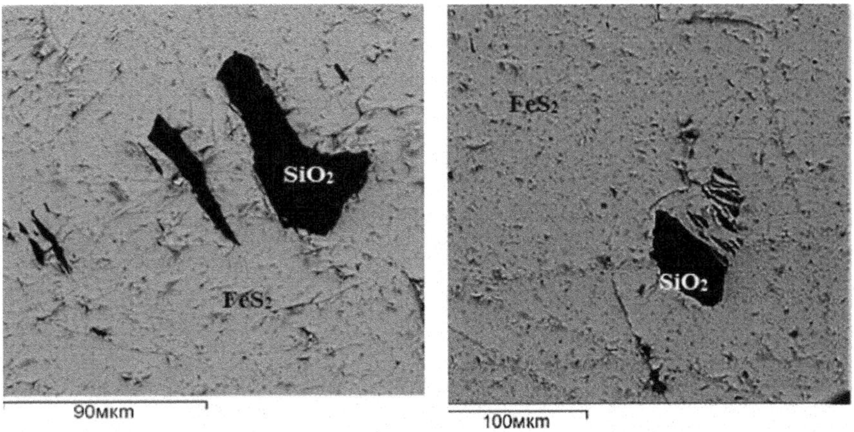

Fig. 3 The SEM image of quartz inclusions in pyrite

Table 2 Chemical composition of ilmenite

Mineral	Major element content					Sum
Ilmenite	O	Ti	Mg	Fe	Co	100.00
	30.57	31.91	0.97	36.42	0.13	

However, it is known that titanium precipitates from acidic solutions in places where environmental pH is locally increased.

Ilmenite composition is shown in Table 2.

Elements included in ilmenite are constant components of magmatic rocks. This fact once more confirms the suggestion that their presence in the pyrite inclusions in brown coals is due to the presence of igneous rocks, which affect the content of mineral nourishment of peats.

Conclusions

To sum up the results of our study, we may conclude that pyrite formation occurred through several stages. The pyrite aggregates filling the cavities and fractures in coals are the most widespread. These aggregates are formed because of penetration of iron containing solution with the help of sulfate reducing bacteria. Inclusions of quartz and ilmenite are associated with the conditions of mineral nourishment of peats as well as with the composition of quartz containing host rocks and traps that neighbor the coals.

References

Finkelman R.B. Modes of occurrence of trace elements in coal. U.S. Geol. Surv. Open File Rep. 1981; 81–99, 322 p.

Geologiya mestorozhdenii uglya i goryuchikh slantsev SSSR. Tom 8. M.: Nedra; 1964. [The Geology of Coal and Pyroshale Depositories of USSR. Vol. 8. Moscow: Nedra; 1964. (In Russ.)].

Hlatshwayo T.B., Matjie R.H., Li Z., Ward C.R. Mineralogical characterization of Sasol feed coals and corresponding gasification ash constituents. Energy Fuels. 2009;23:2867–2873. https://doi.org/10.1021/ef8010806.

Il'enok S.S., Arbuzov S.I. Mineral'nye formy redkikh elementov v uglyakh i zolakh uglei Azeiskogo mestorozhdeniya Irkutskogo ugol'nogo basseina. Izvestiya Tomskogo politekhnicheskogo universiteta. Inzhiniring georesursov. 2016; 327(2): 6–20. [Ilenok S.S., Arbuzov S.I. Mineral modes of rare elements in coals and ashes of coals of Azey deposit in Irkutsk coal basin. Bulletin of the Tomsk Polytechnic University. Geo Assets Engineering. 2016; 327(2): 6–20. (In Russ.)].

Kizilshtein L.Ya., Nastavkin A.V. Iron sulfides in mudstones within the carbonaceous sequence of Donets basin. Lithology and Mineral Resources. 2003; 38(1): 31–35.

Matjie R.H., Li Z., Ward C.R., French D. Chemical composition of glass and crystalline phases in coarse coal gasification ash. Fuel. 2008; 87:857–869. https://doi.org/10.1016/j.fuel.2007.05.050.

Matjie R.H., French D., Ward C.R., Pistorius P.C., Li Z. Behaviour of coal mineral matter in sintering and slagging of ash during the gasification process. Fuel Process. Technol. 2011;92: 1426–1433. https://doi.org/10.1016/j.fuproc.2011.03.002.

Yudovich Ya.E., Ketris M.P. Tsennye elementy-primesi v uglyakh. Ekaterinburg: UrO RAN; 2006. [Yudovich Ya.E., Ketris M.P. Valuable trace elements in coal. Ekaterinburg: UrO RAN [The Urals Division of the Russian Acad. Sci.]; 2006. (In Russ.)].

Dimensions of Atoms in a Crystal: Delusions Versus Reality

Nikolai N. Eremin

Abstract In the review, different systems of atomic radii in a crystal are considered: atomic, ionic, metallic, tetrahedral, crystal (bonded) and their applicability boundaries. Present-day methodical errors in the use of these systems are shown.

Keywords Atomic radius · Crystal radius · Ionic radius · Electron density distribution · Educational crystal-chemical designer

Introduction

The relevance of this topic is due to the fact that, even in the 21-st century, numerous attempts to improve and refine the system of radii of atoms and ions have expanded the scope of their application and made them more universal or, on the contrary, more specific. However, there are many arguments in favor of those concepts that consider the size of an atom in a given valence state as a function of its environment in a crystal, i.e. proceed from the premise that the radius is not a definite constant of the atom. In fact, during the atom transition from a free state to that in a crystal, all its properties (including dimensions) change radically. These changes are caused by the formation of a chemical bond and the interaction of all the atoms in crystal structure with each other.

Discussion

After the discovery of Laue (1912) over a number of years, several tens of crystals were studied by X-ray diffraction analysis, which allowed W. L. Bragg to determine the radii of atoms in crystals as early as in 1920. But in its system, the dimensions

N. N. Eremin (✉)
Department of Geology, Lomonosov Moscow State University, Moscow, Russia
e-mail: neremin@geol.msu.ru

S. Votyakov et al. (eds.), *Minerals: Structure, Properties, Methods of Investigation*, Springer Proceedings in Earth and Environmental Sciences,
https://doi.org/10.1007/978-3-030-00925-0_6

of the electronegative atoms looked considerably smaller in comparison with those of the electropositive elements, which contradicted the idea of the ionic model.

Therefore, Bragg's system of atomic radii was abandoned for several decades. The idea of atomic radii was reconstructed many years later, when the system of Slater atomic radii appeared (Slater 1964). It makes sense to use this system in cases when atoms are connected by covalent or metallic bonds. The accuracy of this system is not very high, but it is possible to achieve higher accuracy by applying the principle of additivity to a narrower group of compounds and constructing a system of radii for it only. Thus, systems of covalent, tetrahedral and metallic radii were obtained.

The electron density distribution in ionic crystals differs significantly from that in covalent or metallic ones; it is characterized by the overlap density shift to a more electronegative atom, and also by the presence of a minimum of the electron density along the bonding line.

In the 1920s, there were methods for determining the radius of ions in a crystal, using some assumptions of a theoretical or semiempirical nature (A. Lande, L. Pauling, V. M. Goldschmidt). Remarkable was the close coincidence of independent systems of ionic radii. This fact made a great impression on contemporaries and subsequent generations of scientists. So, there was a persistent idea in the world that the concept of "ion radius" reflected some objective reality. Gradually, the point of view of the first ionic radii systems creators, who considered them only as some effective quantities, was forgotten. Pauling proclaimed that "… the electron distribution function in the ion has no definite boundary… Therefore, it is impossible to assign to the ion some characteristic size".

At present, the most substantiated and detailed is the ionic radii system of Shannon and Prewitt published first in 1969 (Shannon and Prewitt 1969) and repeatedly refined subsequently on the basis of numerous experimental data for ionic compounds. They estimated the average error of their ionic radii system for interatomic distances predicting about 0.02 Å in ionic crystals. This system has gained so wide popularity and scope that it is considered one of the citation leaders in the modern crystal-chemical literature.

Unfortunately, nowadays, the presence of numerous systems of radii of atoms and ions in crystals leads to serious methodological errors and confusion. For example, rather popular among the school chemical laboratories constructor "The nanoscale model of the crystal structures" Pantuxov et al. registers the size of atoms in all crystal structures without exception only by Slater's atomic radii. Obviously, it is permissible only for simple substances, covalent and metallic structures. For a large number of inorganic crystals (oxides, silicates, phosphates, sulfates, etc.), this approach is wrong and even dangerous, because the actual sizes of atoms and ions in such essentially ionic structures differ significantly from their sizes in simple substances. Visual perception of the developed model by students (for example MgO-periclase) from the authors' library (Fig. 1), can instill in the school audience the idea that the cation (magnesium) in the structure is substantially larger than the anion (oxygen). Nevertheless, the great amount of theoretical and experimental

Fig. 1 Crystal structure of MgO periclase model construction with the help of the educational designer "The nanoscale model of the crystal structures" instills in students incorrect ideas about the relative sizes of atoms in the crystal

material accumulated by crystal chemistry during the 20th and early 21st centuries makes it possible to confidently claim the opposite.

Is there a way to experimentally determine the size of atoms in a crystal nowadays? Of course, yes! Due to the fact that X-rays scatter on electrons, the precise X-ray diffraction analysis can also provide information on the electron density distribution, including the valence electron distribution in the crystal. The distance from the atomic nucleus to the electron density minimum is often called "crystal radius". Another synonym for this word is the term "bonded radius" (Gibbs et al. 1992). Note that it was Gibbs and his co-authors who conducted the most intensive and comprehensive studies of the problem of the real size of atoms in molecules and crystals at the turn of the century. This scientific group used two independent approaches in parallel: theoretical analysis of the results of precise X-ray electron density determination of crystals and *ab-initio* calculations of the electronic structure of molecular fragments (or crystal clusters).

From Figs. 2 and 3, representing the total electron density maps of SiO_2 coesite and NaCl halite, it is easy to determine effective crystal radii from distinct electron density minima. As can be seen, the crystal radii are not equal to the atomic and ionic radii. Moreover, they differ in different directions in the crystal (especially for coesite). Note that the electron density distribution investigations often revealed deviations of the ions in the crystal from sphericity.

The deviations are particularly strong for the ions of transition metals, which can be explained by the crystal field influence on the *d*-orbitals symmetry, as well as the complex nature of the *d*-orbitals splitting in the Coulomb field of ligands, and by the significant deviation of the chemical bond character from the purely ionic one. Numerous experiments have shown that even in one crystal it is possible to detect different sizes of an atom in various bonds and coordination polyhedra. For example, Besancon in 1982 showed that the radii of oxygen and sulfur can vary by up to 0.3 Å in the same structure.

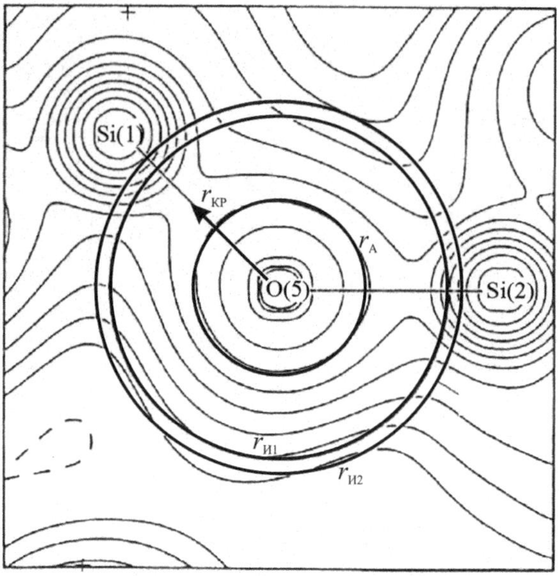

Fig. 2 Total electron density map of SiO_2 coesite according to (Gibbs et al. 1992). r_{KP}—crystal radius O, r_A atomic radius, $r_{И1}$ and $r_{И2}$—ionic radii (in different systems)

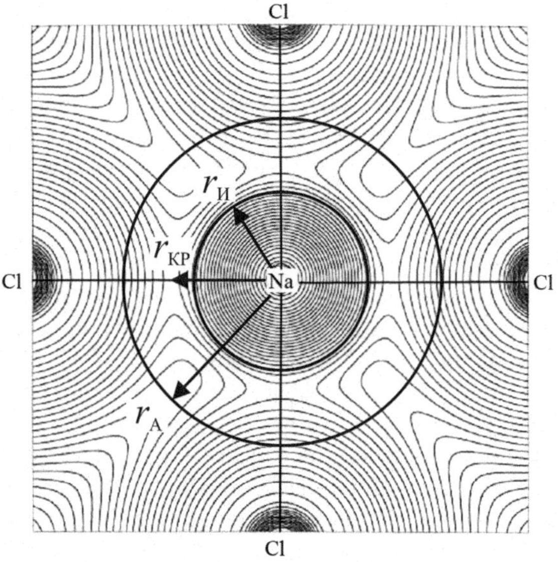

Fig. 3 Total electron density map of halite NaCl according to (Gibbs et al. 1992). The arrows mark: r_{KP}—crystal radius of Na, r_A—atomic radius and $r_И$—ionic radius

Conclusions

The question arises whether the idea of atomic radii should be preserved, if there is no doubt now that the size of the atom in heteroatomic compounds can be somewhere between two limiting values: its radius for a hypothetical purely ionic state (ionic radius) and covalent (or metallic) radius. The answer to this question is obviously positive, because the real interatomic distance depends little on the choice of this or that system (in the first approximation, it does not depend).

The total and complete transition to the variable radii concept will be associated with the loss of the most attractive side of the idea of using dimension systems for the ability to predict many thousands or even millions of distances between atoms in crystals using a relatively small set of numbers (on the order of the number of chemical elements).

Therefore, up to this point, A.E. Fersman's statement remains true: "… whatever the physical meaning of the ion radii… they are of great practical importance as quantities with which it is easy and simple to operate both in crystal chemistry and in geochemistry…".

It is necessary, of course, to be aware of the limits of the applicability of such methods and not to introduce in them an intrinsic physical meaning or to require from them an accuracy, which they are unable to provide by their nature.

References

Besancon P. Determination des rayons loniques absolus et des charges ioniques relatives darts les structures cristallines. Acta Cryst. 1982; B38: 2379–2388.

Gibbs GV et al. Bonded and promolecule radii for molecules and crystals. Amer. Miner. 1992; 77: 741–750.

Pantuxov CV et al. Crystal structure library. https://vk.com/club143236905.

Shannon RD., Prewitt CT. Effective ionic radii in oxides and fluorides. Acta Cryst. 1969; B25: 925–946.

Slater JC. Atomic Radii in Crystals. Journal of Chemical Physics 1964; 41: 3199–3204.

Determination of Major (Ca, P) and Trace (Na, Mg, Al, Si, K, Fe) Elements in Biogenic Apatite Using Inductively Coupled Plasma Atomic Emission Spectroscopy (ICP-AES)

Anastasiya K. Fokina, Daria Kiseleva and Nadezhda V. Cherednichenko

Abstract The major and trace elements in fossil phosphate biominerals are able to retain information both about the environment during the life of the organism and the conditions, time and scale of diagenetic changes in the palaeoenvironmental signal during its burial. The study of the macro- and microelement composition of bone and dental tissues of humans and animals is extremely important in biomedical, ecological, archaeological, palaeoecological, palaeontological and taphonomical studies. Inductively coupled plasma atomic emission spectroscopy (ICP-AES) and mass spectrometry (ICP-MS) are widely applied in the analysis of the bioapatite elemental composition. In the present article the methodology for determining trace (Na, Mg, Al, Si, K, Fe) as well as major (Ca, P) elements by ICP-AES in biogenic apatite samples of bone and dental tissue of fossil and contemporary humans and animals has been described. The sample dissolution scheme is proposed using HNO_3 and H_2O_2 upon heating that provides solid apatite transfer into solution and its separation from the organic matrix with minimal contamination from the reagents used. Its accuracy and correctness is confirmed by the NIST SRM 1400 standard reference material of bone ash. Using the developed technique, a number of samples of contemporary human bone tissue, as well as fossil bone and dental tissue of humans and animals from the Bronze Age fortified settlements and burials of the Southern Urals have been studied.

Keywords Bioapatite · Bone and teeth · Major and trace element composition · ICP-AES

A. K. Fokina (✉) · D. Kiseleva · N. V. Cherednichenko
A.N. Zavaritsky Institute of Geology and Geochemistry, Ural Branch of the Russian Academy of Sciences, 15 Vonsovskogo Street, Ekaterinburg, Russia
e-mail: asiy.fokina93_93@mail.ru

D. Kiseleva
e-mail: kiseleva@igg.uran.ru

A. K. Fokina
Ural Federal University, 19 Mira Street, Ekaterinburg, Russia

S. Votyakov et al. (eds.), *Minerals: Structure, Properties, Methods of Investigation*, Springer Proceedings in Earth and Environmental Sciences,
https://doi.org/10.1007/978-3-030-00925-0_7

Introduction

Human and animal hard tissues (bones, teeth) consist of mineral and organic phases: enamel contains up to 2 wt. % of organic matter, dentin up to 30 wt. % and the bone tissue up to 35 wt. % (Pan and Fleet 2002). The mineral component of hard tissues consists of carbonate-hydroxyapatite, the apatite-group mineral of biogenic origin. Its unique ability to accumulate trace impurities from the environment is provided by the wide isomorphic capacity of the apatite structure, which is manifested through the substitution of calcium for two- and trivalent cations of impurities, and the phosphate and hydroxyl-halogen groups for other anionic groups, as well as by its non-stoichiometric calcium-deficient composition and micro- and nanosized crystals (Pan and Fleet 2002).

The study of the macro- and microelement composition of bone and dental tissues of humans and animals is extremely important in biomedical, ecological, archaeological, palaeoecological, palaeontological and taphonomical studies (Elliott 2002). The major and trace elements in phosphate biominerals are able to retain information, firstly, about the environmental parameters during the organism's life, such as temperature, the consumed food and water composition as well as its place in the trophic (food) chain, and, secondly, about the conditions, time and scale of diagenetic changes in the palaeoenvironmental signal (Reynard and Balter 2014).

For example, calcium (25–36 wt.%) and phosphorus (11–18 wt.%) are the structure forming elements of biogenic phosphate of bone and dental tissue, and the degree of its deviation from stoichiometric composition can be estimated using their molar ratio (Ca/P ratio equal to 1.67 corresponds to synthetic hydroxyapatite) (Gross and Berndt 2002), while sodium, potassium, and magnesium (tenths of a percent) are the most significant electrolytes in the body's liquid media and iron incorporated into the tissues at even lower levels (hundredths of a percent) is the most important essential microelement (Burguera and Burguera 2009).

Due to its high sensitivity, a wide dynamic range of the calibration curve and the ability to determine many elements simultaneously, atomic spectroscopy and mass spectrometry methods have found wide application in the analysis of bioapatite elemental composition, in particular inductively coupled plasma mass spectrometry and atomic emission spectrometry (ICP-MS and ICP-AES). ICP-MS is an unrivalled method for determining bioapatite trace composition (when element concentrations do not exceed 100 μg/g (Burguera and Burguera 2009)), but in order to determine higher concentrations it is necessary to employ the repeated dilution of the sample, which leads to analytical error increase. In this case, the ICP-AES method is preferable, since it allows for the determination of higher concentrations without dilution. In addition, ICP-AES makes it possible to analyse such elements as Ca, Fe, K that appear challenging for quadrupole ICP-mass-spectrometry due to argon isobaric interferences in a rather wide concentration range (from hundredths to tens of percent by weight) (Kiseleva et al. 2017).

Nevertheless, the successful application of the mentioned methods greatly depends on an appropriately chosen and effective dissolution method, which provides quantitative transfer of solid bioapatite sample into the solution, as well as the removal of its organic constituent. Thus, the problem can be solved by selecting an optimal dissolution technique, which involves the choice of a suitable solvent and decomposition conditions. Various methods of open and closed (in autoclaves) decomposition by means of highly oxidising acids or their mixtures, i.e. HNO_3 nitric acid (sometimes combined with H_2O_2 hydrogen peroxide) or $HClO_4$ perchloric acid, usually with the complementary effect of heating or microwave radiation, are used for this purpose (Burguera and Burguera 2009). Thus, for example, an open decomposition using HNO_3 with H_2O_2 added under heating was used for the dissolution of conodont elements (fluorapatite), as well as human dental tissues, while $HClO_4$ was used for layer-by-layer dissolution of human tooth enamel followed by transition into nitrates (Votyakov et al. 2007; Ron' et al. 2012). A mixture of H_2O_2+HNO_3+HF was used for microwave decomposition of bone remains of Quaternary small mammals extracted from the sediments in the karst cavities of the Urals (Smirnov et al. 2009).

The aim of the study was to develop a methodology for determining trace Na, Mg, Al, Si, K, Fe as well as major Ca, P elements by ICP-AES in biogenic apatite samples of fossil and contemporary human and animal bone and dental tissue. To achieve the goal, the following tasks were set:

1. To select a sample dissolution scheme that would ensure the quantitative transfer of biogenic apatite into solution and its separation from the organic matrix with minimal contamination from the reagents used, and analyse the major and trace elements by ICP-AES and ICP-MS;
2. To select the operating parameters of the ICP-AES spectrometer for the analysis of Na, Mg, Al, Si, K, Fe, Ca, P;
3. To validate the developed methodology using the reference material;
4. To perform the decomposition and ICP-AES analysis of contemporary and fossil bone and dental samples using the developed technique.

Materials and Methods

Samples. A standard reference material of bone ash (NIST SRM 1400) was used to test the dissolution procedure and ICP-AES analysis. The samples of contemporary human bone tissue (provided by the clinic of the Ural State Medical University), as well as the samples of fossil bone and dental tissue of humans and animals from the Bronze Age fortified settlements and burials of the Southern Urals (provided by the Institute of History and Archaeology and the Institute of Plant and Animal Ecology, UB RAS) were investigated.

Sample preparation. The samples were prepared and measured in the 6 and 7 ISO Class cleanroom facility of the IGG UB RAS. Ultrapure deionised water (18.2 MΩ cm) was used at all stages of the analysis. The labware used was made of either PTFE or polypropylene. Nitric acid was further purified by sub-boiling distillation (Berghof, Germany). Hydrogen peroxide H_2O_2 did not make a significant contribution to the blanks.
For the elemental analysis, NIST SRM 1400 as well as bone and dental samples with masses 50–100 µg were previously dried during 2 h at 105 °C and placed into 50 cm^3 polypropylene containers. A mixture of 14M HNO_3 and H_2O_2 was used to dissolve the samples. 0.5 ml of 14M HNO_3 and 0.1 ml of H_2O_2 were added to the sample and left in a fume hood for 24 h. The covers were not closed tightly for the gas evolution reaction to be completed. After 24 h, the samples were heated on a water bath for 30 min. The vapour evolution was observed, which indicated partial removal of organic matter. In order to completely remove the organic component, the samples were cooled in a fume hood and boiled on a water bath for another 30 min after adding 15 ml of 0.5% HNO_3. After this, the samples were cooled down to room temperature, diluted to volume of 50 cm^3 with ultrapure water and proceeded for measurement. In the case of subsequent combined measurement of the elemental composition by ICP-AES and ICP-MS, indium (an element of the internal standard for ICP-MS) with a content of 10 µg/l was added to the containers. For measurements of calcium and phosphorus, the final solutions were diluted 200 times.

Elemental composition measurement. The measurements were carried out by the PerkinElmer Optima 8000 ICP-AES spectrometer. Optimised operating parameters for the Optima 8000 ICP-AES spectrometer are presented in Table 1.

The determined elements with the analytical lines selected are presented in Table 2.

PerkinElmer multi-element standard solutions with different concentrations were used to obtain the calibration curves: Al, Fe, K, Mg, Na and Si—0.1; 1; 10 mg/l; P and Ca—1; 10; 100 mg/l.

Table 1 Operating parameters for the Optima 8000 ICP-AES spectrometer

Parameter	Parameter value
RF Power, W	1200
Sample gas, l/min	0.7
Auxiliary gas flow, l/min	0.2
Plasma gas flow, l/min	8
Sample flow, ml/min	1.5
Spectrum registration	By area
Plasma observation	Radial/axial
Nebulizer	Cross-flow
Spray Chamber	Scott chamber (Ryton)

Table 2 Analytical lines for ICP-AES determination of elements in biogenic apatite

Analyte	Analytical line, nm		
	1	2	3
Al	394.401	396.153	
Fe	238.204	259.939	
K	766.490	769.896	
Mg	279.077	285.213	
Na	588.995	589.592	
Si	251.611	252.851	288.158
P	213.617	214.914	
Ca	315.887	317.933	

Results and Discussion

The results of graphical comparison between the measured and certified concentrations in NIST SRM 1400 standard reference material of bone ash are indicated in Fig. 1.

As can be seen from Fig. 1, the measured concentration values are in a good agreement with the certified values. Thus, the developed technique provides for the quantitative transfer of biogenic apatite into solution and separation from the organic matrix, reduces cross-contamination, and is suitable for the subsequent

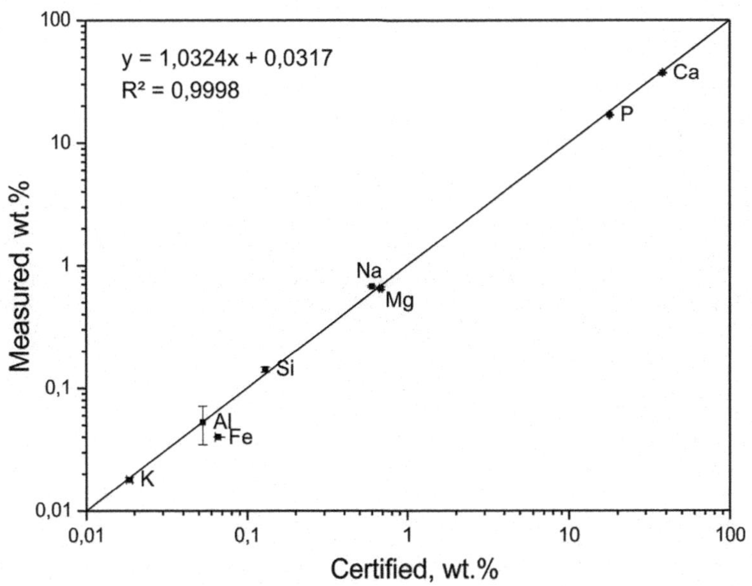

Fig. 1 The results of graphical comparison between the measured and certified elemental concentrations in NIST SRM 1400 standard reference material of bone ash

determination of the major and trace element composition by ICP-AES and ICP-MS.

Using the developed technique, a number of samples of contemporary human bone tissue, and those of fossil bone and dental tissue of humans and animals from the Bronze Age fortified settlements and burials of the Southern Urals have been studied. The results have given new data in the field of biomedicine and archaeology.

Conclusions

The methodology for determining Na, Mg, Al, Si, K, Fe as well as Ca and P by ICP-AES in biogenic apatite samples of fossil and contemporary human and animal bone and dental tissue has been developed. Its accuracy and correctness has been confirmed by the NIST SRM 1400 standard reference material of bone ash. The technique can be used to analyse bone tissue of contemporary humans, as well as fossil bone and dental tissues of humans and animals from fortified settlements and burial grounds of the Southern Urals in the Bronze Age.

Acknowledgements The work was carried out at the UB RAS "Geoanalitik" Center for Collective Use within the state assignment of IGG UB RAS № AAAA-A18-118053090045-8, supported in part by RFBR (project No. 17-05-00618).

References

Burguera JL, Burguera M. Recent on-line processing procedures for atomic samples for determination of trace elements by atomic spectrometric methods. Spectrochimica Acta Part B. 2009;64:451–458.

Elliott JC. Calcium phosphate biominerals. In: Kohn MJ, Rakovan J. & Hughes JM, editors. Reviews in Mineralogy and Geochemistry, vol. 48: Phosphates: Geochemical, geobiological, and materials importance. Washington, DC: Mineralogical Society of America; 2002. p. 427–453.

Gross KA, Berndt CC. Biomedical application of apatites. In: Kohn MJ, Rakovan J. & Hughes JM, editors. Reviews in Mineralogy and Geochemistry, vol. 48: Phosphates: Geochemical, geobiological, and materials importance. Washington, DC: Mineralogical Society of America; 2002. p. 633–672.

Kiseleva DV, Cherednichenko NV, Deryugina LK, Pinigina MA. Vozmozhnosti metoda atomno-emissionnoi spektroskopii s induktivno svyazannoi plazmoi (ICP-AES) pri analize gornykh porod i mineralov. Trudy Instituta geologii i geokhimii im. akademika A.N. Zavaritskogo. 2017;164: 299–302. [Kiseleva DV, Cherednichenko NV, Deryugina LK, Pinigina MA. Possibilities of inductively coupled plasma atomic emission spectroscopy (ICP-AES) in the analysis of rocks and minerals. Proceedings of the A.N. Zavaritsky Institute of Geology and Geochemistry. 2017;164:299–302. (in Russ.)].

Pan Y, Fleet M. Compositions of the Apatite-group Minerals: Substitution mechanisms and controlling factors. In: Kohn MJ, Rakovan J. & Hughes JM, editors. Reviews in Mineralogy and Geochemistry, vol. 48: Phosphates: Geochemical, geobiological, and materials importance. Washington, DC: Mineralogical Society of America; 2002. p. 13–50.

Reynard B, Balter V. Trace elements and their isotopes in bones and teeth: Diet, environments, diagenesis, and dating of archeological and paleontological samples. Palaeogeogr Palaeoclimatol Palaeoecol. 2014; 416:4–16.

Ron' GI, Votyakov SL, Mandra YuV, Kiseleva DV. Morfologicheskie struktury tverdykh tkanei zubov cheloveka. Ekaterinburg: UGMA; 2012. [Ron' GI, Votyakov SL, Mandra YuV, Kiseleva DV. Morphological structures of human dental hard tissues. Ekaterinburg: UGMA; 2012. (in Russ.)].

Smirnov NG, Votyakov SL, Sadykova NO, Kiseleva DV, Shchapova YuV. Fiziko-khimicheskie kharakteristiki iskopaemykh kostnykh ostatkov mlekopitayushchikh i problema otsenki ikh otnositel'nogo vozrasta. Chast' 1. Termicheskii i mass-spektrometricheskii elementnyi analiz. Ekaterinburg: Goshchitskii; 2009. [Smirnov NG, Votyakov SL, Sadykova NO, Kiseleva DV, Shchapova Yu.V. Physicochemical characteristics of fossil bone remains of mammals and the problem of assessing their relative age. Part 1. Thermal and mass spectrometric element analysis. Ekaterinburg: Goschitsky; 2009. (in Russ.)].

Votyakov SL, Kiseleva DV, Rozanov AYu, Cherednichenko NV, Ushatinskaya GT. Mul'tielementnyi mass-spektrometricheskii mikroanaliz v issledovaniyakh biomineral'nykh obrazovanii. Litosfera. 2007;1:123–137. [Votyakov SL, Kiseleva DV, Rozanov AYu, Cherednichenko NV, Ushatinskaya GT. Multielement mass-spectrometric microanalysis in the studies of biomineral formations. Lithosphere (Russia). 2007;1:123–137 (in Russ.)].

Sulfide Minerals in Ordinary Chondrites Chelyabinsk, Northwest Africa 869 and Gao-Guenie

Svetlana S. Hontsova, Elena M. Maksimova, Evgeniya V. Petrova, Irina A. Danilenko, Igor A. Nauhatsky and Grigoriy A. Yakovlev

Abstract A comparative study of sulfide minerals in ordinary chondrites Chelyabinsk, Northwest Africa 869 and Gao-Guenie was carried out using X-ray diffraction, optical microscopy and scanning electron microscopy. Unit cell parameters were obtained for minerals: troilite, pyrrhotite, and pentlandite. A small amount of copper was revealed in LL chondrite Chelyabinsk, while its content in the metal and sulfide grains of H and L chondrites was below detection limit. Unit cell parameters for sulfides from the Chelyabinsk and Northwest Africa 869 chondrites were obtained.

Keywords Ordinary chondrites · Sulfide minerals · X-ray diffraction · Scanning electron microscopy

Introduction

Sulfur is one of the most abundant elements in the Solar system materials. Sulfide minerals are present in a wide range of meteorites with varied metamorphic histories (Schrader and Lauretta 2010). The troilite and pyrrhotite are the main sulfide phases in meteorite, which sometimes occur in combination with other sulfides, such as pyrite, pentlandite and chalcopyrite (Rubin 1997).

Ordinary chondrites are considered to be almost undifferentiated substance, which retains information about the early processes of the solar system matter

S. S. Hontsova (✉) · E. M. Maksimova · I. A. Nauhatsky
V.I. Vernadsky Crimean Federal University, 295007 Vernadsky ave. 4, Simferopol
Russian Federation
e-mail: sgoncova@gmail.com

E. M. Maksimova
e-mail: maksimovaem@cfuv.ru

E. V. Petrova · I. A. Danilenko · G. A. Yakovlev
Ural Federal University, 620002 Mira Str. 19/5, Ekaterinburg, Russian Federation
e-mail: evgeniya.petrova@urfu.ru

S. Votyakov et al. (eds.), *Minerals: Structure, Properties, Methods of Investigation*, Springer Proceedings in Earth and Environmental Sciences,
https://doi.org/10.1007/978-3-030-00925-0_8

formation. This group of meteorites is divided into three subgroups H, L, LL, which differ in the content of iron and other siderophile elements (H > L > LL), and in relation of the amount of oxidized iron to metal (LL > L > H) (Dodd 1981).

Materials and Methods

In this paper the structural and elemental were studied composition of sulfide minerals of the three ordinary chondrites of different groups:

1) Chelyabinsk (light lithology), LL5, S4, W0, (Ruzicka et al. 2015).
2) Northwest Africa 869 L4-6, S3, W1, (Connolly et al. 2006).
3) Gao–Guenie, H5, (Grossman 1999).

The phase composition in meteorite samples was investigated by X-ray diffraction (XRD) using diffractometer with copper monochromatic radiation by powder diffraction technique.

The surface morphology of the samples was investigated using inverted optical microscope Axiovert 40 MAT and scanning electron microscope FE-SEM ZEISS ΣIGMA VP with an energy dispersive X-ray analysis system INCA Energy (Oxford Instruments). The samples were cut and sections were polished with diamond pastes according to the standard method.

Results and Discussion

The XRD analysis has revealed the following sulfide phases in ordinary chondrites Chelyabinsk and Northwest Africa 869: troilite FeS, pyrrhotite $Fe_{1-x}S$ and pentlandite $(Fe, Ni)_9S_8$. For the detected sulfide minerals, structural parameters have been determined, Tables 1, 2 and 3. Their structural parameters have been

Table 1 Structural parameters of troilite (P 6(-)2c) in the samples studied

Meteorite	a, Å	c, Å	V, $Å^3$
Chelyabinsk	5.939 ± 0.001	12.034 ± 0.001	367.71 ± 0.13
NWA 869	5.979 ± 0.001	11.725 ± 0.001	363.08 ± 0.04

Table 2 Structural parameters of pyrrhotite (P 6_3/mmc) in the samples studied

Meteorite	a, Å	c, Å	V, $Å^3$
Chelyabinsk	3.429 ± 0.001	5.679 ± 0.001	57.87 ± 0.01
NWA 869	3.427 ± 0.001	5.681 ± 0.001	57.79 ± 0.01

Table 3 Structural parameters of pentlandite (Fm3 m) in the samples studied

Meteorite	a, Å	V, $Å^3$
Chelyabinsk	10.156 ± 0.001	1047.49 ± 0.16
NWA 869	10.163 ± 0.001	1050.01 ± 0.15

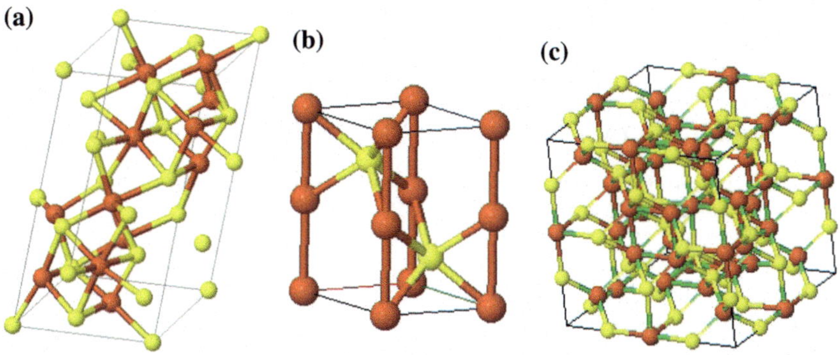

Fig. 1 The unit cells of the sulfide minerals (a) troilite; (b) pyrrhotite; (c) pentlandite. S and Fe atoms are represented by yellow and red spheres, respectively

calculated using the UnitCell program, and the unit cells of the sulfide minerals have been built using the Jmol program, Fig. 1 (Bilbao Crystallographic Server 2018).

It has been found that unit cell of troilite and pyrrhotite from the ordinary chondrite Chelyabinsk is bigger than for those sulfides from the NWA 869, Tables 1–2. At the same time, unit cell parameters for pentlandite from the Chelyabinsk is smaller than for those from the NWA 869, Table 3. It can be due to a higher amount of Ni in the pentlandite from the NWA 869, while pentlandite from the Chelyabinsk contains less amount of impurity: Ni accompanied with Cu, Table 4. This fact may be due to the thermodynamic and kinetic processes of the formation of sulfide crystalline structures. The difference in the structure and

Table 4 Composition of troilite (wt%) in ordinary chondrites

Meteorite	Fe	S	Cu	Ni	Total
Chelyabinsk	62.50	37.50	–	–	100.00
	62.17	37.22	0.61	–	100.00
	60.84	36.48	1.00	1.68	100.00
NWA 869	62.44	37.56	–	–	100.00
	62.88	37.17	–	–	100.05
Gao-Guenie	63.11	36.89	–	–	100.00
	63.32	36.68	–	–	100.00

– Amount is below detection limit

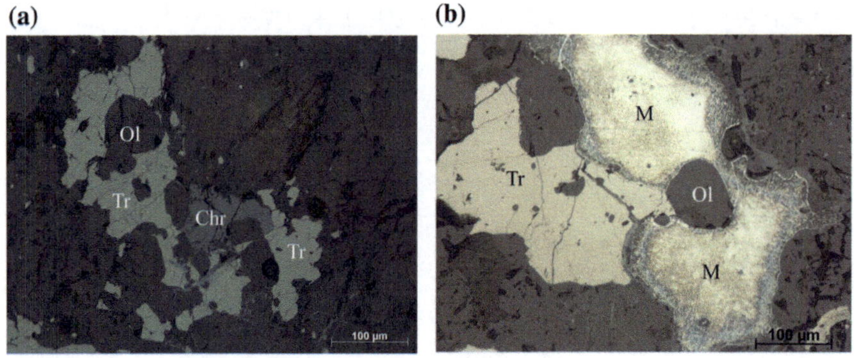

Fig. 2 Troilite inclusions in meteorites (a) Gao-Guenie, (b) NWA 869 (Tr—troilite, M—metal, Ol—olivine, Chr—chromite)

morphology of the sulfide minerals in meteorite fragments is due to the differences in the conditions of their formation.

In all the samples troilite presented both as individual forms and as eutectic with a Fe-Ni alloy, Fig. 2. The troilite composition of NWA 869 and Gao-Guenie without any impurities, Table 4. But the chemical composition of some troilite inclusions in the ordinary chondrite Chelyabinsk slice contains about 1% wt. of Cu. That could be a result from of the lower content of the metal phase in the Chelyabinsk LL5 material in comparison with L and H chondrites, that have a higher amount of metallic iron. Therefore, the less content of metal phase is presented in meteorite the richer in Cu that metal phase.

In all the samples pentlandite is observed as sections in the troilite, Fig. 3. For some metal-troilite grains of Chelyabinsk and Gao-Guenie pentlandite inclusions

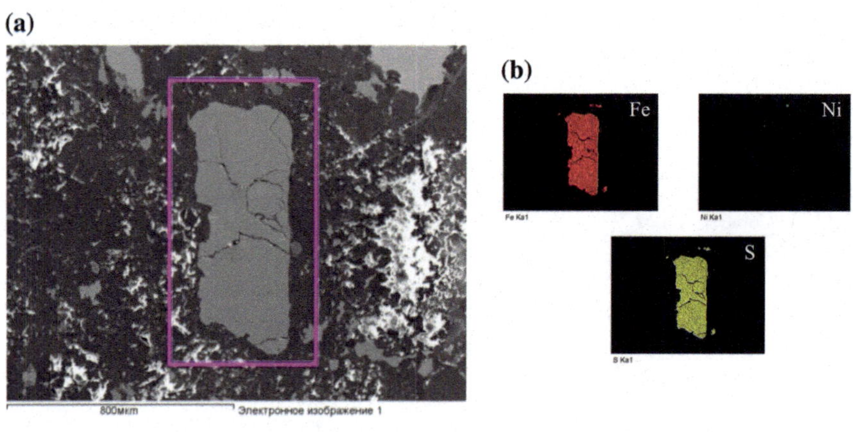

Fig. 3 The structure of the sulfide grain in ordinary chondrite NWA 869 (a) back-scattered electron image; (b) the spatial distribution of Fe, Ni, S within the analyzed area, respectively

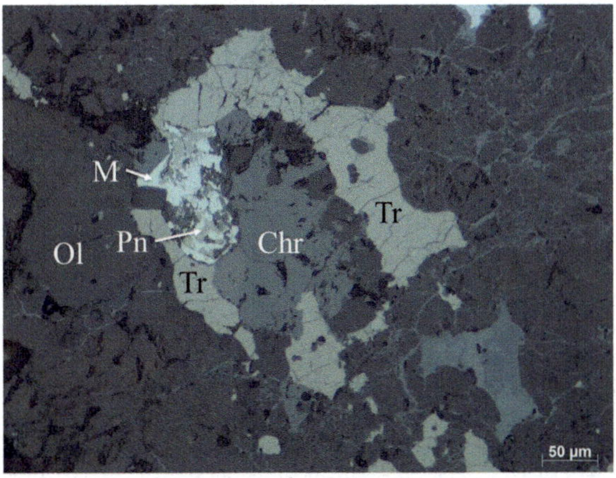

Fig. 4 Metal-sulfide grain in ordinary chondrite Gao-Guenie (Tr—troilite, M—metal, Ol—olivine, Chr—chromite, Pn—pentlandite)

Table 5 Chemical composition of pentlandite (wt%) in the meteorites

Meteorite	Fe	S	Ni	Cu	Total
Chelyabinsk	57.37	28.65	8.26	5.71	99.99
NWA 869	46.68	35.01	18.31	–	100.00
Gao-Guenie	50.63	33.82	15.55	–	100.00

were observed in the metal-sulfide eutectic zones, Fig. 4. According to the classification of minerals (Shushkun et al. 1974), discovered pentlandite is the glandular pentlandite. For the pentlandite from the ordinary chondrite Chelyabinsk as well as for the troilite in Chelyabinsk, copper was registered, Table 5 (Hontsova et al. 2018). Apparently, this is due to the diffusion of copper within the troilite with a subsequent concentration in inclusions of pentlandite.

Conclusions

The investigation mineral composition of ordinary chondrites can give information about of the origin and development of the Solar System.

Formations of glandular pentlandite in the regional part of the troilite grains in the studied samples may be associated with its of nucleation at the latest stages of crystallization at low temperatures. A small amount copper in the chemical composition of pentlandite from the Chelyabinsk ordinary chondrite may be due to the diffusion of copper within troilite grains in the pentlandite inclusions.

The unit cells of sulfide minerals have various values in the different chemical groups of ordinary chondrites. This may be due to the conditions of their formation.

Acknowledgments The authors are grateful to V.I. Nekrasov and to V.I. Grokhovsky for the samples. This work was supported in part by the Development Program of the Crimean Federal University under the project "Academic Mobility of Young Scientists of Russia—AMMUR" in the Ural Federal University, by the Ministry of Education and Science (Project # 5.3451.2017/4.6), by the Act 211 of the Government of the Russian Federation, agreement no. 02.A03.21.0006.

References

Bilbao Crystallographic Server. University of the Basque Country, Biscay http://www.cryst.ehu.es/cryst/visualize.html Accessed 30 Apr 2018.

Connolly HC., Zipfel J., Grossman JN., Folco L., Smith C., Jones RH., et al. The Meteoritical Bulletin, No. 90. Meteorit Planet Sci. 2006; 41(9): 1383–1418.

Dodd R. Meteorites: a Petrologic-chemical Synthesis. NY: Cambridge University press; 1981.

Grossman JN. The Meteoritical Bulletin, No. 83. Meteorit Planet Scie. 1999; 34: 169–186.

Hontsova SS., Danilenko IA., Yakovlev GA., Petrova EV., Maksimova EM., Nauhatsky IA. Morphology of sulfide minerals in some ordinary chondrites. AIP Conference Procceedings. 2018; 2015(1): 020029-1–020029-5. https://doi.org/10.1063/1.5055102.

Rubin AE. Mineralogy of meteorite groups. Meteorit Planet Sci. 1997; 32: 231–247.

Ruzicka J., Grossman A., Bouvier C., Herd DK., Agee CB. The Meteoritical Bulletin, No. 102. Meteorit Planet Sci. 2015; 50: 1662.

Schrader DL., Lauretta DS. High-temperature experimental analogs of primitive meteoritic metal-sulfide–oxide assemblages. Geochim Cosmochim Acta. 2010; 74(5): 1719–1733. https://doi.org/10.1016/j.gca.2009.11.030.

Shushkun NN., Karpenkov AM., Kulagov JeA., Mitenkov GA. O klassifikacii mineralov gruppy pentlandita. Doklady Akademii Nauk SSSR 1974; 217(1): 194–197. [Shishkin NN., Karpenkov AM., Kulagov EA., Mitenkov GA. On the classification of minerals pentlandite group. Doklad of the Acad of Sci of the USSR. 1974; 217(1): 194–197. (In Russ.)].

Fracture Micromechanism of the Hierarchically Organized Biocomposite (Dentin of the Human Teeth)

Anna V. Kabanova, Larisa P. Kiselnikova, Zaoli Zhang and Peter E. Panfilov

Abstract Cracks in the thin samples of teenage dentin were examined on the nanoscale (TEM) and microscale (optical microscopy in the transmission light). Cracks in the bioorganic dentin matrix of the teenage dentin on the nanoscale behave like a crack in viscoelastic media, where the plastic zone develops ahead of a crack tip. There are pore-like satellite cracks in the narrow strip ahead of a microcrack tip. These satellite cracks have the tendency to merge with the main crack. The character of crack growth does not depend on the age of teenage dentin, while the angle at the crack tip is governed by the filler of the bioorganic dentin matrix. The presence of both collagen fibres and inorganic component in the dentin matrix does not change the type of its deformation behavior but plays the role of strengthening elements in a viscoelastic media. The mechanism of crack growth on the nanoscale is similar to the mechanism of microcrack development.

Keywords Dentin · Cracks · Fracture

A. V. Kabanova (✉) · P. E. Panfilov
Ural Federal University, Mira 19, Yekaterinburg, Russia
e-mail: gerbers13@mail.ru

P. E. Panfilov
e-mail: peter.panfilov@urfu.ru

L. P. Kiselnikova
A. I. Yevdokimov Moscow State University of Medicine and Dentistry, Vuchetich St. 9a, Moscow, Russia
e-mail: lpkiselnikova@mail.ru

Z. Zhang
Erich Schmid Institute for Materials Science, Jahnstrasse 12, Leoben, Austria
e-mail: zaoli.zhang@oeaw.ac

S. Votyakov et al. (eds.), *Minerals: Structure, Properties, Methods of Investigation*, Springer Proceedings in Earth and Environmental Sciences,
https://doi.org/10.1007/978-3-030-00925-0_9

Introduction

The design of bioinspired materials or biomimetics is a modern trend in materials science (Wegst et al. 2014; Naleway et al. 2015). The choice of a living tissue for the copying is the cornerstone for such composites (Barthelat 2007). Dentin possessing a multilevel hierarchical structure is the hard basis of human teeth. It consists of bioorganic and inorganic components in a volume ratio of 1:1 or a weight ratio of 7:3. Biological tissues have two basic functions. The first function, which could be considered as biological, is nucleation and formation of a biological tissue and its further functioning, including regeneration, in the life cycle of an organism. The second, strength function is supporting the shape of organs of an organism. In comparison with bone, human dentin possesses more simple structure inasmuch as it has no regenerative function. Dentin exhibits contradictory deformation behavior: it behaves like a filled polymer or a viscoelastic solid under uniaxial compression, while its behavior is brittle under tension (Zaytsev et al. 2012; Zaytsev and Panfilov 2014; Panfilov et al. 2016). On the microscopic scale, dentin fracture under tension behaves similar to such viscoelastic solid as plastic metal, because the main crack in dentin grows by merging with the satellite cracks (Nalla et al. 2003; Kruzic et al. 2003, 2005).

The aim of the present paper is to analyze the driving force of this apparent contradiction when this hard tissue demonstrates both viscoelastic and brittle behavior.

Experimental Procedure

Ten intact human permanent premolars of 13–18-year-old patients, which have been extracted according to the medical diagnosis and the Ethics Protocol of Moscow State Medical University (Russia), are used in experiments. Teenage dentin is usually considered as the basic state of this hard tissue inasmuch as the ageing process has not begun. The samples for TEM thin foils were cut normally to the main axis of a tooth from the middle parts of a tooth with the help of a diamond disc saw under water irrigation (thickness of samples was 1 mm). Abrasive papers with different grain sizes were used for mechanical thinning of the samples up to the thickness of 0.15 mm. The samples were chemically thinned by the "window" technique in a jet of concentrated H_3PO_4 acid for 1 h. After that, the foils were washed in the flowing water and dried on air. Thin foils were glued on the standard copper cartridges for the Gatan™ tensile holder that helped prevent the torsion of dentin sample under an electron beam in the TEM column (Panfilov et al. 2017). A TEM/STEM JEOL2100 operating at 200 kV, equipped with an image-side C_S-corrector and an image filter (Tridiem™), was used. The alignment of the C_S-corrector was performed using the CEOS software, based on the aberration measurements deduced from Zemlin tableaus. All TEM images shown here were recorded on a 2k × 4k pixel CCD camera at the resolution of 12k ~ 60k using a very short acquisition time, to keep electron current density as small as possible. No obvious burning of dentin samples was observed in the TEM column due to small electron dose used for imaging. Cracks on the microscopic scale or microcracks in thin samples of teenage dentin were examined with the help of an optical microscope in the transmission light (Panfilov et al. 2017).

Results and Discussion

The morphology of thin foils of teenage dentin was examined on the nanoscale. The foil represents the stack of single layers with the thickness of 100 nm. Hence, human dentin has a layered morphology, as it occurs according to dentinogenesis (Arana-Chavez and Massa 2004). A diffuse halo on the electron diffraction means that the sample is in the amorphous state. There is no inorganic phase in the thin foils. It seems that orthophosphoric acid dissolute calcium hydroxyapatite present in teenage dentin during the thin foil preparation. Therefore, it may be concluded that the object of this TEM study is the bioorganic matrix of the teenage dentin. There are concentric cells having the size of 10–100 nm in every single layer. The boundaries between the cells have the width of about 10 nm. No fine structure was observed inside the boundaries between the dentin cells of 13–14-year-old humans (Fig. 1a). However, the examination of thin dentin foils of 16–18-year-old humans has shown that the boundaries between the cells in the single layer of the dentin bioorganic matrix contain the collagen fibres or they possess a thin structure (Fig. 1b).

Cracks in the dentin thin foils have appeared during the preparation of samples. There are notch-like defects at the edges of thin foils of 13–14 year old dentin on the nano-scale (Fig. 2). They have a V-shape and a blunted tip; the angle at their tips is 20°–30°; their length is 50–100 nm. There is a light strip ahead of some

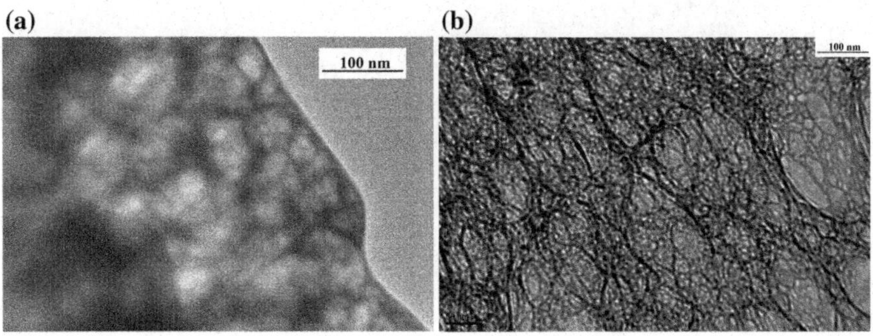

Fig. 1 Microstructure of teenage dentin: **a** the cell structure of 13–14-year-old dentin, **b** collagen fibers

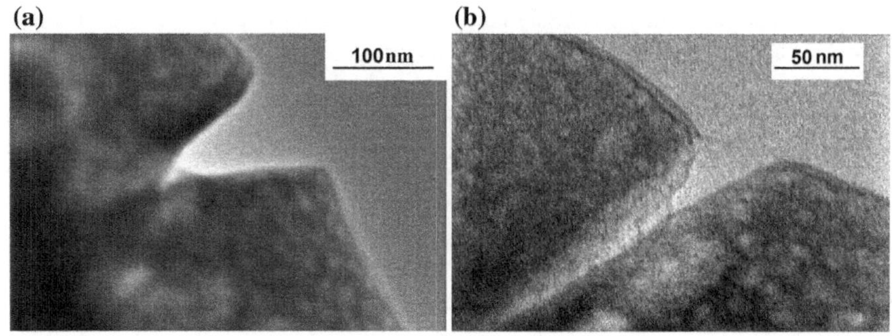

Fig. 2 Notch-like defects (cracks) at the edges of thin foils of a 13–14-year-old dentine

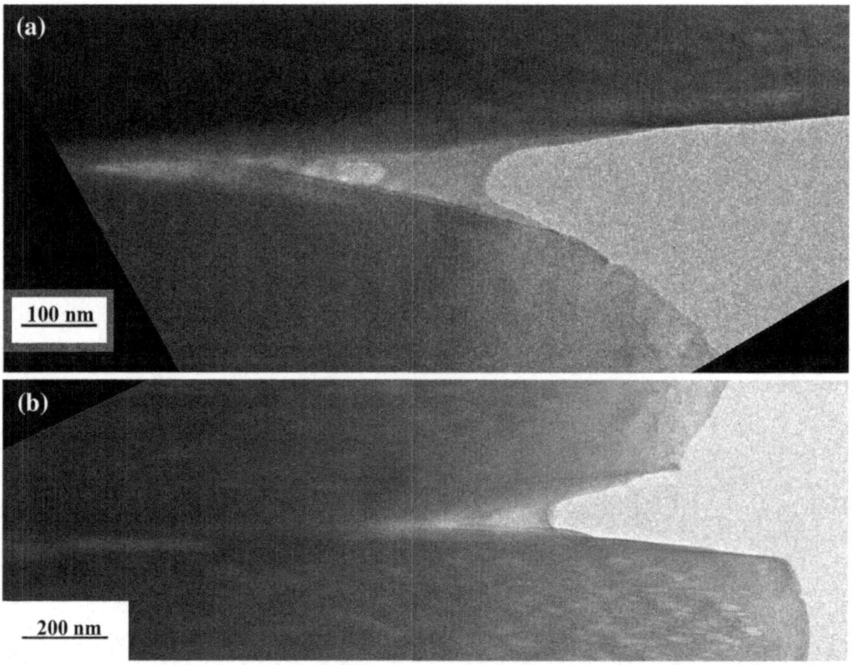

Fig. 3 Cracks in a teenage dentin foil of a 16–18-year-old human: **a** the main crack with satellite cracks at the crack tip and **b** the crack with a localized deformation region

notch-like defects. The angle at the crack tip in thin foils of 16–18 year old dentin decreases down to 10°–15°. The edge of the neighbouring single layer has an elliptic profile that shows the deformation localization in the crack vicinity. There is a light strip containing pore-like satellite cracks ahead the crack tip in thin foils of 16–18 year old dentin (Fig. 3a, b). Such mechanism of crack growth is similar to the fracture of a viscoelastic solid such as a polymer film or a plastic metal. Hence,

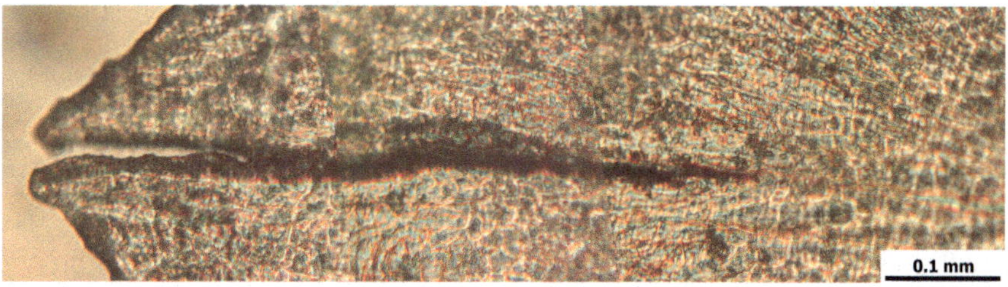

Fig. 4 The main crack with localization region in a teenage dentin foil (optical microscopy, transmission geometry)

the light narrow strip can be qualified as a plastic zone near a crack tip. The angle at the crack tip decreases with an increase of the age of teenage dentin. This effect is caused by collagen fibres in the boundaries between the cells, which play the role of a strengthening element in the bioorganic dentin matrix.

Cracks on the microscopic scale in thin samples of teenage dentin appear at the sites where samples have been bent locally. This is where the deformation is localized. A microcrack in a thin sample of teenage dentin is shown in Fig. 4. It has a V-shape and the angle at the crack tip is less than 5°. There are pore-like satellite cracks in the narrow strip ahead of the microcrack tip. These satellite cracks have the tendency to merge with the main crack. The same type of crack growth is observed in viscoelastic polymer films (Argon 2013). Hence, it may be concluded that the mechanism of microcrack development is similar to the mechanism of crack growth on the nanoscale. The decrease of the angle at the microcrack tip as distinct from the cracks on the nanoscale can be associated with the presence of hydroxyapatite crystallites that play the role of strengthening elements in the samples.

Conclusions

The findings give the basis for the conclusion that the bioorganic dentin matrix is a viscoelastic media, where mechanisms for the accommodation of an elastic energy such as elastic/inelastic deformation and crack growth are active. The deformation behavior of the human dentin is governed by the competition between these channels of stress accommodation. The presence of both collagen fibres and inorganic components in the dentin matrix does not change the type of its deformation behavior, but plays the role of the strengthening elements in a viscoelastic media.

Cracks in human dentin behave like cracks in a viscoelastic media. They appear in the sites of the localization of irreversible deformation, whereas the plastic zone containing pore-like satellite cracks develops ahead of crack tip.

References

Arana-Chavez VE, Massa LF. Odontoblasts: the cells forming and maintaining dentine. The International Journal of Biochemistry and Cell Biology. 2004; 36: 1367–1373. https://doi.org/10.1016/j.biocel.2004.01.006.

Argon A.S. The Physics of Deformation and Fracture of Polymers. Cambridge University Press; 2013. https://doi.org/10.1017/CBO9781139033046.

Barthelat F. Biomimetics for next generation materials. Phil. Trans. R. Soc. A. 2007; 365: 2907–2919. https://doi.org/10.1098/rsta.2007.0006.

Kruzic JJ, Nalla RK, Kinney JH, Ritchie RO. Crack blunting, crack bridging and resistance-curve fracture mechanics in dentin: effect of hydration. Biomaterials. 2003; 24: 5209–5221. https://doi.org/10.1016/S0142-9612(03)00458-7.

Kruzic JJ, Nalla RK, Kinney JH, Ritchie RO. Mechanistic aspects of in vitro fatigue-crack growth in dentin. Biomaterials. 2005; 26: 1195–1204. https://doi.org/10.1016/j.biomaterials.2004.04.051.

Naleway SE, Porter MM, McKittrick J, Meyers MA. Structural Design Elements in Biological Materials: Application to Bioinspiration. Adv. Mater. 2015; 27(37): 5455–76. https://doi.org/10.1002/adma.201502403.

Nalla RK, Kinney JH, Ritchie RO. Effect of orientation on the in vitro fracture toughness of dentin: the role of toughening mechanism. Biomaterials. 2003; 24: 3955–68. https://doi.org/10.1016/S0142-9612(03)00278-3.

Panfilov P, Zaytsev D, Antonova OV, Alpatova V, Kiselnikova LP. The Difference of Structural State and Deformation Behavior between Teenage and Mature Human Dentin. International journal of biomaterials. 2016; 2016. https://doi.org/10.1155/2016/6073051.

Panfilov P, Kabanova A, Guo J, Zhang Z. Transmission electron microscopical study of teenage crown dentin on the nanometer scale. Materials Science & Engineering C. 2017; 71: 994–998. https://doi.org/10.16/j.msec.2016.11.016.

Panfilov PE, Kabanova AV, Borodin IN, Guo J, Zang Z. Fracture Mechanism of Coronal Teenage Dentin. Russian Metallurgy (Metally). 2017; 10: 879–883. https://doi.org/10.1134/S0036029517100172.

Wegst U, Hao Bai, Saiz E, Tomsia A, Ritchie RO. Bioinspired structural materials. Nature Materials. 2014; 14: 23–36. https://doi.org/10.1038/nmat4089.

Zaytsev D, Grigoriev S, Panfilov P. Deformation behavior of human dentin under uniaxial compression. International Journal of Biomaterials. 2012; 2012. Article ID 854539, 8 pages. https://doi.org/10.1155/2012/854539.

Zaytsev D, Panfilov P. Deformation behavior of human dentin in liquid nitrogen: A diametral compression test. Materials Science and Engineering C. 2014; 42: 48–51. https://doi.org/10.1016/j.msec.2014.05.011.

Bedrock Phase Composition of Kamenny Ambar and Konoplyanka Bronze Age Fortified Settlements and Neplyuevsky Burial Ground (Southern Urals)

Sofia V. Karpova, Daria Kiseleva, Anastasiya D. Ryanskaya, Olga L. Galakhova, Evgeny S. Shagalov and Lyudmila N. Koryakova

Abstract Among the main archaeological sites of the Late Bronze Age are the fortified settlements of Kamenny Ambar and Konoplyanka and Nepluyevsky burial ground, located in the basin of the Karagaily-Ayat river (Chelyabinsk region, Southern Urals). The studies of cultural heritage objects collected from these sites are impossible without a comprehensive study of the surrounding geological and ecological environment, representing some reference points or background values for comparison. The region under consideration appears to be poorly studied, particularly regarding isotopic and geochemical characteristics of complex interactions between the underlying rocks and soils, groundwater and plants. The present work is devoted to the study of phase composition of rocks characterising the surrounding geological environment in the area of the major Bronze Age archaeological monuments by X-ray diffraction. Using the SiroQuant software, the phase composition of bedrocks characterising the surrounding geological environment in the region of the major Bronze Age archaeological sites of the Southern Urals is was determined.

Keywords Bronze Age archaeological sites · Southern Urals · Bedrock phase composition · X-ray diffraction

---

S. V. Karpova (✉)
Ural Federal University, 19 Mira Street, Ekaterinburg, Russia
e-mail: svkarpo@mail.ru

D. Kiseleva · A. D. Ryanskaya · O. L. Galakhova · E. S. Shagalov
A.N. Zavaritsky Institute of Geology and Geochemistry of the Ural Branch of the Russian Academy of Sciences, 15 Vonsovskogo Street, Ekaterinburg, Russia
e-mail: kiseleva@igg.uran.ru

L. N. Koryakova
Ural Branch, Russian Academy of Sciences, Institute of History and Archaeology, 16 S. Kovalevskoy Str., Ekaterinburg, Russia

S. Votyakov et al. (eds.), *Minerals: Structure, Properties, Methods of Investigation*, Springer Proceedings in Earth and Environmental Sciences,
https://doi.org/10.1007/978-3-030-00925-0_10

Introduction

The main archaeological monuments of the Southern Urals of the late Bronze Age are the fortified settlements of Kamenny Ambar and Konoplyanka (21–17 centuries BC) as well as Nepluyevsky burial ground (19–16 centuries BC), located in the basin of the Karagaily-Ayat river (Chelyabinsk region) (Krause and Koryakova 2013; Sharapova et al. 2014, 2016).

An extensive collection of material items of cultural heritage including fossil bone and dental fragments of humans and animals, both domestic (domestic dog—*Canis familiaris*, cow—*Bos taurus*, sheep—*Ovis aries*, horse—*Equus caballus*) and wild (moose—*Alces alces*, etc.), as well as freshwater fish (pike—*Esox lucius*, perch—*Percafluviatilis*, crucian—*Carassius carassius*) along with soil, ancient tools and objects made of metal, stone, ceramics, as well as the sources of natural raw materials for their manufacturing is gathered from these monuments. This material is very promising for the analysis of palaeodiets due to the fact that there are both human bones and the remains of a whole series of potential sources of animal nutrition; it also can be very useful for obtaining the data about the way of life and migration of the population and domestic animals on the basis of isotopic characteristics as well as for the detection of isotopic-geochemical indicators in the ancient products of metallurgical processing and ceramics for determining the source of mineral raw materials in antiquity.

Such studies of cultural heritage items are impossible without a comprehensive study of the surrounding geological and ecological environment and its isotopic-geochemical characteristics, representing some reference points or background values for further comparison. For example, the isotope ratio $^{87}Sr/^{86}Sr$ of a particular geological region remains unaltered when it incorporates bone (dental) tissues of humans and animals from the underlying rocks through the soil and food chain while calcium is replaced by strontium in the hydroxyapatite crystal lattice (Ericson 1985; Copeland et al. 2008). Likewise, the ratio of stable isotopes in archaeological artefacts (for example, $^{207}Pb/^{206}Pb$, $^{65}Cu/^{63}Cu$, etc.) plays the role of the most stable and information-saturated natural genetic labels.

The region under consideration is represented by two granitoid massifs—the Nepluyevsky and Varshavsky plutons. The terrigenous rocks of the Lower-Middle-Ordovik Rymnik Suite, which are exposed to the west of the granitoids, represent their host rocks (Tevelev et al. 2006). To the east, in the territory of archaeological monuments, there are carbonaceous-clay phyllite schists of the Early Carboniferous age, and to the south there is a hyperbasite massif. Though some isotopic data on the Nepluyevsky and Varshavsky granitoid massifs is available (Rb-Sr age (Tevelev et al. 2006)), this region appears to be poorly studied, especially in the field of isotopic-geochemical characteristics of a complex system of interaction between the underlying rocks and soils (as a medium for burial of the cultural heritage objects), groundwater, as well as living organisms (from bacteria formed during the decay of organic matter after the death of organisms, to plants growing on the surface).

The present work is devoted to the study of phase composition of rocks characterising the surrounding geological environment in the area of the major Bronze Age archaeological monuments of the Southern Urals for further isotopic-geochemical research.

Materials and Methods

Eight samples of rocks collected in the areas of the major Bronze Age archaeological monuments of the Southern Urals—Kamenny Ambar and Konoplyanka (21–17 centuries BC) and burial ground Neplyuevsky (19–16 centuries BC) (Chelyabinsk region)—were studied (Fig. 1, Table 1).

Sample preparation. Before the analysis, the samples were ground by a crusher; in order to obtain a finer sample, a roll crusher was used, and finally the samples were ground manually in a jasper mortar.

Phase composition determination. The mineral composition of multicomponent mixtures can be determined most correctly by means of a full-profile X-ray phase analysis. The full-profile X-ray phase analysis by Rietveld is based on computer fitting of the calculated X-ray diffractograms of mixtures to experimental ones based on minimising the "calculation-experiment" deviation when varying a set of

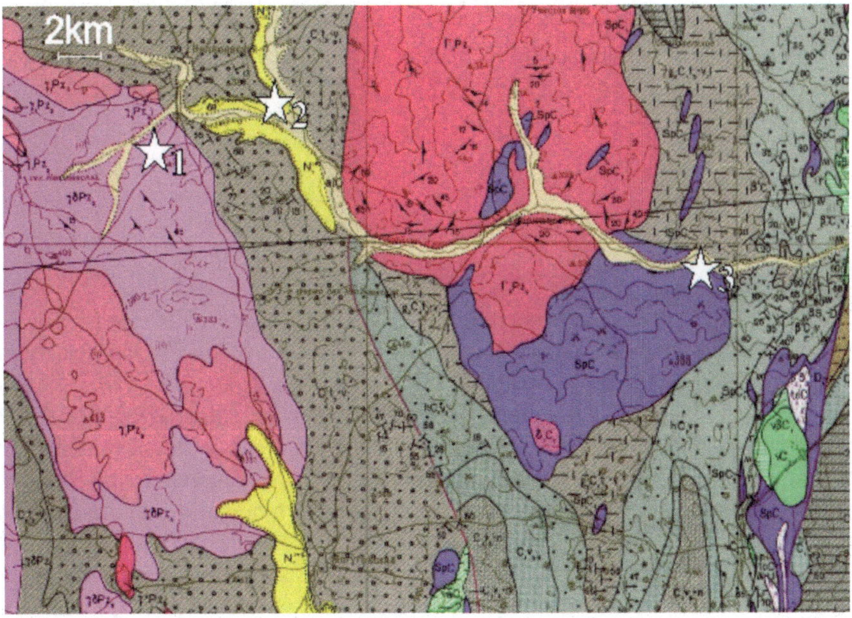

Fig. 1 Places of rock sampling: 1—Neplyuevsky, 2—Konoplyanka, 3—Kamenny Ambar

Table 1 The results of quantitative X-ray phase analysis of the studied rock samples

№	Sample name	Rock	GPS Coordinates	Composition
1	B4-1	Granitoid	52°51′20 ″N 60°21′48 ″E	Quartz 27% Plagioclase (albite) 47% Muscovite 14% Potassium feldspar (microcline) 8% Chlorite 4%
2	B4-2	Granitoid	52°51′20 ″N 60°21′48 ″E	Quartz 20% Mica (biotite) 30% Plagioclase 24% Potassium feldspar (orthoclase) 11% Apatite 5% Chlorite 4% Epidote 6%
3	B4-3	Granitoid	52°51′20 ″N 60°21′48 ″E	Quartz 30% Plagioclase (albite) 42% Potassium feldspar (microcline) 20% Muscovite 8%
4	B9-1	Phyllite	52°49′58 ″N 60°32′50 ″E	Quartz 34% Clinochlor 11% Muscovite (2M1 - polytype) 38% Paragonite 17%
5	B14-1	Serpentinite	52°47′6 ″N 60°22′43 ″E	Serpentine 100%
6	B19-1	Serpentinite	52°47′58 ″N 60°23′27 ″E	Serpentine 92% Chromospinelide 5% Dolomite 3%
7	K8-1	Quartzite schist	52°51′18 ″N 60°13′30 ″E	Quartz 97% Muscovite 1% Plagioclase 2%
8	NP-4	Granite	52°51′54 ″N 60°6′27 ″E	Quartz 39% Potassium feldspar (microcline) 27% Plagioclase (albite) 28% Muscovite 6%

variables (Taylor 2004; Ryanskaya et al. 2014). One of the most powerful contemporary software products implementing the Rietveld method for phase analysis is the SiroQuant (Sietronics, Australia) (Taylor 2004).

The experimental diffractograms were obtained using Shimadzu XRD-7000 X-ray diffractometer. The operation conditions were as follows: filtered copper radiation in the 3–70° range of 2Θ Bragg angles, recording speed 1°/min, sample weight about 2 g. The samples had a diameter of 25 mm and a thickness of not more than 2.5 mm. When preparing the samples, the sample surface was flat and strictly parallel to the border of the cuvette, avoiding, if possible, the formation of undesirable texture along the crystallite faces during pressing. A preliminary qualitative X-ray phase analysis was performed by the basic reflexes using the

Powder Diffraction File-2 database according to the method (MI № 88-16360-119-01.00076-2011). A quantitative full-profile X-ray phase analysis was performed using the SiroQuant software (Sietronics, Australia) according to Ryanskaya et al. (2014).

Results

The results of quantitative X-ray phase analysis of the studied rock samples are given in Table 1.

Conclusions

Using the SiroQuant software, the phase composition of a number of rocks characterising the surrounding geological environment in the region of the major archaeological monuments of the Southern Urals of the Bronze Age has been determined.

Acknowledgements The work was carried out at the UB RAS "Geoanalitik" Center for Collective Use; the results of mineralogical study of the rocks were obtained within the state assignment of IGG UB RAS № AAAA-A18-118053090045-8; the results of historiographical interpretation were obtained within the RSF grant No. 16-18-10332.

References

Copeland SR, Sponheimer M, le Roux PJ, Grimes V, Lee-Thorp JA, de Ruiter DJ, Richards MP. Strontium isotope ratios ($^{87}Sr/^{86}Sr$) of tooth enamel: a comparison of solution and laser ablation multicollector inductively coupled plasma mass spectrometry methods. Rapid Communications in Mass Spectrometry. 2008; 22: 3187–3194.

Ericson JE. Strontium isotope characterization in the study of prehistoric human ecology. J. Hum. 1985; 14: 503–514.

Krause R, Koryakova L, editors. Multidisciplinary research of the Bronze Age settlements in the Southern Transurals (Russia). Bonn; Verlag Dr. Rudolf Habelt GmbH; 2013.

MI № 88-16360-119-01.00076-2011. Gornye porody i mineraly. Identifikatsiya fazovogo sostava s ispol'zovaniem difraktometra XRD-7000 firmy "Shimadzu". Ekaterinburg: IGG UrO RAN; 2011. [MI No. 88-16360-119-01.00076-2011. Rocks and minerals. The identification of the phase composition using the Shimadzu XRD-7000 diffractometer. Ekaterinburg: IGG UB RAS; 2011. (in Russ.)].

Ryanskaya AD., Shchapova YuV., Gulyaeva TYa., Galakhova OL, Petrishcheva VG, Gorbunova NP, Tatarinova LA. Polnoprofil'nyi rentgenodifraktsionnyi analiz fazovo-mineral'nogo sostava porod-kollektorov nefti i gaza s ispol'zovaniem programmy SiroQuant (na primere iskusstvennykh smesei). EZhEGODNIK 2014 - Trudy IGG UrO RAN; 2015; 162: 267–275. [Ryanskaya AD, Shchapova YuV, Gulyaeva TYa, Galakhova OL,

Petrischeva VG, Gorbunova NP, Tatarinova LA. Full-profile X-ray diffraction analysis of the phase-mineral composition of the reservoir rocks of oil and gas using the SiroQuant software (by the example of artificial mixtures). EZhEGODNIK 2014 – Collection of works of IGG UB RAS. 2015; 162: 267–275. (in Russ.)].

Sharapova SV, Krauze R, Molchanov IV, Shtobbe A, Soldatkin NV. Mezhdistsiplinarnye issledovaniya poseleniya Konoplyanka: predvaritel'nye rezul'taty. Vestnik NGU. Seriya: Istoriya, filologiya. 2014; 13(3): 101–109. [Sharapova SV, Krause R, Molchanov IV, Shtobbe A, Soldatkin NV. Interdisciplinary studies of the Konoplyanka settlement: preliminary results. Vestnik Novosibirsk State University. Series: "History and Philology". 2014; 13(3): 101–109. (in Russ.)].

Sharapova SV, Gol'eva AA, Koryakova LN, Krauze R, Luaie Zh. Detskie pogrebeniya mogil'nika Neplyuevskii v Yuzhnom Zaural'e (predvaritel'nye rezul'taty). Materialy V mezhdunarodnoi nauchnoi konferentsii "Ekologiya drevnikh i traditsionnykh obshchestv", noyabr' 7–11, 2016, Tyumen', Russia, ss. 70–73. [Sharapova SV, Gol'eva AA, Koryakova LN, Krauze R, Luaie Zh. Children's burials of the Neplyuevsky burial ground in the Southern Trans-Urals (preliminary results). Materials of the V International Scientific Conference "Ecology of ancient and traditional societies", November 7–11, 2016, Tyumen, Russia, pp. 73–78. (in Russ.)].

Taylor JC. Rietveld made easy: a practical guide to the understanding of the method and successful phase quantifications. Canberra, Australia: Sietronics Pty Limited; 2004.

Tevelev AV, Kosheleva IA, Popov VS, Kuznetsov IE, Osipova TA, Pravikova NV, Vostretsova ES, Gustova AS. Paleozoidy zony sochleneniya Vostochnogo Urala i Zaural'ya. Trudy laboratorii skladchatykh poyasov (vypusk 4). Pod red. prof. Nikishina A.M.M.: Geologicheskii f-t MGU; 2006. [Tevelev AV, Kosheleva IA, Popov VS, Kuznetsov IE, Osipova TA, Pravikova NV, Vostretsova ES, Gustova AS. Paleozoids of the joint zone of the Eastern Urals and the Trans-Urals. In: Nikishin AM, editor. Proceedings of the Laboratory of folded belts (issue 4). Moscow: Geological Faculty of Moscow State University; 2006. (in Russ.)].

About the Features of Exsolution Phenomena of NaCl-KCl Solid Solutions from Arsenatnaya Fumarole (Tolbachik Volcano, Kamchatka)

Dmitry A. Khanin and Valery M. Chubarov

Abstract The fumarolic formations of GTFE (The Great Tolbachik Fissure Eruption of 1975–76) slag cones are excellent polygons for studying the peculiarities of mineral formation processes at high temperatures (up to 500 °C) and atmospheric pressure. Halite and sylvine are the most common occurring minerals formed from the gas phase. Their aggregates from Tolbachik fumaroles are characterised by structural heterogeneity. The samples from different sections of Arsenatnaya fumarole located on the Second cone of the Northern Breakthrough of the GTFE varying in temperature and in mineral associations were selected for the study. Their investigation by scanning electron microscopy (SEM) was carried out after four months—the period, after which the final exsolution of the NaCl-KCl solid solution took place. For the NaCl-KCl system from Arsenatnaya fumarole, NaCl:KCl ratio increases while the temperature decreases. The most "pure" halite is observed in the secondary and, probably, redeposited crusts from the low-temperature parts of the fumarole. A mixture of KCl and NaCl is observed in the intermediate zone between the hottest and the coldest parts of the fumarole. For sylvine from the mid-temperature zones of Arsenatnaya fumarole, a rubidium impurity is characteristic (up to 0.2%wt).

Keywords Fumaroles · Kamchatka · The Great Tolbachik Fissure Eruption · NaCl-KCl solid solution · Exsolution structures

D. A. Khanin (✉) · V. M. Chubarov
Institute of Volcanology and Seismology FEB RAS, 9 Piip Boulevard, Petropavlovsk-Kamchatskiy, Russia
e-mail: mamontenok49@yandex.ru

D. A. Khanin
Department of Geology, Lomonosov Moscow State University, 1 Leninskiye Gory, Moscow, Russia

S. Votyakov et al. (eds.), *Minerals: Structure, Properties, Methods of Investigation*, Springer Proceedings in Earth and Environmental Sciences,
https://doi.org/10.1007/978-3-030-00925-0_11

Introduction

The fumarolic formations of GTFE [The Great Tolbachik Fissure Eruption of 1975–76 (Fedotov 1980)] slag cones are excellent polygons for studying the peculiarities of mineral formation processes at high temperatures (up to 500 °C) and atmospheric pressure and allow the results of these processes to be directly observed. Here, halite and sylvine are the most common occurring minerals formed from the gas phase. The formations comprised by them, as was noted in a number of works (Vergasova 1983; Naboko 1947; Piip 1956; Serafimova 1979), refer to the products of direct deposition from a variety of volcanic exhalations. The structural heterogeneity of halite-sylvine aggregates from Tolbachik fumaroles was first characterised by Filatov and Vergasova (2002) on a material derived from the fumaroles of various lava flows and slag cones of GTFE.

Materials and Methods

The samples from different sections of Arsenatnaya fumarole located on the Second cone of the Northern Breakthrough of the GTFE varying in temperature and in mineral associations were selected for the study. Their investigation by scanning electron microscopy (SEM) was carried out after four months—the period after which, according to the data (Safonov and Tsygankov 2011), the final exsolution of the NaCl-KCl solid solution takes place.

Results and Discussion

Basically, the chlorides under consideration form transparent skeletal crystals up to 5 mm across the cubic habit (1Ks-5 samples from the hottest zone with a temperature of up to 500 °C), along with brushes and crystalline crusts. Due to the occurrence of gas-liquid inclusions, the crystals often acquire a milky white colour. In addition, solid milk-white crusts (samples 6 and 7) covering the slag pieces, which were in contact with atmospheric moisture, were collected from the areas with a low temperature (up to 200 °C). The internal structure of crystals and crusts is very similar—in the reflected light, the presence of two phases is clearly visible, which according to their composition correspond to KCl and NaCl, clearly distinctive under a scanning electron microscope in the backscattered electron mode (Fig. 1).

These phases form irregular allocations up to a few millimetres in diameter. On the fresh cleavage of 1Ks, 2,3,4,5 samples the inclusions of smaller precipitates of NaCl and KCl in the halite-sylvine aggregate are observed. They represent the rectangular allocation from 2 to 5 μm wide and practically do not differ in composition from the main mass (Fig. 2).

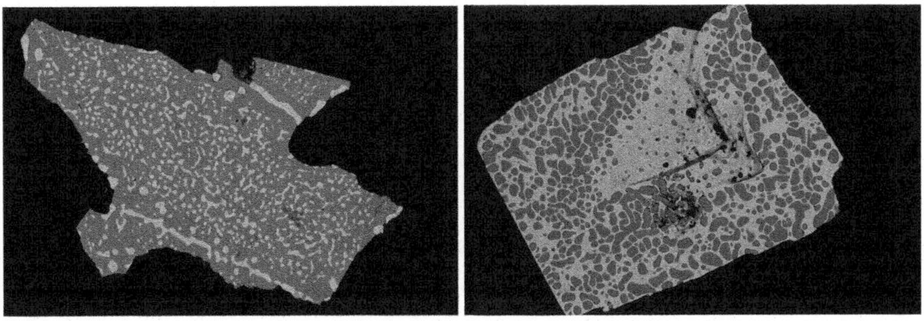

Fig. 1 The exsolution structure of the NaCl-KCl solid solution. Back-scattered electron image. Dark gray is NaCl, light gray is KCl. The field of view is 2 mm

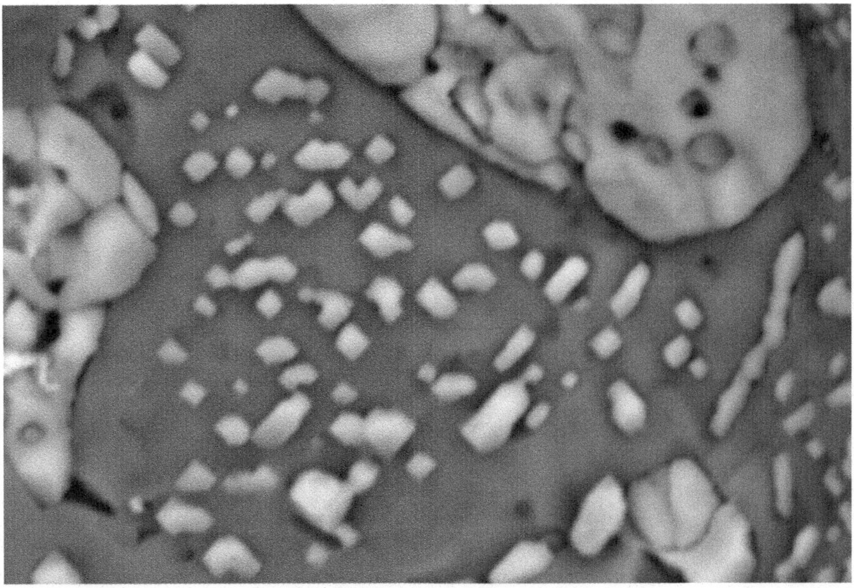

Fig. 2 Inclusions of KCl crystals in NaCl, back-scattered electron image. Dark gray is NaCl, light gray is KCl. The field of view is 70 μm

Table 1 KCl and NaCl ratio, vol%

Sample No.	1Ks	1	2	3	4	5	6	7
NaCl	50	100	30	50	70	10	100	95
KCl	50	–	70	50	30	90	–	5

Table 2 Concentration intervals of the main components in the exsolution products of the NaCl-KCl solid solution in samples from different zones

No.	1	1ks[a]	1ks	2[a]	2	3[a]	3	4[a]	4	5[a]	5	6	7[a]	7
Na	39.1–40.1	3.1–5.1	35.1–37.7	3.6–4.0	37.0–38.6	4.09–5.84	37.2–37.5	1.9–6.3	35.9–39.6	1.0–5.4	33.6–39.0	39.1–40.0	1.5–2.8	37.5–39.0
K	0.2–0.8	46.4–48.9	2.0–7.2	47.8–48.8	2.0–2.8	45.3–47.7	3.5–3.8	45.6–50.8	1.8–5.9	45.9–51.1	1.3–7.0	Up to 0.2	48.6–50.9	1.2–1.5
Cl	60.3–61.7	48.0–49.0	58.3–60.5	48.1–48.6	59.1–60.0	48.3–49.0	58.1–60.6	47.4–49.6	59.3–60.4	46.8–49.3	57.4–61.4	60.3–61.5	47.4–48.5	59.2–61.0

[a]The compositions correspond to Na-containing sylvine

The phase ratios of KCl and NaCl for samples from different zones of Arsenatnaya fumarole differ. They are represented in Table 1. The average contents of the main components in different phases are given in Table 2.

Conclusions

In general, for the NaCl-KCl system from Arsenatnaya fumarole, the NaCl:KCl ratio increases while the temperature decreases. The most "pure" halite (practically without potassium) is observed in the secondary crusts and probably redeposited ones from the low-temperature parts of the fumarole. It is also observed in the form of individual crystals collected from the zone with the highest temperature, where, in addition to the halite, hematite is present. A mixture of KCl and NaCl is observed in the intermediate zone between the hottest and the coldest parts of the fumarole. For sylvine from the mid-temperature zones of Arsenatnaya fumarole, a rubidium impurity is characteristic (up to 0.2%wt).

Acknowledgements The authors are grateful to I. V. Pekov for consultations during the preparation of this article. The study was supported by the RSF grant 14-17-00048.

References

Fedotov SA, red. Bol'shoe treshchinnoe Tolbachinskoe izverzhenie (BTTI, Kamchatka, 1975–1976). M.: Nauka; 1980. [Fedotov SA, editor. The Great Tolbachik Fissure Eruption (GTFE, Kamchatka, 1975–1976). Moscow: Nauka; 1980. (in Russ.)]

Filatov S.K., Vergasova L.P. Protsessy raspada i gomogenizatsii dvoinykh solei NaCl-KCl iz inkrustatsii fumarol po dannym termorentgenografii. Vulkanologiya i seismologiya. 2002;5:25–31. [Filatov SK, Vergasova LP. The processes of exsolution and homogenization of NaCl-KCl double salts from fumarole incrustations according to thermoradiography. Journal of Volcanology and Seismology. 2002;5:25–31. (in Russ.)]

Naboko SI. Izverzhenie Bilyukaya, pobochnogo kratera Klyuchevskogo vulkana v 1938 godu. Tr. Lab. vulkanologii i Kamchatskoi vulkanol. stantsii. 1947;5:134. [Naboko SI. The eruption of Bilyukay, a secondary crater of Klyuchevsky volcano in 1938. Proceedings of Laboratory of volcanology and Kamchatka volcanological station. 1947;5:134. (in Russ.)]

Piip BI. Klyuchevskaya sopka i ee izverzhenie v 1944–1945 gg. i v proshlom. Tr. Lab. vulkanologii. 1956;11:309. [Piip BI. Klyuchevskaya Sopka and its eruption in 1944–1945 and in the past. Proceedings of Laboratory of volcanology and Kamchatka volcanological station. 1956;11:309. (in Russ.)]

Safonov V.V., Tsygankov V.N. Svoistva tverdykh rastvorov sistemy NaCl-KCl. Zhurnal neorganicheskoi khimii. 2011;56(7):1179–1181. [Safonov VV, Tsygankov VN. The properties of solid solutions of the NaCl-KCl system. Journal of Inorganic Chemistry. 2011;56(7):1179–1181. (in Russ.)]

Serafimova EK. Mineralogiya vozgonov vulkanov Kamchatki. M.: Nauka; 1979. [Serafimova E. K. Mineralogy of sublimations of Kamchatka.volcanoes. Moscow: Nauka; 1979. (in Russ.)]
Vergasova L.P. Ob inkrustatsii fumarol lavovykh potokov v effuzioeksplozivnyi period BTTI. Vulkanologiya i seismologiya. 1983;6:75–87. [Vergasova L.P. On the incrustation of the fumaroles of lava flows in the effusion-explosive period of the GTFE. Journal of Volcanology and Seismology. 1983;6:75–87. (in Russ.)]

Localized Shock-Produced Melting in Meteorites

Natalia R. Khisina, Elizaveta A. Pankrushina and Albert M. Abdrakhimov

Abstract Local regions of shock-induced melting in meteorite Elga are represented by stratified schreibersite-oxide reaction rims developed around silicate inclusions during the primary shock event, and by melt pockets produced in silicate inclusions during a subsequent shock event. Melt pockets have been investigated with EMPA, SEM, TEM and Raman spectroscopy. Melt pockets manifest themselves by abundant schreibersite-dominated and silicate-dominated emulsions resulting from phosphide-silicate liquid immiscibility in shock-produced mixed melts. The first finding of phosphide-siderite liquid immiscibility in one of melt pockets is indicative of the extraterrestrial shock-induced origin of siderite in Elga.

Keywords Meteorites · Shock metamorphism · Melt pockets · Extraterrestrial siderite · Raman spectroscopy · Electron microprobe analyses · Transmission electron microscopy · Scanning electron microscopy

Introduction

Meteorites are the main source of information about chemical composition, mineralogy, lithology and shock history of the rocks on their parent bodies, i.e., on asteroids and planets. Throughout a long pre-history in the solar system, cosmic objects repeatedly collided with one another, and shock waves that were generated in collided bodies, produced in turn shock effects in meteorites. Melt pockets and melt veins represent local regions of shock-induced melting that are often associated with pre-existing vugs in meteoritic rocks (Beck et al. 2007; Walton and Shaw 2009;

N. R. Khisina (✉) · A. M. Abdrakhimov
Vernadsky Institute of Geochemistry and Analytical Chemistry of Russian Academy of Sciences, Kosygin St., 19, 119334 Moscow, Russia
e-mail: khisina@gmail.com

E. A. Pankrushina
Zavarytsky Institute of Geology and Geochemistry Ural Branch of Russian Academy of Sciences, Ekaterinburg, Russia

S. Votyakov et al. (eds.), *Minerals: Structure, Properties, Methods of Investigation*, Springer Proceedings in Earth and Environmental Sciences,
https://doi.org/10.1007/978-3-030-00925-0_12

Kuchka et al. 2017). Numerous data indicate that shock-induced localized melting in meteorites might be related to former pores and fissures in an unload rock. The magnitude of the temperature increase at the material interface is caused by different effects: shock heating due to the passage of shock wave closer of the gap, and impedance difference at the interface. Increased shock temperature and shock pressure at the interfaces between the target materials of different density (strong impedance contrast) were estimated in numerical simulation experiments (Heider and Kenkmann 2003), and the temperature increase at open fissures can be especially high because impedance contrasts of empty space and rock are extreme (Heider and Kenkmann 2003). The shock process in a porous material involves melting at the grain/vug boundaries and jetting of melt into pores, thus resulting in the appearance of melt pockets and melt veins in meteorites (Kiefer 1975). Shock mechanisms of the formation of localized melted areas, such as melt pockets and melt veins imply a different origin for the melt: (1) injection of externally generated melts into the target rock along cracks and fractures and (2) in situ melting of a target rock by the void space collapse. The microstructure of regions with a local increase in pressure and temperature due to shock wave propagation through an inhomogeneous rock is documented by experiments (Kenkmann et al. 2000).

Shock melt pockets and veins are observed in ordinary chondrites and Martian meteorites (Tomkins et al. 2013; Kuchka et al. 2017); recently melt pockets were found in Nechaevo IIE iron meteorite (Van Roosbroek et al. 2017). Here we give the description of melt pockets in meteorite Elga that relates to IIE irons containing silicate inclusions.

Materials and Methods

Meteorite Elga relates to silicate-bearing iron meteorites (IIE clan) that consist of rounded silicate inclusions embedded into FeNi matrix. Shock-produced zones in Elga manifest themselves by stratified schreibersite-oxide reaction rims around silicate inclusions (Fig. 1) and by products of the late shock-induced transformation of the rims that are represented by breccias and melt pockets (Fig. 2). The mineralogy of the shock-produced zones was investigated with EMPA, SEM, and Raman spectroscopy.

Raman spectra of siderite, merrillite and Fe, Ni-oxides in the rim and breccias were measured using Senterra (Bruker GmbH) with lazer power 5 mW, diameter of beam at the sample 3 μm, excited wave length 532 nm. Raman microprobe measurements of individual micrometer-sized siderite precipitates in the melt pocket were performed using a Horiba Jobin Yvon LabRam-HR (high resolution) 800 system equipped with Olympus BX-FM optical microscope, a grating with 1800 grooves per millimetre, and Si-based, Peltier thermo-electrically cooled charge-coupled device (CCD) detector. Spectra were excited using the He–Ne 632.8 nm emission line (2 mW measured behind the microscope objective). With the system operated in confocal mode and 100× objective (numerical aperture 0.9),

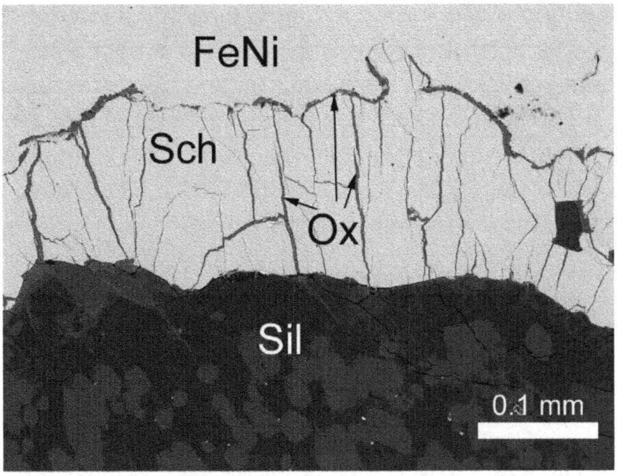

Fig. 1 Schreibersite-oxide rim around silicate inclusion in Elga meteorite. Sch—schreibersite, Ox—Fe, Ni-oxides, Sil—silicate inclusion, Me—FeNi metal

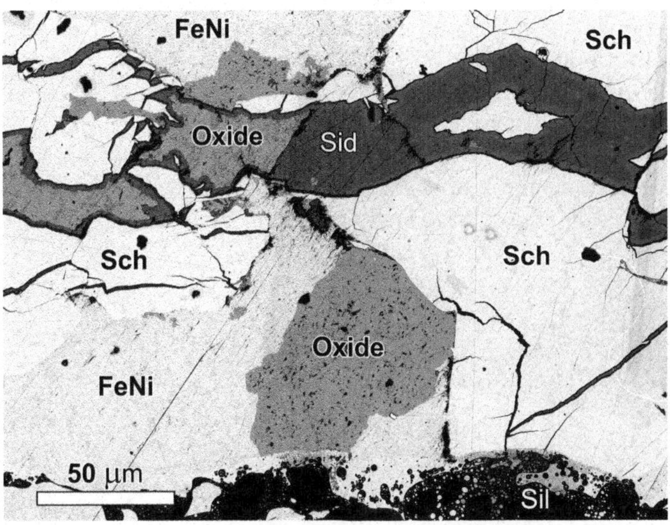

Fig. 2 Brecciated zone on the border between silicate inclusion and crack

the lateral resolution was better than 2 μm (acquisition time 150 s, accumulation number 10).

Phosphides, sulphides and FeNi metal were analyzed using SX 100 Cameca with acceleration voltage 15 kV and beam current 10 nA; the reference materials employed were synthetic FeS and Ni, and meteoritic $(FeNi)_3P$. Quantitative chemical analyses of siderite, merrillite and Fe, Ni-oxides were carried out in the wavelength-dispersion mode, using a Superprobe JEOL JXA-8230 operated under the following conditions: 20 kV acceleration voltage and 2×10^{-8} A beam current. A focused beam was used to analyze minerals and a defocused beam (diameter 10 μm) for the analysis of glass. Both samples and reference materials were analyzed at the same conditions. The reference materials employed were: apatite (Ca, P), $ScPO_4$ (P), dolomite (Mg, Ca), siderite (Fe, Mn). The counting time of the signal at the peak for each element was 30 s, and that for the background from each side was 15 s. A ZAF matrix correction was applied.

Results and Discussion

Well pronounced shock-produced zones in Elga manifest themselves by (i) the double-layered schreibersite-oxide rims at the silicate inclusion/host metal boundaries (Fig. 1), (ii) breccias (Fig. 2), and (iii) melt pockets in the shock-transformed zone inside a large silicate inclusion (Figs. 3, 4, and 5). The Sch-Ox rims around silicate inclusions have been found both in the marginal (Khisina et al. 2017; Teplyakova et al. 2012) and in the central (Sharygin 2017) parts of the meteorite. The melt pockets and breccias contain fragments of destroyed Sch-Ox rims, thus they resulted from a late shock transformation of the already formed schreibersite-oxide rims. This observation has led to a conclusion that at least two separated in time collisions happened in Elga "life" (Khisina et al. 2017). The former one resulted in the appearance of double-layered rims stratified into schreibersite and siderite-bearing oxide(s) layers and developed around silicate inclusions or along the cracks in the metal matrix (Fig. 1). The latter event

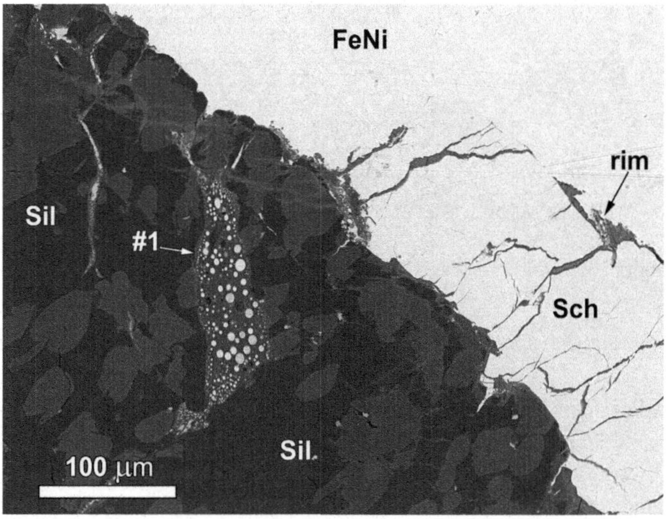

Fig. 3 Melt pocket #1 in silicate inclusion

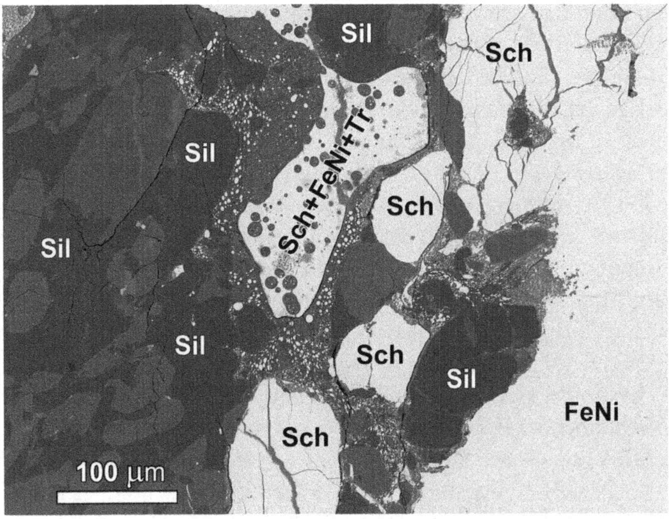

Fig. 4 Melt pocket #2 in silicate inclusion

Fig. 5 Melt pocket #3 in silicate inclusion

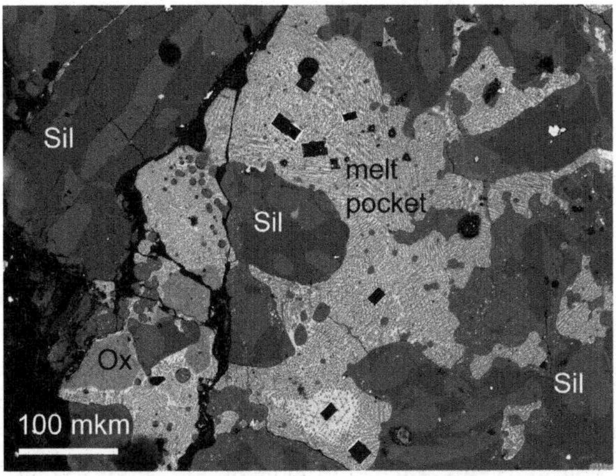

manifests itself by the rim fragmentation accompanied by the formation of the breccia consisted of phosphide, oxide, metal and siderite clasts (Fig. 2), and by melting of phosphide and oxide matter with their transfer from the silicate/FeNi boundary to the interior of a silicate inclusion to form melt pockets in the shock-produced zone (Figs. 3, 4, and 5). Shock-produced siderite is localized in shock-produced zones (Khisina et al. 2017).

The shock-produced zone extending from the boundary into the interior of the silicate inclusion manifests itself by several regions of local melting with the schreibersite-oxide rim fragments injected in them. Three types of melt pockets occur inside this zone. Melt pockets #1 (Fig. 3) and #2 (Fig. 4) are on the border of a silicate inclusion and schreibersite-oxide rim, whereas the melt pocket #3 occurs far from the silicate/FeNi boundary in the depth of shock-transformed zone (Fig. 5).

The BSE image (Fig. 3) demonstrates that the melt pocket #1 was developed along a crack in the silicate inclusion. A passage of a shock wave through the crack was accompanied by melting of an adjacent silicate. Inside the melt pocket #1 there are a lot of schreibersite globules less than 1 μm in size embedded into silicate glass. Obviously, the crack was formed before the shock, and the molten schreibersite rim portion was injected by a shock wave into the crack. Under post-shock cooling the liquid phosphide droplets were exsolved due to liquid immiscibility from the mixed silicate-phosphide melt.

A specific feature of the melt pocket #2 (Fig. 4) is the presence of numerous brecciated fragments represented by schreibersite and schreibersite-FeNi alloy transferred from the outside of silicate inclusion, i.e. from the silicate/metal boundary. Schreibersite-FeNi clasts are homogeneous on the special resolution levels of EMPA and SEM. Silicate glass in the melt pocket #2 has MgO-Al_2O_3-Na_2O-K_2O-CaO-SiO_2 composition. Small silicate globules composed by a glass of a similar composition occur inside the schreibersite-FeNi clasts, but they were never observed inside pure schreibersite clasts. Boundaries of the melt pocket #2 are decorated by numerous very small schreibersite-troilite globules embedded in silicate glass. These patterns are similar to metal-dominated emulsions, i.e. immiscible silicate melt globules within metal melt, and silicate-dominated emulsion, the opposite, as was observed in ordinary chondrites (Tomkins et al. 2013). The lack of silicate globules inside the schreibersite clasts confirms that pure schreibersite have not been involved in the melting process and opposite to schreibersite-FeNi fragments, the clasts of pure schreibersite have been transported by the shock wave into the silicate inclusion without melting. We conclude that the temperature of shock-induced heating during the formation of the melt area #2 was lower than the melting temperature of schreibersite but higher than the temperature of schreibersite-troilite-FeNi eutectic. Thus, shock-induced heating during the formation of the melt pocket #2 was lower than those of #1.

The melt pocket #3 inside the silicate inclusion (Fig. 5) is formed by shock-molten fragments of the stratified schreibersite-oxides-silica rim transferred by the same late shock wave from the silicate-metal boundary. The size of the #3 is about 120 μm; smaller fragments are dispersed in the surrounding silicate substance. The #3 has an octopus-like shape with tentacles extended along cracks into surrounding silicate. In the area around the melt pocket #3 a specific rotation type texture formed by elongated and curved pyroxene crystals in the feldspar silicate glass is observed (Fig. 5). Such a rotation texture looks like "floating" deformed pyroxene crystals embedded in turbulently disturbed feldspar silicate melt. The border of the #3 is marked by very small schreibersite globules embedded in the silicate glass; in turn, silicate globules occur inside schreibersite matrix of the #3. It can be seen that the largest of the silicate globules contain nanosized schreibersite globules inside, thus exhibiting a secondary phase separation. On a macroscopic level, the microstructure of the melt pocket #3 looks like continuous parallel rows of precipitates in a matrix, which at first glance resemble dendritic type microstructures. Dendritic growth is known to proceed during crystallization from overcooled melts. However, SEM/BSE images (Fig. 6a) and elemental maps (Fig. 6b) demonstrate instead a system of rounded precipitates of 1–2 μm in size, isolated from each other and uniformly distributed in the matrix, which is consistent with a phase separation by nucleation and growth in a liquid immiscible melt (Van Roosbroek et al. 2017; Shepilov et al. 2007). This kind of phase separation is restricted to the region between the binodal and the spinodal of a miscible gap at the phase diagram of immiscible melts. In other words, the microstructure of the #3 has been formed at temperatures much higher than that of crystallization temperature and the phase separation proceeded during cooling of the initially homogeneous melt.

SEM/EDS elemental mapping (Fig. 6b) and Raman spectrum (Fig. 6c) as well as previous TEM elemental mapping, EDS/TEM analyses and SAED patterns

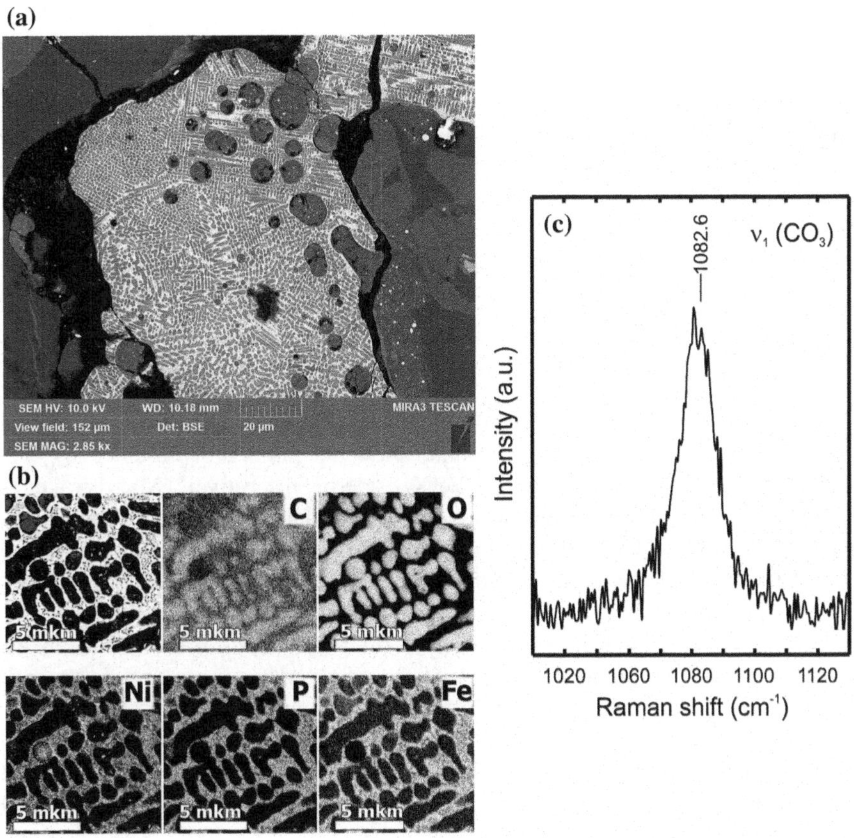

Fig. 6 Fragment of the melt pocket #3. Microstructure (**a**) and elemental mapping (**b**) of siderite precipitates in schreibersite matrix of the melt pocket #3. **c** Raman spectrum of individual siderite grain

(Khisina et al. 2017; Teplyakova et al. 2012) testify that rounded precipitates in the matrix are composed of siderite. Raman spectra from individual rounded precipitate reveal the band of siderite at 1082.6 cm^{-1} (Fig. 6c). Electron diffraction patterns from rounded inclusions show ring-shaped, weak and diffuse siderite reflections indicating that siderite occurs as nano-crystalline aggregates. Siderite inclusions in schreibersite often contain areas with small precipitates of Ni-phosphide, elemental carbon and SiO_2, and they often have SiO_2-rich rims.

The secondary minerals of extraterrestrial origin, formed as a result of the impact, occur in the shock-produced zones in Elga meteorite: $(Fe,Ni)_3P$ schreibersite, FeS troilite, NiS millerite, $NaMgCa_9(PO_4)_7$ merrillite, $FeCO_3$ siderite, Fe_2O_3 hematite, and Ni magnetite. We conclude that Fe, Ni, P, S, C and oxygen have been involved in shock transformation process, wherein each of these elements occurs in different valence states: Fe^0, Fe^{2+} and Fe^{3+}; Ni^0 and Ni^{2+}; P^0 and P^{5+}; C^0 and C^{4+}. Shock-induced redox reactions proceed obviously at the contacts of silicate inclusion and FeNi host. The schreibersite forming rims at the boundaries between silicate inclusions and FeNi host indicate that the system might approximate a phosphide-phosphate buffer partly through the reduction reaction at high temperatures. The bulk equilibrium was not reached in the system, however, because of a very short duration of the impact. The results indicate the heterogeneity of pressure, temperature and fO_2 distribution in shocked material in a case of shock wave propagation through heterogeneous rock. A high-temperature shock-induced melting in meteorite Elga might be caused by the presence of mechanically weakened areas and vugs (open spaces) in the target rock.

Conclusion

Melt pockets in Elga are represented by phosphide-silicate, silicate-phosphide and siderite-phosphide emulsions that contain globular particles represented by $(Fe,Ni)_3P$ schreibersite, silicate glass, and $FeCO_3$ siderite, respectively. Microstructures of melt pockets in silicate inclusions exhibit both sphericity and monodispersity of particles, as well as partial ordering in the particle arrangement. Such patterns are typical for the nucleation-and-growth stage of binodal liquid-liquid phase separation (Shepilov et al. 2007).

Our data demonstrate the first finding of siderite in shock melted zones in IIE iron meteorites. The liquid-liquid siderite-phosphide phase separation in the melt pocket #3 proves an extraterrestrial shock-produced origin of siderite in Elga.

The extraterrestrial source of CO_2 in Elga is still debated. Siderite is only found in highly fractured marginal part of Elga. The presence of mechanically weakened areas and vugs (open spaces) in marginal parts of the meteorite might trigger local melting and provide trapping of atmospheric gas (Beck et al. 2007). The laboratory shock emplacement of CO_2, N_2 and noble gases into basalt (Wiens and Pepin 1988) showed high emplacement efficiencies for CO_2, on average, by a factor of 3.2 greater than those of N_2 and noble gases in a sample shocked to 60 GPa. There are a lot of evidences for the presence of trapped atmospheric gases in Martian meteorites (Beck et al. 2007; Treiman et al. 2000; Walton et al. 2007; Bogard et al. 1986; Carr et al. 1985). We can't exclude that in case of Elga, the atmospheric CO_2 gas could be implanted into silicate melt due to the impact event. If so, then we can speculate about a "Martian transit" for Elga.

Acknowledgements Authors are thankful to A. A. Shiryaev, N. G. Zinovieva and A. Burmistrov for their help in conducting research. This research was supported by the Program 7 of ONZ RAN.

References

Beck P, Ferroir T, Gillet P. Shock-induced compaction, melting and entrapment of atmospheric gases in Martian meteorites. Geophys Res Letters. 2007;34:L01203. https://doi.org/10.1029/2006gl028141.2007

Bogard DD, Hörz F, Johnson PH. Shock-implanted noble gases: an experimental study with implications for the origin of Martian gases in shergottite meteorites. Proceedings: 17th Lunar and Planetary Science Conference. J. of Geophysical Research. 1986;91:E99-E114. https://doi.org/10.1029/jB091;B13p00E99

Carr RH, Grady MM, Wright IP, Pillinger CT. 1985. Martian atmospheric carbon dioxide and weathering products in SNC meteorites. Nature, 314,248–250. https://doi.org/10.1038/314248a0

Heider N and Kenkmann T. Numerical simulation of temperature effects at fissures due to shock loading. Meteoritics and Planetary Sci. 2003;38:1451–1460. https://doi.org/10.1111/j.1945-5100.2003.tb00250.x

Kiefer SW. From regolith to rock by shock. The Moon. 1975;13:301–320. https://doi.org/10.1007/BF00567522

Kenkmann T., Hornemann U., Stöffler D. Experimental generation of shock-induced pseudo-tachylytes. Meteoritics and Planetary Sci. 2000;35:1275–1290. https://doi.org/10.1111/j.1945-5100.2000.tb01516.x

Khisina NR, Teplyakova SN, Wirth R, Senin VG, Averin AA, Shiryaev AA. Carbon-bearing phases in shock-induced melt zones of the Elga meteorite. Geochemistry International. 2017;55(4):317–320. https://doi.org/10.1134/S0016702917040036

Kuchka CR, Herd CDK, Walton EL, Guan Y, Liu Y. Martian low-temperature alteration minerals in shock melt pockets in Tissint: constraints on their preservation in shergottite meteorites. Geochimica et Cosmochimica Acta. 2017;210:228–246. https://doi.org/10.1016/j.gca.2017.04.037

Sharygin VV. Meteorite Elga: mineralogy of inclusion with silicate-natrophosphate immiscibility. 200th Anniversary Meeting of the Russian Mineralogical Society, Sankt-Petersburg, Russia. Conference papers. 2017;2:349–351

Shepilov MP, Kalmykov AE, Sycheva GA. Liquid-liquid phase separation in sodium borosilicate glass: ordering phenomena in particle arrangement. Journal of Non-Crystalline Solids. 2007;353:2415–2430. https://doi.org/10.1016/j.jnoncrysol.2007.03.019

Teplyakova SN, Khisina NR, Artemov VV, Vasil'ev AL. Nanomineralogiya dendritnych vklyuchenii v zheleznom meteorite Elga. Zapiski RMO. 2012;141(2):42–52. [Teplyakova SN, Khisina NR, Artemov VV, Vasil'ev AL. Nanomineralogy of dendritic inclusions in the Elga iron meteorite. Proceedings of Russian Mineralogical Society. 2012;141(2):42–52. (In Russ.)])

Tomkins AG, Weinberg RF, Schaefer BF, Langendam A. Disequilibrium melting and melt migration driven by impacts: implications for rapid planetesimal core formation. Geochimica et Cosmochimica Acta. 2013;100:41–59. https://doi.org/10.1016/j.jgca.2012.09.044

Treiman AH, Gleason JD, Bogard DD. The SNC meteorites are from Mars. Planetary and Space Science. 2000;48:1213–1230

Van Roosbroek N, Hamann C, McKibbin S, Greshake A, Wirth R, Pittarello L, et al. Immiscible silicate liquids and phosphoran olivine in Netschaevo IIE silicate: analogue for planetesimal core-mantle boundaries. Geochimica et Cosmochimica Acta. 2017;197:378–395. https://doi.org/10.1016/j/gca.2016.10.042

Walton EL, Kelley SP, Spray JG. Shock implantation of Martian atmospheric argon in four basaltic shergottite: a laser probe $^{40}Ar/^{39}Ar$ investigation. Geochimica et Cosmochimica Acta. 2007;70:1059–1075. http://doi.org/10.1016/j.gca.2016.10.042

Walton EL and Shaw CSJ. Understanding the textures and origin of shock melt pockets in martian meteorites from petrographic studies, comparisons with terrestrial mantle xenoliths, and experimental studies. Meteoritics and Planetary Sci. 2009;44:55–76. https://doi.org/10.1111/j/1945-5100.tb00717.x

Wiens RC and Pepin RO. Laboratory shock emplacement of noble gases, nitrogen and carbon dioxide into basalt, and implications for trapped gases in shergotites EET 79001. Geochimica et Cosmochimica Acta. 1988;52:295–307. https://doi.org/10.1016/0016-7037(88)90085-3

Formation Stages of Calc-Alkaline Granites in the Western Transbaikalian Granitoid Province: LA-ICP-MS U–Pb Age Data on Detrital Zircons from Modern Sediments

Valentin B. Khubanov, Andrey A. Tsygankov and Tatyana N. Antsifirova

Abstract The paper presents LA-ICP-MS U–Pb age data on detrital zircons from modern sediments, whose source was the calc-alkaline granites of the Barguzin complex (Western Transbaikalian granitoid province). The greatest number of zircons is late Paleozoic with two age peaks: ~290 and ~315 Ma.

Keywords Calc-alkaline granites · Western transbaikalian granitoid province · LA-ICP-MS U–Pb dating · Detrital zircons

Introduction

The Western Transbaikalian granitoid province with a total area of more than 200 thousand km^2 hosts hundreds or thousands of granitoid plutons, see Fig. 1. Five intrusive suites composed of different geochemical types of granitoids were emplaced throughout the region (Tsygankov et al. 2010, 2017; Litvinovsky et al. 2011). In order of formation they are: (1) the Barguzin suite consisting of calc-alkaline auto- and allochthonous biotite granites; (2) the Chivyrkui suite consisting of high-potassium quartz monzonites, quartz syenites and gabbroids; (3) the Zaza suite comprising transitional from calc-alkaline to alkaline granite and quartz syenite; (4) the shoshonitic Low Selenga suite made up of monzonite, syenite and alkali rich microgabbro (5) the Early Kunalei suite of alkaline (alkali feldspar) and peralkaline syenite and granite.

More than two-thirds of the province is represented by calc-alkaline granites of the Barguzin suite. Over the past two decades, a significant number of absolute dates have been obtained for these granites (Tsygankov et al. 2007, 2010, 2017; Yarmolyuk et al. 1997; Kotov 2012; Khubanov et al. 2016). Isotope data of

V. B. Khubanov (✉) · A. A. Tsygankov · T. N. Antsifirova
Geological Institute of SB RAS, 6a Sakh'yanova, 670047 Ulan-Ude, Russia
e-mail: khubanov@mail.ru

S. Votyakov et al. (eds.), *Minerals: Structure, Properties, Methods of Investigation*, Springer Proceedings in Earth and Environmental Sciences,
https://doi.org/10.1007/978-3-030-00925-0_13

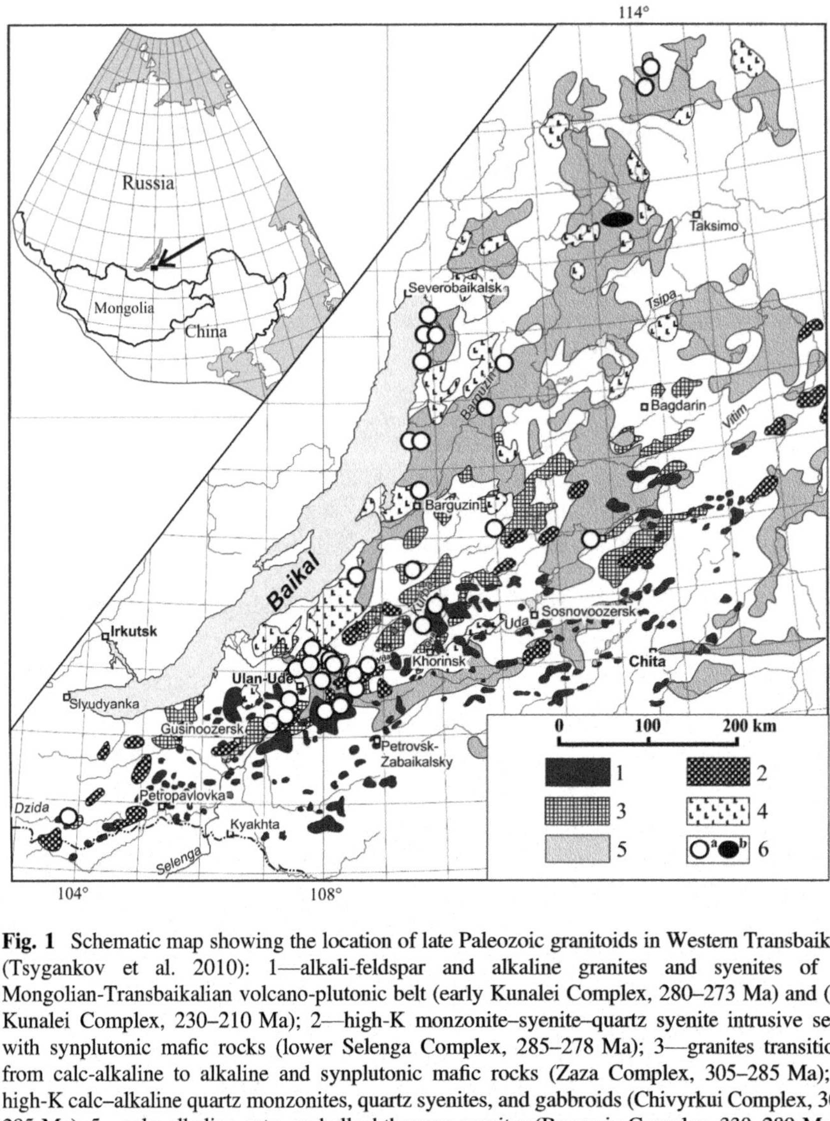

Fig. 1 Schematic map showing the location of late Paleozoic granitoids in Western Transbaikalia (Tsygankov et al. 2010): 1—alkali-feldspar and alkaline granites and syenites of the Mongolian-Transbaikalian volcano-plutonic belt (early Kunalei Complex, 280–273 Ma) and (late Kunalei Complex, 230–210 Ma); 2—high-K monzonite–syenite–quartz syenite intrusive series with synplutonic mafic rocks (lower Selenga Complex, 285–278 Ma); 3—granites transitional from calc-alkaline to alkaline and synplutonic mafic rocks (Zaza Complex, 305–285 Ma); 4—high-K calc–alkaline quartz monzonites, quartz syenites, and gabbroids (Chivyrkui Complex, 305–285 Ma); 5—calc–alkaline auto- and allochthonous granites (Barguzin Complex, 330–280 Ma); 6—sampling sites for U–Pb zircon dating: a—igneous zircons (literature data), b—detrital zircons (Angarakan river basin)

individual granite plutons suggest that emplacement of granites occurred in the interval of 330–280 Ma in two separate stages (Fig. 2). However, because of the impossibility of dating each of the many hundreds of plutons, the question of the interval and stages of the Late Paleozoic calc-alkaline granitoid magmatism is debatable (Tsygankov et al. 2017; Kotov 2012). The age analysis of the zircons from the sediments of the catchment areas, where the outcrops of the rocks of the Barguzin complex significantly dominate, can be one of the tools for estimating the interval and stages of calc-alkaline granitoid magmatism. The sampling of detrital zircons from these sediments can be indicative of the age of the sources of demolition in a large area, encompassing a large number of the late Paleozoic granite plutons of the Barguzin complex, emerging on the surface.

Materials and Methods

The Angarakan river basin in the Northern Muya Range in the northern part of Western Transbaikalia is one of the territories composed mainly of calc-alkaline granites of the Barguzin complex. In this area, among the calc-alkaline allochthonous and autochthonous granites, there are rare occurrences of only Precambrian metamorphosed rocks, which form the roof pendants for granite massifs. Detrital zircons were derived from the gray heavy concentrate from the river medium-grained arkose sand of the Middle Angarakan region.

U–Pb isotopic dating for zircons was carried out using Laser Ablation Inductively Coupled Plasma Mass Spectrometry (LA-ICP-MS) (Khubanov et al. 2016; Buyantuev et al. 2017) at the Research Resource Center "Analytical center for the mineralogical-geochemical and isotope research", GIN SB RAS (Ulan-Ude). It should be noted that LA-ICP-MS is the most accessible and simple in sample preparation express U–Th–Pb isotope-geochronological method. As an external reference material, zircon standard samples 91,500 were measured (1065 Ma) (Wiedenbeck et al. 1995), as a key sample—standard samples Plešovice (337 Ma) (Sláma et al. 2008) and GJ-1 (610 Ma) (Jackson et al. 2004). The relative errors in the measurement of isotope ratios in the key samples varied within: for $^{208}Pb/^{232}Th$, $^{207}Pb/^{206}Pb$, $^{206}Pb/^{238}U$ and $^{207}Pb/^{235}U$ varied within 1–2.5%. The relative error of the weighted average concordant ages of zircons Plešovice and GJ-1, identified by

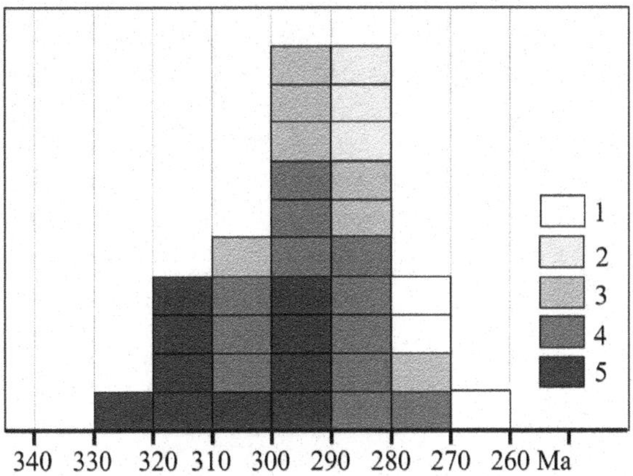

Fig. 2 Histogram showing the distribution of the U–Pb zircon dates of late Paleozoic magmatic rocks in Western Transbaikalia (Tsygankov et al. 2017) granitoid complexes: 1—early Kunalei, 2—lower Selenga; 3—Zaza; 4—Chivyrkui; 5—Barguzin

the LA-ICP-MS method, was less than 2% of their nominal age value. During the analysis, isotope ratios in the zircons were measured without any preliminary selection by size or morphology. The processing of the measurements was carried out using GLITTER (Griffin et al. 2008), ISOPLOT (Ludwig et al. 2008) and DensityPlotter software (Vermeesch et al. 2012). In the interpretation only age assessments, discordance D (1) not exceeding 10% was taken into account. Histograms and relative probability curves for the zircons older than 1 billion years were constructed by $^{207}Pb/^{206}Pb$ age values with an error of 2σ, and for the zircons younger than 1 billion years old, they were calculated by $^{206}Pb/^{238}U$ with an error of 2σ.

$$D = 100\left(\frac{\frac{^{207}\mathrm{Pb}}{^{235}\mathrm{U}}\,age}{\frac{^{206}\mathrm{Pb}}{^{238}\mathrm{U}}\,age} - 1\right), \quad (1)$$

Results and Discussion

206 zircons from a sample of sand from the Middle Angarakan region were analyzed, 135 zircons with D <10% were selected for chronological assessments. Considering that in the catchment area of the river Angarakan the calc-alkaline granites of the Barguzin complex are the dominant source of sedimentary material, the age relations of detrital zircons most probably reflect the age position of these granites. Figure 3 shows the largest number of zircons of the late Paleozoic (68 zircons or 50% of the sample), with two age peaks recorded: ~290 and ~315 Ma. It should be noted that the age range and their peaks are consistent with the duration and two age maxima of the formation of Barguzin granites identified during the dating of magmatic zircons and shown in Fig. 2 (Tsygankov et al. 2017). Twenty-nine zircons (21%) are middle Paleozoic with a peak of 424 Ma. At the same time, Mesozoic and Cambrian (except for two zircons with the ages of 504 and 518 Ma) ages, which are common for detrital zircons in current sediments in the southern part of the Western Transbaikalia, are absent (Khubanov et al. 2017; Ivanov et al. 2016). A part of the zircons (35 grains, 26%) are of Neoproterozoic age with a peak of 780 Ma. Single zircons have ages: 1100, 1400, 1760, 1825 Ma. These ancient ages and especially distinct peaks for the Middle Paleozoic (~425 Ma) and Neoproterozoic (780–770 Ma) reflect the age of the basement and protoliths of the late Paleozoic calc-alkaline granitoid magmatism in the northern part of the Western Transbaikalia.

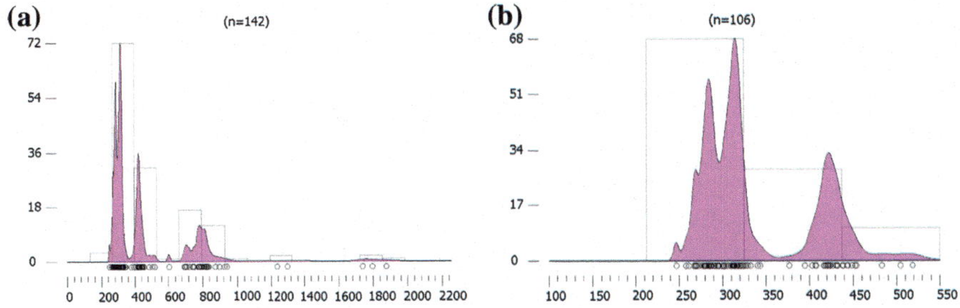

Fig. 3 $^{206}Pb/^{238}U$ age histogram and relative probability density plot of the detrital zircon ages from Angarakan river: **a** all ages; **b** only Paleozoic ages

Conclusions

A significant part of the Western Transbaikalia granitoid province is composed of calc-alkaline granites of the Barguzin complex.

U–Pb LA-ICP-MS isotope-geochronological study of detrital zircons from the current Angarakan river sediments, the sourceland of which was the granites of the Barguzin complex in the northern part of the province, has shown that the greatest number of them are of late Paleozoic age, and they form two distinct age peaks (~290 and ~315 Ma). These data are consistent with the two age maxima of the formation of the Barguzin granites established by the predecessors during the dating of magmatic zircons that together implies the formation of these granites during two stages that are closely related in time.

The middle Paleozoic (~425 Ma) and Neoproterozoic (780–770 Ma) ages of zircons, as well as older single zircons with ages of 1100, 1400, 1760, 1825 Ma reflect the age of the basement and protoliths of the late Paleozoic calc-alkaline granitoid magmatism in the northern part of the Western Transbaikalia.

Acknowledgements The studies were carried out with the financial support of the RFBR grant No. 17-05-00275a and the comprehensive program of fundamental scientific research of the SB RAS "Carboniferous-Permian-Triassic granitoid batholiths and basite-ultrabasite complexes of the southern frame of the North Asian Craton: age limits, numerical models of formation, metallogeny" №II.1.39.

References

Buyantuev MD, Khubanov VB, Vrublevskaya TT. U-Pb LA-ICP-MS dating of zircons from subvolcanics of the bimodal dyke series of the Western Transbaikalia: Technique, and evidence of the Late Paleozoic extension of the crust. Geodynamics & Tectonophysics. 2017;8:369–384. https://doi.org/10.5800/GT-2017-8-2-0246

Griffin WL, Powell WJ, Pearson NJ, O'Reilly SY. GLITTER: Data reduction software for laser ablation ICP–MS. In: Sylvester PJ, editor. Laser Ablation ICP-MS in the Earth. Mineralogical Association of Canada short course series; 2008. P. 204–207

Ivanov AV, Demonterova EV, Reznitsky LZ, Barash IG, Arzhaninikov SG, et al. Catastrophic outburst and tsunami flooding of Lake Baikal: U–Pb detrital zircon provenance study of the Palaeo-Manzurka megaflood sediments. International Geology Review. 2016;58:1818–1830. https://doi.org/10.1080/00206814.2015.1064329

Jackson SE, Pearson NJ, Griffin WL, Belousova EA. The application of laser ablation-inductively coupled plasma-mass spectrometry to in situ U–Pb zircon geochronology. Chemical Geology. 2004;211:47–69. https://doi.org/10.1016/j.chemgeo.2004.06.017

Khubanov VB, Buyantuev MD, Tsygankov AA. U–Pb dating of zircons from PZ_3 –MZ igneous complexes of Transbaikalia by sector-field mass spectrometry with laser sampling: technique and comparison with SHRIMP. Russian Geology and Geophysics. 2016;57:190–205. https://doi.org/10.1016/j.rgg.2016.01.013

Khubanov VB, Tsygankov AA, Burmakina GN, Buyantuev MD, Burdukovskii VV. Vozrast tsirkonov iz sovremennykh rechnykh osadkov (reki Selenga, Muya, Angarakan) Zapadnogo Zabaikal'ya: k otsenke prodolzhitel'nosti i stadiinosti pozdnepaleozoiskogo magmatizm. Materialy soveshchaniya « Geodinamicheskaya evolyutsiya litosfery Tsentral'no-Aziatskogo podvizhnogo poyasa (ot okeana k kontinentu) » . 2017, Irkutsk: Institut zemnoi kory SO RAN, s.286–287. [Khubanov VB, Tsygankov AA, Burmakina GN, Buyantuev MD, Burdukovsky VV. The age of zircons from modern river sediments (the Selenga River, the Moya River, the Angarakan River) of the Western Transbaikalia: to an assessment of the duration and stages of Late Paleozoic magmatism. Proceedings of the meeting "Geodynamic evolution of the lithosphere of the Central Asian mobile belt (from the ocean to the continent)". 2017, Irkutsk. Institute of the Earth's Crust SB RAS, pp. 286–287 (in Russia)]

Kovach VP, Sal'nikova EB, Rytsk EY, Yarmolyuk VV, Kotov AB, et al. 2012, The time length of formation of the Angara-Vitim batholite: Results of U-Pb geochronological studies. Doklady Earth Sciences. 2012;444:553–558. https://doi.org/10.1134/S1028334X12050133

Litvinovsky BA, Tsygankov AA, Jahn BM, Katzir Y, Be'eri-Shlevin Y. Origin and evolution of overlapping calc-alkaline and alkaline magmas: The Late Palaeozoic post-collisional igneous province of Transbaikalia (Russia). Lithos. 2011;125:845–874 https://doi.org/10.1016/j.lithos.2011.04.007

Ludwig KR. User's Manual for Isoplot 3.70: A Geochronological Toolkit for Microsoft Excel. Berkeley Geochronology Center, Berkeley; 2008

Sláma J, Košler J, Condon DJ, Crowley JL, Gerdes A, Hanchar JM, et al. Plešovice zircon – A new natural reference material for U-Pb and Hf isotopic microanalysis. Chemical Geology. 2008;249:1–35. https://doi.org/10.1016/j.chemgeo.2007.11.005

Tsygankov AA, Burmakina GN, Khubanov VB, Buyantuev MD. Geodynamics of Late Paleozoic Batholith-Forming Processes in Western Transbaikalia. Petrology. 2017;25:396–418. https://doi.org/10.1134/S0869591117030043

Tsygankov AA, Litvinovsky BA, Jahn BM, Reichow MK, Liu DY, et al. Sequence of magmatic events in the Late Paleozoic of Transbaikalia, Russia (U-Pb isotope data). Russian Geology and Geophysics. 2010;51:972–994. https://doi.org/10.1016/j.rgg.2010.08.007

Tsygankov AA, Matukov DI, Berezhnaya NG, Larionov AN, Posokhov VF, et al. Late Paleozoic granitoids of western Transbaikalia: magma sources and stages of formation. Russian Geology and Geophysics. 2007;48:120–140. https://doi.org/10.1016/j.rgg.2006.12.011

Vermeesch P. On the visualisation of detrital age distributions. Chemical Geology. 2012;312–313:190–194. https://doi.org/10.1016/j.chemgeo.2012.04.021

Wiedenbeck M, Allé P, Corfu F, Griffin WL, Meier M, Oberli F, et al. Three natural zircon standards for U–Th–Pb, Lu–Hf, trace element and REE analysis. Geostandards Newsletter. 1995;19:1–23. https://doi.org/10.1111/j.1751-908X.1995.tb00147.x

Yarmolyuk VV, Budnikov SV, Kovalenko VI, Antipin VS, Goreglyad AV, et al. Geochronology and geodynamic setting of the Angara–Vitim batholith. Petrology. 1997;5:401–414

Physical and Chemical Characteristics of Pathogenic Tooth Pulp Calcifications

Daria Kiseleva, Evgeny S. Shagalov, Maria V. Zaitceva, Elizaveta A. Pankrushina, Sergey G. Sustavov and Natalia M. Spivak

Abstract Denticles (pulp stones) are formed of dentin or dentin-like tissue as a result of the crown and root pulp mineralisation. The etiological factors of pathogenic mineral formation in the pulp have not yet been studied. The aim of the study is to determine the physico-chemical characteristics of pulp stones (denticles) from patients of the Ural region. Due to the small size of the samples, the local analytical methods were used for the study: SEM-EDS, LA-ICP-MS and Raman microspectroscopy. According to X-ray phase analysis and Raman spectroscopy, the phase composition of denticle mineral component is carbonate-hydroxyapatite. For the first time, the information on the organic denticle component corresponding to collagen has been obtained by Raman spectroscopy. Elemental analysis by SEM-EDS and LA-ICP-MS has indicated the presence of a number of essential (sodium, magnesium, potassium, chlorine, sulfur, zinc) and toxic (lead, barium) elements in the composition of pulp stones. The trace element composition of denticles may indicate the development of inflammatory processes and the state of oxidative stress in the pulp, leading to the initiation of a pathogenic mineralisation reaction.

Keywords Tooth pulp · Pathogenic mineralisation · Denticles · Element and phase composition · Carbonate-hydroxyapatite · Collagen

---

D. Kiseleva (✉) · E. S. Shagalov · M. V. Zaitceva · E. A. Pankrushina
A.N. Zavaritsky Institute of Geology and Geochemistry of the Ural Branch of the Russian Academy of Sciences, 15 Vonsovskogo Street, Ekaterinburg, Russia
e-mail: Kiseleva@igg.uran.ru

S. G. Sustavov
Ural State Mining University, 85 Khokhryakova Street, Ekaterinburg, Russia

N. M. Spivak
Ural State Medical University, 3 Repina Street, Ekaterinburg, Russia

S. Votyakov et al. (eds.), *Minerals: Structure, Properties, Methods of Investigation*, Springer Proceedings in Earth and Environmental Sciences,
https://doi.org/10.1007/978-3-030-00925-0_14

Introduction

Denticles (pulp stones) are formed of dentin or dentin-like tissue as a result of crown and root pulp mineralisation. Such calcifications are found during routine radiographic examinations of the oral cavity and are quite widespread among all age groups (Berès et al. 2016).

The etiological factors of pathogenic mineral formation in the pulp have not yet been studied. The proposed mechanisms include the development of denticles around damaged pulp tissue (e.g., decaying cells, blood clots, or collagen fibres) and calcium phosphate crystal depositing within mineralising cells (Luukko et al. 2011).

Most often, denticles are formed without any obvious symptoms while an inflammatory process with pronounced pain starts only when the pulp neurovascular bundle is squeezed by these new formations.

There are several shapes of pulp stones: free (located directly in the pulp), adherent (merged with the tooth wall) and interstitial (arising from free or adherent denticle fouling with new layers of secondary dentin) (Johnson and Bevelander 1956). Depending on the structure, one can distinguish true or dentin-like (with dentinal tubules), false (structureless localised aggregates of calcified matter) and diffuse (usually found near vessels) denticles (Sayegh and Reed 1968).

To date, there is very little data on the physico-chemical characteristics of denticles (its composition, structure and properties) in the literature (both Russian and foreign); it is only known that their mineral phase consists of carbonate-hydroxyapatite (Berès et al. 2016; Trowbridge et al. 1996). Berès et al. (2016) tried to reveal the crystallographic structure and elemental composition of the mineral phase of pulp stones using X-ray diffraction analysis and EDS scanning electron microscopy.

The aim of the study was to determine the physico-chemical characteristics of pulp stones (denticles) of patients in the Ural region.

Materials and Methods

Two samples of pulp stones (denticles), extracted from patients during their dental treatment (provided by the clinic of the Ural State Medical University) were studied.

SEM images were obtained using JSM-6390LV (JEOL) scanning electron microscope with the accelerating voltage of 20 kV; the samples were sputtered with carbon. The elemental composition was determined using the Oxford X-ray energy dispersive spectrometer (EDS). The trace element composition was determined using NexION 300S quadrupole mass spectrometer with New Wave Research NWR 213 laser ablation system. Laser operating parameters: energy 3 J/cm^2, frequency

10 Hz, spot diameter 50 μm. MAPS-4 reference material (pressed calcium phosphate) was used for instrument calibration.

Debyegrams were obtained using URS-55 X-ray powder diffractometer (RKD chamber, FeKα + β, U = 30 kV, I = 10 mA) at the Ural State Mining University.

Raman spectra were collected by LabRam HR Evolution Raman spectrometer in the spectral range 300–4000 cm^{-1}; excitation laser 633 nm; lens 50×; grating 600 g/mm.

Results and Discussion

SEM images of the investigated samples are shown in Fig. 1.

As can be seen from Fig. 1, the denticles had a heterogeneous porous, parallel-fibrous, tangled-fibrous structure. The internal structure was traced in the large pores and channels; the sites with an increased content of organic matter as well as substantially mineralised areas were observed. Dentin-like structure was not observed. According to the results of EDS analysis, the mineralisation was represented by biogenic apatite of variable composition, with thin films of chlorine-sulphur-siliceous-alumina composition. The elevated levels of sodium, magnesium, potassium, aluminum and silicon were found, as well as zinc and strontium at the separate analysis points. The anions were represented by chlorine and sulphur in high concentrations. Sodium, potassium, magnesium and chlorine are usually the most important electrolytes in the liquid media of the body; sulphur is an essential component of blood and neural tissue, collagen organic matrix, as well as ameloblasts and odontoblasts (the cells producing tooth enamel and dentin) (J. L. Burguera and M. Burguera 2009). Some trace elements (Cr, Zn, B, etc.), so-called essential, are necessary for the body to function properly. On the contrary, a number of elements (Al, Sr, etc.) do not play a definite role in the life of the organism, while some of them (Pb, Cd, Hg, etc.) are toxic even in small amounts (Reeder et al. 2006). The EDS analysis data are generally confirmed by the results of LA-ICP-MS in the denticles studied: zinc content is 0.1–0.13%, strontium 185–380 ppm, barium 14–23 ppm, lead 7–14 ppm, boron 4–7 ppm. An increased content of zinc and copper in denticles was noted earlier (Berès et al. 2016); the authors linked this fact to the possible increased activity of Zn, Cu-containing enzyme-antioxidant referred to as superoxide dismutase (SOD) under the oxidative stress caused by inflammatory processes in the pulp, which in turn facilitated the initiation of a pathogenic mineralisation reaction. The presence of lead in pulp stones can also indirectly indicate the development of an inflammatory process in the pulp, as lead is highly neurotoxic and can provoke inflammation (Chibowska et al. 2016); even its smallest concentration can lead to negative consequences. In general, the presence of heavy elements (strontium, barium, and lead) in pulp stones (and, consequently, in the pulp) can be a reflection of an unfavourable ecological situation in a megapolis or even the entire Ural industrial region.

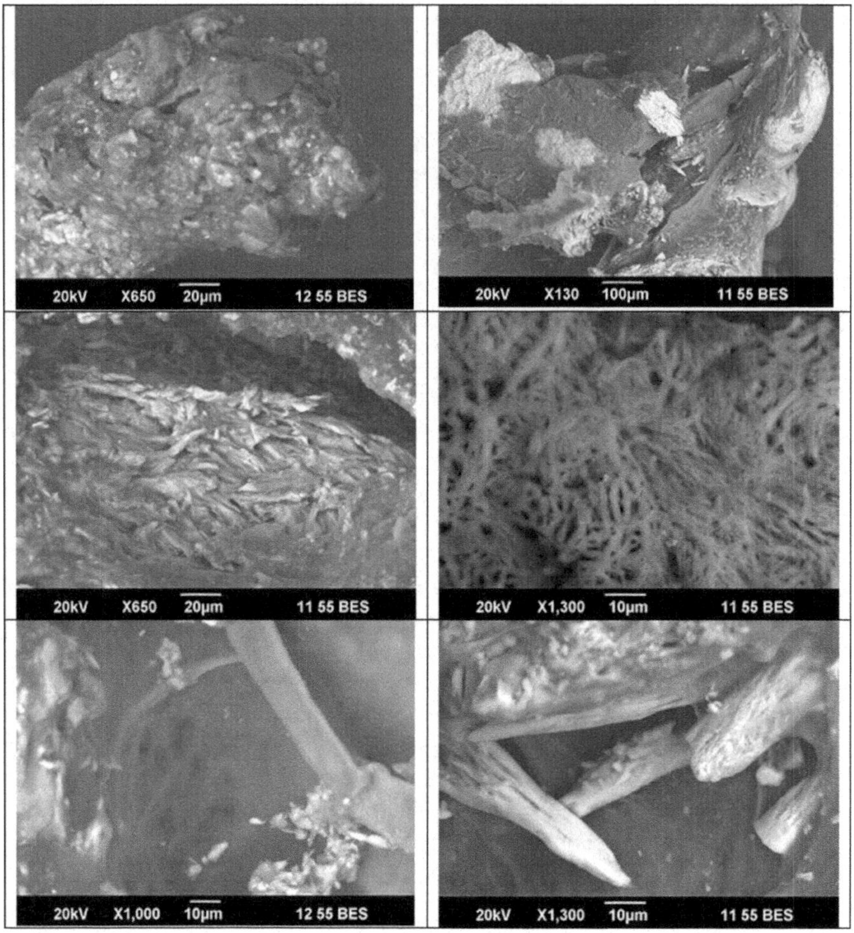

Fig. 1 SEM images of the denticles

According to the results of X-ray phase analysis, the mineral component of a denticle corresponds to poorly structured carbonate-hydroxyapatite characterised by broadened major peaks.

The Raman spectra of the samples studied are characterised by high continuous background, probably due to the luminescence of organic compounds (Sa et al. 2012). Figure 2 demonstrates background subtracted Raman spectra of the denticle (sample No. 1) (1–4 various analysis points on its surface). The registered spectral bands are arising from the vibrations of the carbonate-hydroxyapatite mineral component: ν_1 PO_4^{3-} symmetric stretching vibration ($\sim$960 cm^{-1}), ν_4 PO_4^{3-} bending vibrations (430 and 580 cm^{-1}), and ν_1 stretching vibrations of B-type carbonate ions replacing the phosphate ions in the apatite lattice (1065–1070 cm^{-1}) (band assignment according to Kirchner et al. (1997)). The organic matrix

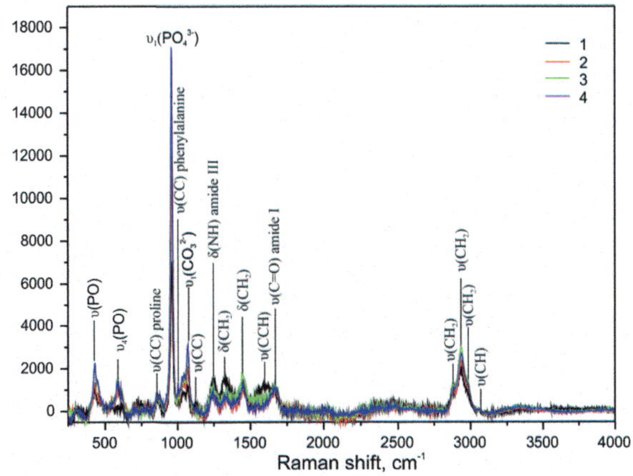

Fig. 2 Raman spectra of the investigated denticle sample: 1–4 various analysis points on the denticle surface

vibrations are also clearly manifested: the vibrations of carbon bonds in the proline, hydroxyproline and phenylalanine aminoacids; NH and C = O bonds in type I and III amides (band assignment according to Kirchner et al. (1997)). All the observed vibrations of the protein matrix correspond to the collagen fibrillar protein, which is the main organic component of tooth enamel and dentin.

Conclusions

The physico-chemical characteristics of denticles (pulp stones) extracted from patients of the Ural region have been determined. Due to the small size of the samples, the following local analytical methods have been used for the study: SEM-EDS, LA-ICP-MS, Raman microspectroscopy. According to X-ray phase analysis and Raman spectroscopy, the phase composition of denticle mineral component has been determined as carbonate-hydroxyapatite. The data obtained are consistent with the literature data. For the first time, the information on the organic component of denticle corresponding to collagen has been obtained by Raman spectroscopy.

Elemental analysis by SEM-EDS and LA-ICP-MS has indicated the presence of a number of essential (sodium, magnesium, potassium, chlorine, sulfur and zinc) and toxic (lead, barium) elements in the composition of pulp stones, provided that the data on the trace element composition by LA-ICP-MS has been obtained for the first time. The trace element composition of denticles may indicate the development of inflammatory processes and the state of oxidative stress in the pulp, leading to the initiation of a pathogenic mineralisation reaction.

Acknowledgements The work was carried out at the UB RAS "Geoanalitik" Center for Collective Use within the IGG UB RAS № AAAA-A18-118053090045-8.

References

Berès F, et al. Comparative physicochemical analysis of pulp stone and dentin. Journal of Endodontics. 2016;42(3):432–438.

Burguera JL, Burguera M. Recent on-line processing procedures for atomic samples for determination of trace elements by atomic spectrometric methods. Spectrochimica Acta Part B. 2009;64:451–458.

Chibowska K, Baranowska-Bosiacka I, Falkowska A, Gutowska I, Goschorska M & Chlubek D. Effect of lead (Pb) on inflammatory processes in the brain. International Journal of Molecular Sciences. 2016;17(12):2140.

Johnson PL, Bevelander G. Histogenesis and histochemistry of pulpal calcification. J Dent Res. 1956;35:714–22.

Kirchner MT, Edwards HGM, Lucy D., Pollard AM. Ancient and modern specimens of human teeth: a Fourier transform Raman spectroscopic study. Journal of Raman Spectroscopy. 1997;28:171–178.

Luukko K, Kettunen P, Fristad I, et al. Structure and functions of the dentin-pulp complex. In: Cohen S, Burns RC, editors. Pathways of the Pulp. 10th ed. St Louis: Mosby; 2011. p. 494–497.

Reeder RJ, Schoonen MAA, Lanzirotti A. Metal speciation and its role in bioaccessibility and bioavailability in medical mineralogy and geochemistry. In: Sahai N and Schoonen MAA, editors. Reviews in Mineralogy and Geochemistry, vol. 64: Medical mineralogy and geochemistry. Washington, DC: Mineralogical Society of America 2006. p. 59–114.

Sa Y, Chen D, Liu Y, et al. Effects of two in-office bleaching agents with different pH values on enamel surface structure and color: An in situ vs. in vitro study. J Dent. 2012;40(SUPPL. 1): e26–e34.

Sayegh FS, Reed AJ. Calcification in the dental pulp. Oral Surg Oral Med Oral Pathol. 1968;25:873–82.

Trowbridge HO, Tewart JCB, Shapiro IM. Evaluation of indurated, diffusely calcified human dental pulp. Proceedings of the International Conference of Dentin/Pulp Complex. Tokyo: Quintescence Publishing; 1996:297.

Raman and IR Spectroscopy of Na_2O–SiO_2–GeO_2 System

Nadezhda M. Korobatova, Mikhail V. Shtenberg, Tatyana N. Ivanova and Olga N. Koroleva

Abstract Germanosilicate glasses of compositions (mol%): $(100 - x)(33Na_2O \cdot 67SiO_2) \cdot xGeO_2$, where x = 20, 25, 30, 35 and 40 were studied using Raman and IR spectroscopy. Structural units existing in glasses were determined. The transition of silicon-containing anionic sheet structures to three-dimensional network with the formation of Si–O–Si bridge atoms was observed when germanium oxide was added to sodium disilicate.

Keywords Raman · Infrared · Spectroscopy · Germanosilicate glasses

Introduction

Germanosilicate systems are of interest for solving a number of issues of chemistry, geochemistry, materials science, etc. Scientists began to explore them in the eighties of the last century by means of Raman (Verweij 1979a, b; Potter et al. 1992; Bykov et al. 2011; Furukawa and White 1991) and IR spectroscopy (Jiao et al. 2017), XRD (Smets and Lommen 1981), and NMR (Du et al. 2007). In addition, the authors (Osaka et al. 1986) measured the density and elasticity of germanosilicate glasses. Since the most promising area for studying germanium-containing systems is the structure and properties of glasses doped with rare-earth elements, many modern publications are devoted to the study of this type of glasses (Santos et al. 2010a, b; Narendrudu et al. 2015, 2017; Ventura et al. 2015; Cao et al. 2018). The presence in the system of two analogical glass-forming cations gives a particular interest to the study. The system contains five- and six-coordinated atoms

N. M. Korobatova (✉) · M. V. Shtenberg · T. N. Ivanova · O. N. Koroleva
Institute of Mineralogy, Ural Branch of RAS, Ilmensky Reserve,
456317 Miass, Chelyabinsk Region, Russia
e-mail: n.korobatova@yandex.ru

O. N. Koroleva
South Ural State University, Branch in Miass, Kalinin Street 37,
456304 Miass, Chelyabinsk Region, Russia

S. Votyakov et al. (eds.), *Minerals: Structure, Properties, Methods of Investigation*,
Springer Proceedings in Earth and Environmental Sciences,
https://doi.org/10.1007/978-3-030-00925-0_15

[Ge(V) and Ge(VI)] in addition to four-coordinated silicon and germanium atoms. In the study of sodium and potassium metasilicate systems, it has been concluded that there are chains of metagermanosilicate in glasses consisting of SiO_4 and GeO_4 tetrahedra, which are likely to be mixed randomly (Verweij 1979a, b). Strong polarized peaks at about 1100 and 870 cm^{-1} assigned to the vibrations of XO_4 tetrahedra with one non-bridging oxygen atom (NBO) have been found in the Raman spectra of alkaline digermanosilicate glasses (Verweij 1979a, b). In addition, a weak band of about 945 cm^{-1}, according to the authors, indicates a small amount of XO_4 tetrahedra with two NBO (X = Si, Ge). Glass compositions $Na_2O \cdot (2 - x) SiO_2 \cdot GeO_2$ (x = 0–2) and $Na_2O \cdot (1 - x) SiO_2 \cdot GeO_2$ (x = 0–1) have been studied using Raman spectroscopy (Furukawa and White 1991). The authors have determined that alkali-germanate glasses are not fully equivalent to the corresponding silicate glasses due to the possibility of transition Ge(IV) to Ge(VI). Their results show that the spectra and the method of crystallization of glasses with a high alkali content are similar to the corresponding silicates, whereas low-alkaline glasses are different. Adding a small amount of alkali to GeO_2 results in the net formation, possibly with a six-coordinated Ge(VI), without the formation of an NBO. Glasses and melts of germanosilicate with the sodium oxide content of 33 mol% and that of germanium oxide of 20–40 mol% were studied by Raman spectroscopy in the temperature range from 20 to 1100 °C (Bykov et al. 2011) According to the authors, the disordered net of germanosilicate glass is formed by sheet anionic groups of SiO_4 and GeO_4 tetrahedra with one non-bridging oxygen atom (Q^3—units), and the melt consists of separate anions—fragments of these layers. The uniform depolymerization occurs during the interaction of oxide modifier (Na_2O) with glass-forming oxides (SiO_2 and GeO_2) and the resulting non-bridging oxygen atoms are statistically distributed between the silicate and germanate structural units.

Materials and Methods

Three-component glasses of the Na_2O–SiO_2–GeO_2 system were synthesized from analytically pure amorphous SiO_2 and GeO_2, and carbonate Na_2CO_3. The batch was carefully mixed in a mortar with alcohol, dried, and melted in a platinum crucible in a Silit furnace at a temperature below 1473 K until the melt was totally homogenized and then air hardened. The structure of the germanosilicate glasses as a function of the composition system with a mass content of GeO_2 from 20 to 40 mol% was investigated by Raman and IR spectroscopy. A Labram iHR 320 Raman spectrometer with an Olympus BX41 microscope was used for recording the Raman spectra. The IR transmission spectra of glasses were recorded on a Nicolet 6700 Thermo Scientific Fourier spectrometer by the standard technique (as KBr pellets). The baseline correction procedure was performed for all recorded spectra. The spectra were processed using the OMNIC Thermo Nicolet software.

Results and Discussion

This work is devoted to the study of germanosilicate glasses of compositions (mol%): $(100 - x)(33Na_2O \cdot 67SiO_2) \cdot xGeO_2$, where x = 20 (20Ge), 25 (25Ge), 30 (30Ge), 35 (35Ge), 40 (40Ge). Figures 1 and 2 demonstrate the Raman and IR spectra of germanosilicate glasses with GeO_2 content up to 40 mol%.

The same bands (~555, 862 and 1090 cm^{-1}), which differ in intensity, can be observed in comparison of 20Ge Raman spectra (Fig. 1) and $33\%Na_2O \cdot 67\%SiO_2 + 20\%GeO_2$ (16.6Ge) (Bykov et al. 2011). The band observed in the Raman spectra of glasses in the 1090 cm^{-1} region is due to vibrations of the ending groups in the Q^3 (Si) structural units (Anfilogov et al. 2005). Along with this band, there is a band with a maximum of 862 cm^{-1} in the high-frequency region of the spectra. This band is due to the vibrations of the GeO_4 tetrahedra with one non-bridging oxygen atom (Furukawa and White 1980; Henderson and Wang 2002), which are germanium analogues of Q^3 (Si) structural units. A band of 555 cm^{-1} is observed due to symmetric valence vibrations of Si–O–Si (Ge) bridges in the low-frequency region. Bands in the region of 600–650 cm^{-1}, which correspond to vibrations of highly coordinated GeO_6 germanium polyhedra–octahedra (Furukawa and White 1980; Kamitsos et al. 1996), are missing in the spectra of germanosilicate glasses.

IR spectra also carry information about the structure of the glasses, complementing the data obtained from the Raman spectra. In particular, the peaks near 609 and 648 cm^{-1} are characteristic of Si–O–Ge stretching vibration (Jiao et al. 2017) and the peak at 460 cm^{-1} is a superposition of anti-symmetrical Si–O–Si bond (Koroleva et al. 2014; Bykov et al. 2001) and symmetrical Ge–O–Ge bond (Ivanova 2013). The bands near 855 and 1055 cm^{-1} are characteristic of Q^3(Ge)

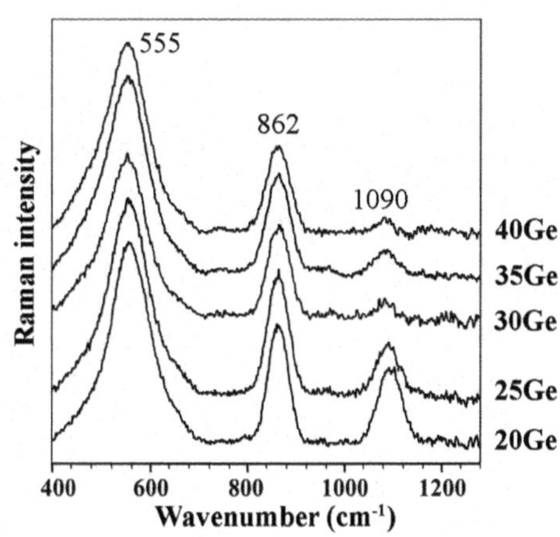

Fig. 1 Raman spectra of 20Ge, 25Ge, 30Ge, 35Ge, 40Ge germanosilicate glasses

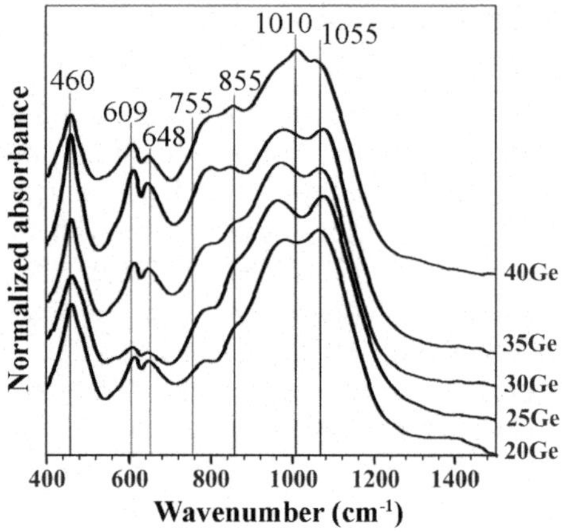

Fig. 2 IR spectra of 20Ge, 25Ge, 30Ge, 35Ge, 40Ge germanosilicate glasses

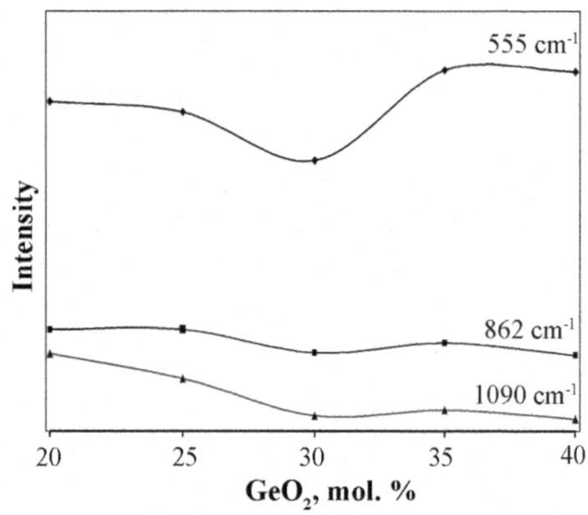

Fig. 3 Intensities of 555, 862 and 1090 cm^{-1} bands from Raman spectra of 20Ge, 25Ge, 30Ge, 35Ge, 40Ge germanosilicate glasses

and Q^3(Si) vibrations, respectively (Jiao et al. 2017; Koroleva et al. 2013, 2014; Bykov et al. 2001; Ivanova 2013). The 755 and 975 cm^{-1} bands can be attributed to TO/LO vibrational modes (Ivanova 2013).

The 1090–1110 cm^{-1} band [Q^3 (Si)] dominates in the Raman spectrum of 16.6Ge glass (Bykov et al. 2001), whereas the 862 cm^{-1} band [Q^3 (Ge)] has a greater intensity in the high-frequency region of the 20Ge spectrum (Fig. 1). Consequently, Q^3 (Si) units dominate in the structure of the 16.6Ge glass, consisting of sheet groups. The number of Q^3 units (Si) decreases with the addition of germanium oxide in the system to the composition of 30Ge. A further increase of germanium content in the glass leads to a decrease of the band intensity at 862 cm^{-1}, whereas the intensity of the band at 555 cm^{-1} increases. The ratios of the intensities of the bands in the Raman spectra are shown in Fig. 3.

Conclusions

It can be concluded that the number of Q^3 (Si) structural units decreases sharply with the addition of germanium oxide in the system up to 30 mol%, and then becomes close to zero. The intensity of the 862 cm^{-1} band, characterizing the vibrations of Q^3 (Ge) structural units, slightly decreases throughout the graph, which can be associated with an increase of the Ge/Na ratio, and, as a consequence, with the increase in glass polymerization. Thus, the transition of silicon-containing anionic sheet structures to the three-dimensional network with the formation of Si–O–Si bridge atoms is observed first, and only then the reduction of non-bridging oxygen atoms in Q^3 (Ge) germanium-oxygen tetrahedra occurs, when germanium oxide is added to sodium disilicate.

Acknowledgements This work was financially supported by the Russian Foundation for Basic Research, projects No. 16-35-60045 and No. 18-05-00079.

References

Anfilogov VN., Bykov VN., Osipov AA. Silicatnie rasplavy, M.: Nauka; 2005. [Anfilogov VN., Bykov VN., Osipov AA. Silicate melts. M.: Nauka; 2005. (In Russ.)]

Bykov VN., Anfilogov VN., Osipov AA. Spectroskopiya i structura silicatnih rasplavov i stekol. Miass: Institute of mineralogy, UrB RAS; 2001. [Bykov VN., Anfilogov VN., Osipov AA. Spectroscopy and structure of silicate melts and glasses. Miass: Institute of mineralogy, UrB RAS; 2001. (In Russ.)]

Bykov VN., Ivanova TN., Koroleva ON. Raman spectroscopy of borosilicate and germanate-silicate glasses and melts. Russ. Metall. 2011;8:719–722. https://doi.org/10.1134/S0036029511080040

Cao W., Huang F., Ye R., Cai M., Lei R., Zhang J., Xu S., Zhang X. Structural and fluorescence properties of Ho^{3+}/Yb^{3+} doped germanosilicate glasses tailored by Lu_2O_3. J. Alloys Compd. 2018;746:540–548. https://doi.org/10.1016/j.jallcom.2018.02.325

Du LS., Peng L., Stebbins JF. Germanosilicate and alkali germanosilicate glass structure: New insights from high-resolution oxygen-17 NMR. J. Non-Cryst. Solids. 2007;353:2910–2918. https://doi.org/10.1016/j.jnoncrysol.2007.05.122

Furukawa T., White, WB. Raman spectroscopic investigation of the structure and crystallization of binary alkali germanate glasses. J. Mater. Sci. 1980;15(7):1648–1662. https://doi.org/10.1007/BF00550581

Furukawa T., White WB., Raman spectroscopic investigation of the structure of silicate glasses. IV. Alkali–silico–germanate glasses. J. Chem. Phys. 1991;95:776–784. https://doi.org/10.1063/1.461084

Henderson GS., Wang HM. Germanium coordination and the germanate anomaly. Eur. J. Mineral. 2002;14(4):733–744. https://doi.org/10.1127/0935-1221/2002/0014-0733

Ivanova TN. Structura germanatnih stekol i rasplavov po dannim kolebatelnoy spektroskopiy. Ekaterinburg: Institute of metallurgy UrB RAS; 2013. [Ivanova TN. The structure of germanate glasses and melts from the data of vibrational spectroscopy. Ekaterinburg: Institute of metallurgy UrB RAS; 2013. (In Russ.)]

Jiao K., Zhang, Z., Xu X., Lv Z., Song J., Lin C., Sun J., He M., Gies H. Synthesis and characterization of germanosilicate molecular sieves: GeO_2/SiO_2 ratio, H_2O/TO_2 ratio and temperature. Dalton Trans. 2017;46(7):2270–2280. https://doi.org/10.1039/C6DT04688F

Kamitsos EI., Yiannopoulos YD., Karakassides MA., Chryssikos, GD., Jain H. Raman and infrared structural investigation of $xRb_2O{\bullet}(1\text{-}x)GeO_2$ glasses. J. Phys. Chem. 1996;100 (28):11755–11765. https://doi.org/10.1021/jp960434+

Koroleva ON., Anfilogov VN., Shatskiy A., Litasov KD. Structure of $Na_2O–SiO_2$ melts as a function of composition: in situ Raman spectroscopic study. J. Non-Cryst. Solids. 2013;375:62–68. https://doi.org/10.1016/j.jnoncrysol.2013.04.060

Koroleva ON., Shtenberg MV., Khvorov PV. Vibrational spectroscopic and X-ray diffraction study of crystalline phases in the $Li_2O\text{-}SiO_2$ system. Russ. J. Inorg. Chem. 2014;59(3):255–258. https://doi.org/10.1134/S0036023614030139

Narendrudu T., Suresh S., Chinna Ram G., Veeraiah N., Krishna Rao D. Spectroscopic and structural properties of Cr^{3+} ions in lead niobium germanosilicate glasses. J. Lumin. 2017;183:17–25. https://doi.org/10.1016/j.jlumin.2016.11.005

Narendrudu T., Suresh S., Yusub S., Kumar AS., Rajyasree C., Rao MV., Kumar VR., Rao DK. Structural investigations of lead germanosilicate glasses doped with Nb_2O_5 by means of spectroscopic and dielectric studies. J. Mol. Struct. 2015;1098:181–190. https://doi.org/10.1016/j.molstruc.2015.05.051

Osaka A., Ariyoshi K., Takahashi K. Network structure of alkali germanosilicate glasses. J. Non-Cryst. Solids. 1986;83:335–343. https://doi.org/10.1016/0022-3093(86)90246-2

Potter BG Jr., Ochoa R., Chen DG., Simmons JH. Raman scattering investigation of $xGeO_2\text{-}(l - x) SiO_2$ planar waveguides. Opt. Lett. 1992;17(19):1349–1351. https://doi.org/10.1364/OL.17.001349

Santos R., Santos LF., Almeida RM., Deubener J., Wondraczek L. Crystallization of niobium germanosilicate glasses. J. Solid State Chem. 2010b;183(1):128–135. https://doi.org/10.1016/j.jssc.2009.10.031

Santos R., Santos LF., Almeida RM. Optical and spectroscopic properties of Er-doped niobium germanosilicate glasses and glass ceramics. J. Non-Cryst. Solids. 2010a;356(44-49):2677–2682. https://doi.org/10.1016/j.jnoncrysol.2010.03.036

Smets BMJ., Lommen TPA. The structure of germanosilicate glasses, studied by X-ray photoelectron spectroscopy. J. Non-Cryst. Solids. 1981;46:21–32. https://doi.org/10.1016/0022-3093(81)90071-5

Ventura M., Monteiro G., Almeida RM., Schwarz R., Santos LF. Germanosilicate glassceramics for non-linear optics. J. Mater. Sci. 2015;50(9):3477–3484. https://doi.org/10.1007/s10853-015-8908-1

Verweij H. Raman study of the structure of alkali germanosilicate glasses (I): sodium and potassium metagermanosilicate glasses. J. Non-Cryst. Solids. 1979a;33:41–53. https://doi.org/10.1016/0022-3093(79)90094-2

Verweij H. Raman study of the structure of alkali germanosilicate glasses (II) lithium, sodium and potassium digermanosilicate glasses. J. Non-Cryst. Solids. 1979b;33:55–69. https://doi.org/10.1016/0022-3093(79)90095-4

Preparation and Properties of Ceramic Materials from Coal Fly Ash

Olga B. Kotova, Grigory V. Ignatiev, Dmitry A. Shushkov, Maria Harja and Maarten A. T. M. Broekmans

Abstract The development of technologies for utilization of coal fly ash, the huge accumulations of which represent a serious environmental threat, is an essential problem. Obtaining of ceramic materials can be a promising way of fly ash utilization. The article presents the results of sintering behavior study of Vorkuta thermal power plant fly ash at 800–1250 °C. Apparent density, water absorption, open porosity, and crushing strength of sintered samples were determined. Experiments have shown that sintering occurs at temperatures of 1050 °C and above. Ceramics obtained at a firing temperature of 1150 °C is characterized by the highest values of apparent density and crushing strength and low values of water absorption and porosity. At 1200 °C and above there is an increase in sample dimensions (foaming) with the formation of internal pores, i.e. a porous ceramics is obtained. It has been shown that the pressure and the sintering temperature affect the characteristics of ceramics: apparent density, water absorption, porosity, and crushing strength. According to its technological characteristics, the obtained materials can be used for the production of building ceramics.

O. B. Kotova (✉) · G. V. Ignatiev · D. A. Shushkov
N. P. Yushkin Institute of Geology of the Komi Science Center of the Ural Branch of the Russian Academy of Sciences, Pervomayskaya St. 54, 167982 Syktyvkar, Russia
e-mail: kotova@geo.komisc.ru

G. V. Ignatiev
e-mail: ignatiev.grigoriy@gmail.com

D. A. Shushkov
e-mail: dashushkov@geo.komisc.ru

M. Harja
"Gheorghe Asachi" Technical University of Iasi, Professor Dimitrie Mangeron Blvd 67, 700050 Iasi, Romania
e-mail: mharja@tuiasi.ro

M. A. T. M. Broekmans
Geological Survey of Norway (NGU), Leiv Eirikssons Vei 39 7040, Trondheim, Norway
e-mail: maarten.atm.broekmans@gmail.com

S. Votyakov et al. (eds.), *Minerals: Structure, Properties, Methods of Investigation*, Springer Proceedings in Earth and Environmental Sciences,
https://doi.org/10.1007/978-3-030-00925-0_16

Keywords Coal fly ash · Ceramics · Industrial waste · Sintering

Introduction

Coal combustion at thermal power plants generates ash and slag, which are among the multi-tonnage industrial waste of the energy industry. The development of technologies for utilization of coal fly ash, occupying large areas and causing damage to the environment, is a very important problem. A promising way of their utilization is the conversion of fly ash into zeolites (Belviso 2018; Yao et al. 2015; Kotova et al. 2016; Shushkov et al. 2018). The technology of ceramic material production can be an alternative way to utilize fly ash. Many articles are devoted to the production of ceramics and glass ceramics from fly ash with the addition of natural raw materials and industrial wastes: quartz, clay, talc, dolomite, tincal and titanium ore wastes (Chandra et al. 2005; Dana et al. 2004; Olgun et al. 2005; Shao et al. 2004; Vilches et al. 2003). However, little attention is paid to the production of ceramics using only fly ash (Erol et al. 2008). The aim of the work is to show the possibility of obtaining ceramics from fly ash without the addition of other components and to find out the conditions for producing ceramics with optimal characteristics of crushing strength and water absorption.

Materials and Methods

For experiments, Vorkuta thermal power plant fly ash with a particle size less than 0.05 mm was used.

To prepare samples, 1.5 g of fly ash were thoroughly mixed with 0.150 ml of distilled water in a mortar. The resulting mixture was pressed at 7, 21 and 35 MPa and an exposure time of 5 min. As a result, cylindrical samples with a diameter of 10 mm were obtained. For pressing fly ash at 35 MPa, 5% polyvinyl alcohol in the volume of 0.500 ml was used instead of water. Before sintering, the compressed samples were dried at 103 °C for 8 h to remove moisture. Firing was carried out in a furnace at temperatures from 800 to 1250 °C (800, 900, 1000, 1050, 1100, 1150, 1200, 1250 °C) for 2 h. The heating rate was 10 °C/min. The samples were cooled down in a furnace.

The mineral composition of fly ash and ceramic materials was determined by Shimadzu XRD-6000 diffractometer (CuK radiation, Ni filter, 30 kV, 30 mA). The powder samples were scanned in the 2θ range between 2 and 70° at a rate of 1°/min and a scanning step of 0.05°. The chemical composition was determined by x-ray fluorescence analysis (XRD-1800 Shimadzu).

Apparent density, open porosity and water absorption were determined according to GOST 2409-95. The crushing strength was determined by a universal testing machine model WDW-T100. The loading speed was 1 mm/min. Electric properties of ceramics were characterized by the impedance spectroscopy (E7-28 admittance analyzer) in the frequency range from 25 Hz to 10 MHz at room temperature. The voltage on the electrodes was 0.5 V.

Results and Discussion

The experiments have shown that sintering *occurs* at temperatures 1050 °C and above. In Fig. 1 samples of initial fly ash and ceramic materials obtained at firing temperatures of 1050–1250 °C are presented. The changes in color and size are visually observed.

Color. The initial ash has a light gray color. Ceramics obtained at 1050–1100 °C become reddish-brown in color because of iron oxides (hematite). At 1150 °C the color becomes brown, at 1200–1250 °C—dark brown (chocolate).

Chemical composition. The components of chemical composition of initial fly ash (mass%) are as follows: SiO_2—59.10; TiO_2—0.85; Al_2O_3—21.03; Fe_2O_3 total —8.39; MnO—0.07; MgO—2.38; CaO—3.80; Na_2O—1.13; K_2O—2.38; P_2O_5—0.15; SO_3—0.57, S_{org}—0.88. In sintered samples the sulfur content decreases, which is related to the burning of the organic part of fly ash (unburned carbon).

Phase transformations. XRD patterns of fly ash and sintered samples can be seen in Fig. 2. In the initial fly ash quartz, mullite, magnetite and hematite have been determined. The broad "hump" (area of increased background) on the diffraction pattern in the area 15–35°2θ indicates the presence of amorphous phase (probably silicate or aluminosilicate glass).

On XRD patterns of the sintered samples there are no peaks of magnetite, which is transformed to hematite. When the firing temperature increases, the intensity of quartz peaks decreases and the intensity of mullite peaks increases. At a temperature of 1200 °C cristobalite occurs.

Linear shrinkage. Figure 3 shows the changes in the size of samples depending on the firing temperature and pressure. In the range of 1100–1150 °C, a decrease in the diameter and height of samples pressed at 7 and 21 MPa was observed. Linear shrinkage in diameter was 7.6–11.6%, in height was 8–12%. Firing at 1200 °C and

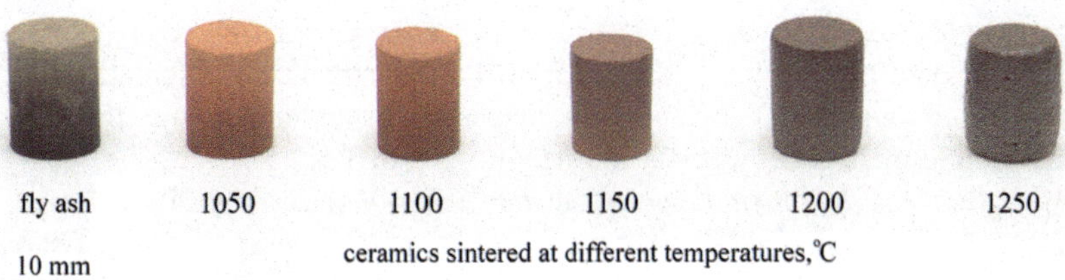

Fig. 1 Initial fly ash and ceramic materials

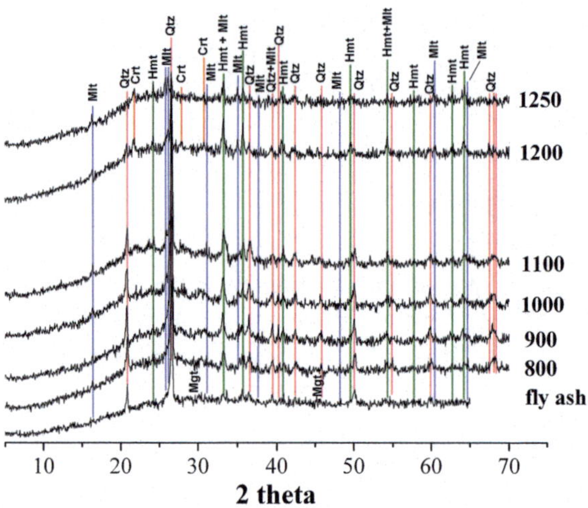

Fig. 2 XRD patterns of fly ash and samples sintered at 800–1250 °C. Mlt—mullite, Qtz—quartz, Mgt—magnetite, Hmt—hematite, Crt—cristobalite

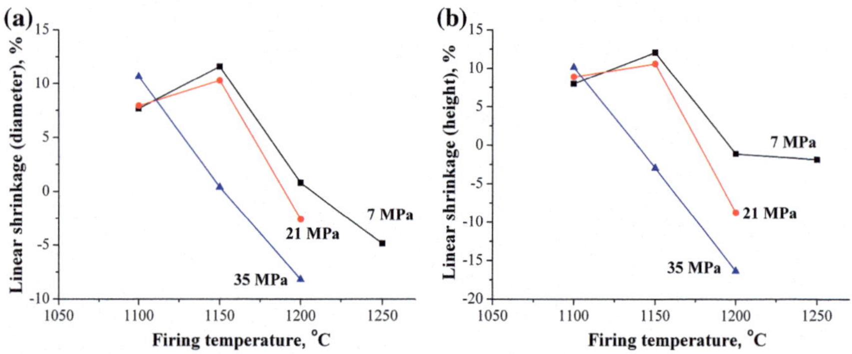

Fig. 3 Effect of firing temperature and pressure on linear shrinkage in diameter (**a**) and height (**b**)

above led to an increase in the size of the sample with the formation of internal pores, i.e. a porous ceramic was obtained. The diameter increased by 2.6–4.8%, the height—by 1.1–8.7%.

The behavior of samples pressed at 35 MPa was somewhat different: shrinkage was observed only at 1100 °C; with an increase in firing temperature, the diameter of the sample increased by 8.1%, the height—by 16.3%.

Apparent density, open porosity, water absorption. In Fig. 4 the results of measurements of apparent density, open porosity and water absorption of samples obtained at different firing temperatures and pressures are presented.

The apparent density of the sample pressed at 7 MPa increased from 1.5 to 2.0 g/cm^3 in the temperature range of 1050–1150 °C. With increasing apparent

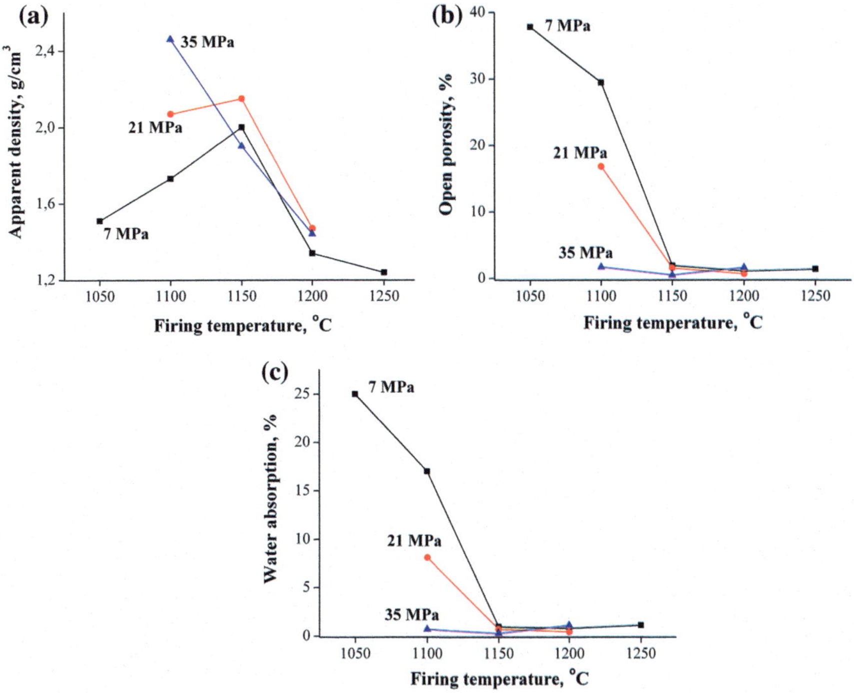

Fig. 4 Effect of firing temperature and pressure on apparent density (**a**), open porosity (**b**) and water absorption (**c**)

density, open porosity and water absorption decreased significantly. Thus, open porosity decreased 20 times (from 38 to 1.9%), water absorption 26 times (from 25 to 0.9%). Firing at 1200 °C and above promoted the formation of internal pores and, accordingly, the decrease of apparent density to 1.2 g/cm^3. The value of the open porosity of porous ceramics was 1.1–1.4%, water absorption –0.8–1.1%.

Pressing at 21 and 35 MPa allowed increasing the apparent density of the ceramics and reducing open porosity and water absorption at a certain firing temperature. Thus, the sample pressed at 35 MPa and a firing temperature of 1100 °C was characterized by a fairly high apparent density (2.46 g/cm^3) and low open porosity and water absorption (0.4 and 0.2%, respectively).

Crushing strength. The results of the crushing strength measurement at different firing temperatures and pressures are shown in Fig. 5. As seen from Fig. 5, the firing temperature significantly affects the crushing strength in contrast to the pressure. Ceramics obtained at a firing temperature of 1150 °C are characterized by the highest values of crushing strength.

Conductivity. The study of conductivity in the frequency range from 100 Hz to 10 MHz has shown (Fig. 6) that ceramics refers to dielectric materials that can be used as an insulator in microwave technology.

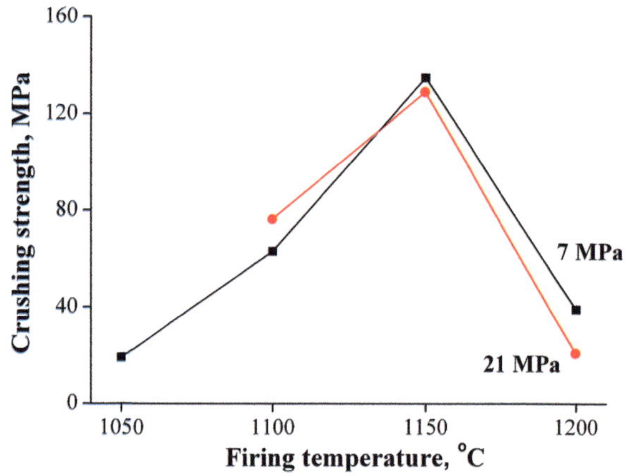

Fig. 5 The effect of firing temperature and pressure on the crushing strength

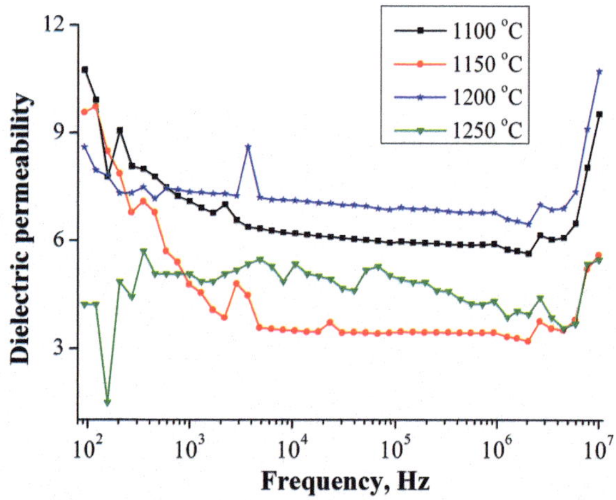

Fig. 6 Ceramics dielectric permeability versus frequency characteristic

Conclusions

Ceramic materials were obtained from the Vorkuta thermal power plant fly ash at firing temperatures of 1050–1250 °C. It was shown that the pressure and the firing temperature affected the characteristics of ceramics: apparent density, water absorption, open porosity, and crushing strength. The ceramic obtained at 1150 °C was characterized by the highest values of apparent density and crushing strength and low water absorption and open porosity. At temperatures of 1200 °C and above, an increase of sample dimensions (foaming) with the formation of internal pores was determined, i.e. a porous ceramics was obtained. According to its technological characteristics, the obtained materials can be used for the production of building ceramics.

Acknowledgements The authors express gratitude to the common use center "Geonauka", Koroleva M.S. (Institute of Chemistry of Komi SC of UB of RAS), Kormschikova Z.I. and Sevbo O.A. (OOO "Vertical") for their help in the analytical work.

The work was carried out according to the research theme "Scientific basis for effective mining, the exploration and development of mineral resources base, development and implementation of innovative technologies, economic-geological zoning of Timan-Northern Ural Region" (GR No. AAAA-A17-117121270037-4).

References

Belviso C. Ultrasonic vs hydrothermal method: Different approaches to convert fly ash into zeolite. How they affect the stability of synthetic products over time? Ultrasonics—Sonochemistry. 2018;43:9–14. https://doi.org/10.1016/j.ultsonch.2017.12.050.

Chandra N., Agnihorti N., Bhasin S., Khan AF. Effect of addition of talc on the sintering characteristics of fly ash based ceramic tiles. J Eur Ceram Soc. 2005;25: 81–88. https://doi.org/10.1016/j.jeurceramsoc.2004.01.004.

Dana K., Das S., Das SK. Effect of substitution of fly ash for quartz in triaxial kaolin-quartz-feldspar system. J Eur Ceram Soc. 2004;24:3169–3175. https://doi.org/10.1016/j.jeurceramsoc.2003.10.008.

Erol M., Kucukbayrak S., Ersoy-Mericboyu A. Characterization of sintered coal fly ashes. Fuel. 2008;87:1334–1340. https://doi.org/10.1016/j.fuel.2007.07.002.

GOST 2409-95. Ogneupory. Metod opredeleniya kazhuscheisya plotnosti, otkrytoi I obschei poristosti, vodopogloscheniya [Refractory. Method of determination of apparent density, open and general porosity, water absorption (In Russ.)].

Kotova OB., Shabalin IN., Shushkov DA., Kocheva LS. Hydrothermal synthesis of zeolites from coal fly ash. Adv in Appl Ceram: Structural, Functional and Bioceramics. 2016;115(3): 152–157. https://doi.org/10.1179/1743676115Y.0000000063.

Olgun A., Erdogan Y., Ayhan Y., Zeybek Development of ceramic tiles from coal fly ash and tincal ore waste. Ceram Int. 2005;31:153–158. https://doi.org/10.1016/j.ceramint.2004.04.007.

Shao H., Liang K., Zhaou F., Wang G., Peng F. Characterization of cordierite-based glass-ceramics prodused from fly ash. J Non-Cryst Solids. 2004;337:157–160. https://doi.org/10.1016/j.jnoncrysol.2004.04.003.

Shushkov DA., Shuktomova II., Rachkova NG., Harja M. Porosity and sorption properties of zeolites synthesized from coal fly ash. Vestnik of Institute of Geology of Komi SC of UB of RAS. 2018;3:31–37. https://doi.org/10.19110/2221-1381-2018-3-32-37.

Vilches LF., Fernandez-Pereira C., Olivars del Valle J., Vale J. Recycling potential of coal fly ash and titanium waste as a new fireproof products. Chem Eng J. 2003;95:155–161. https://doi.org/10.1016/s1385-8947(03)00099-8.

Yao ZT., Ji XS., Sarker PK., Tang JH., Ge LQ., Xia MS., Xi YQ. A comprehensive review on the application of coal fly ash. Earth-Sci Rev. 2015;141:105–121. https://doi.org/10.1016/j.earscirev.2014.11.016.

Platinum Assemblage of Matysken River (Koryak Highlands, Russia): Key Properties and Relation with Bedrock Source

Anton V. Kutyrev and Eugene G. Sidorov

Abstract In this work, we present our first data on the spatial distribution of platinum group elements (PGE) in the minerals of the Matysken massif of Uralian-Alaskan type (Koryak Highlands). The platinum assemblage of the massif is represented by Ir-rich isoferroplatinum rimmed with secondary tetragonal minerals (tulameenite and tetraferroplatinum), native Os and Ir inclusions, minerals of laurite RuS_2—erlichmanite OsS_2 series. Both primary Pt–Fe and Os–Ir–Ru alloys of central parts of the massif dunite core are enriched in Ir relative to the samples from its outermost parts. The reasons for such distribution remain unclear. However, it corresponds with Nekrasov's model, which describes the peculiarities of the composition of PGE minerals in different parts of zonal Uralian-Alaskan type massifs.

Keywords Platinum-group elements · Iridium · Chromite · Uralian-Alaskan type · Dunite · Mafic rocks · Placer

Introduction

One of the main problems related to the massifs of Uralian-Alaskan type is the extremely uneven distribution of platinum group elements (PGE) inside the dunite core. According to the model proposed by Nekrasov et al. (1991) for the placers related to the massifs of Aldan shield (Kondyor massif and others) and developed by other researchers on the basis of Seinav-Galmoenan district placers (Sidorov

A. V. Kutyrev (✉) · E. G. Sidorov
Institute of Volcanology and Seismology FEB RAS, Petropavlovsk-Kamchatsky, Russia
e-mail: anton.v.kutyrev@gmail.com

E. G. Sidorov
e-mail: mineral@kscnet.ru

S. Votyakov et al. (eds.), *Minerals: Structure, Properties, Methods of Investigation*,
Springer Proceedings in Earth and Environmental Sciences,
https://doi.org/10.1007/978-3-030-00925-0_17

et al. 2012), the platinum mineralization at the Uralian-Alaskan type massifs is characterized by the following zonality: in the central parts of the massifs, platinum is generally more Ir- and Fe-rich, while the outer parts bear the platinum, rich in Pd, Rh, and Cu. The assumed sources for these associations are the inner and outer parts of massifs, respectively. For further information on the Uralian-Alaskan type, see Razin (1976), Tolstykh et al. (2000), Augé et al. (2005), Kozlov et al. (2011), Stepanov et al. (2017), Zhmodik et al. (2016) and Johan (2006).

In the Koryak-Kamchatka Platinum Belt, there are several tens of the potentially-platiniferous Uralian-Alaskan massifs. Two of them—the Galmoenan and Seinav massifs—are the sources for economic level placers of the Ledyanoy and Levtyrinvayam rivers, which provided in total more than 60 tons of PGE for 20 years of exploration (Sidorov et al. 2012; Kozlov et al. 2011). Only the subeconomic placers related to other massifs have been found by this moment. One of them is localized at the Matysken river valley and is related to the same-named massif. This massif is the northernmost of the Koryak-Kamchatka platinum belt. Due to its perfect exposure and almost ideal concentrically-zonal structure, it became the optimal ground for studying the principles of PGE distribution in different parts of the massif. In this work, we present our first data on PGE distribution in a specific massif based on the study of lode mineralization and compare this data with placer mineral assemblage.

Materials and Methods

The materials used in this work were collected during the field prospecting works at the Matysken placer, executed by KAMP Ltd. The grains were obtained from alluvium of the Matysken river and chromite veins in dunite. The latter was disintegrated to the particle of 0.1–1.0 mm size and then panned. After panning, they were classified by size (>1.0, 1.0–0.5, 0.5–0.25, 0.25–0.16, 0.16> mm classes) and then the platinum minerals were extracted with the help of binocular lupe. The morphology and composition of the obtained grains were investigated by the VEGA-3 scanning electron microscope with an attached XMax 80 (Oxford Instruments) energy dispersion system (the composition was studied after polishing).

Results and Discussion

The platinum minerals at the Matysken river alluvium are represented by the grains of 0.05–1.50 mm size. Regardless of the distance from the massif, all grains are middle- to well-rounded. The probable explanation of this fact is the transportation of platinum grains in the form of intergrowths with chromite, which may be disintegrated both near and at a distance from the source.

The dominant mineral is isoferroplatinum with a slight excess of Fe and high Ir impurity (up to 7 wt%). An example of the formula: $Fe_{1.14}(Pt_{2.60}Ir_{0.07}Cu_{0.07}Ni_{0.04}Pd_{0.02})_{2.86}$. Numerous mineral inclusions may be found inside, among them native Os, native Ir, laurite RuS_2, erlichmanite OsS_2, kashinite $(Ir, Rh)_2S_3$ and substitution rims of tetraferroplatinum PtFe and, much rarely, tulameenite Pt_2FeCu, chromian spinel and several silicates. The described association is typical for Uralian-Alaskan type massifs and was described in numerous objects, including the Uralian Platinum Belt (Razin 1976; Stepanov et al. 2017; Palamarchuk et al. 2017; Malitch et al. 2017), Koryak Highlands (Sidorov et al. 2012; Tolstykh et al. 2000) and Siberian placer occurrences of an unknown source (Zhmodik et al. 2016) (Table 1).

Lode mineralization is close to the placer, but also includes several additional minerals—sulfides, arsenides, tellurides and unnamed phases. The binary diagram for isoferroplatinum in coordinates of Ir–Fe (Fig. 1a) illustrates the described properties of their composition. The compositions of Pt–Fe minerals from lode deposits fit into the range of placer compositions, but the latter ones are wider. This is quite an expected pattern: the placer represents the whole massif, while the mineral assemblage of the lode represents only the lode itself. This leads us to the first conclusion—the placer represents a broader range of compositions than the sampled parts of the massif. On the other hand, it is evident that in general the compositions of placer-hosted and bedrock-hosted platinum are close.

The comparison of Pt–Fe alloys from different parts of the massif (Fig. 2a) shows that in central parts they are enriched in Ir, which corresponds with Nekrasov's model. However, the concentration of other elements mentioned in the model (Rh, Pd, Cu, Fe) does not depend on the structural position of the studied sample. The diagram in coordinates Ir content in Pt–Fe alloys versus Ir content in Os–Ir–Ru alloys shows several interesting regularities. First, we can see that Os–Ir–Ru alloy compositions also depend on the structural position in the dunite core. Second, a strong correlation between these two factors occurs ($R^2 = 0.79$, partitioning coefficient $Ir_{Pt\text{–}Fe}/Ir_{Os\text{–}Ir\text{–}Ru} = 0.12$). Note that coefficients were counted only for Os-rich hexagonal alloys, while two points of cubic Ir-rich alloys (Fig. 2b) were not taken into account.

Conclusions

1. The association of the Matysken massif and the related placer occurrence fit in most parameters. However, the placer assemblage represents a broader range of compositions than the sampled parts of the massif.
2. The Ir content in coexisting Pt–Fe and Os–Ir–Ru alloys shows a strong correlation. The partitioning coefficient of Ir between these two phases is 0.12.

Table 1 Selected compositions of platinum grains from Matysken river and massif

№	Fe	Cu	Ru	Rh	Pd	Os	Ir	Pt	Sum	Formula
1	7.47	1.32	–	0.45	0.37	–	3.78	86.58	99.97	$(Pt_{2.83}Ir_{0.13}Cu_{0.13}Rh_{0.03}Pd_{0.02})_{3.14}Fe_{0.85}$
2	10.40	–	0.52	1.26	0.91	–	3.43	83.41	99.93	$(Pt_{2.60}Ir_{0.11}Rh_{0.07}Pd_{0.05})_{2.87}Fe_{1.13}$
3	8.85	–	–	–	0.70	–	1.81	87.37	98.73	$(Pt_{2.88}Ir_{0.06}Pd_{0.04})_{2.98}Fe_{1.02}$
4	9.14	–	–	–	0.84	–	1.49	87.95	99.41	$(Pt_{2.86}Ir_{0.05}Pd_{0.05})_{2.96}Fe_{1.04}$
5	8.71	0.44	–	–	–	–	2.45	87.93	99.53	$(Pt_{2.83}Ir_{0.12})_{2.95}Fe_{1.05}$
6	9.29	–	–	–	–	–	3.56	87.74	100.58	$(Pt_{2.91}Ir_{0.04})_{2.95}Fe_{1.04}$
7	–	–	1.19	–	–	78.85	20.20	–	100.23	$Os_{0.78}Ir_{0.20}Ru_{0.02}$
8	–	–	1.53	–	–	83.50	12.83	–	97.87	$Os_{0.84}Ir_{0.13}Ru_{0.03}$
9	–	–	1.82	–	–	74.44	23.51	–	99.77	$Os_{0.74}It_{0.23}Ru_{0.03}$

Analyse 1–6—isoferroplatinum: 1, 2—placer, 3, 4—outer part of the dunite core, 5, 6—its inner part; 7–9—native osmium with Ir-impurity, 7—placer, 8—outer, 9—inner parts of the dunite core

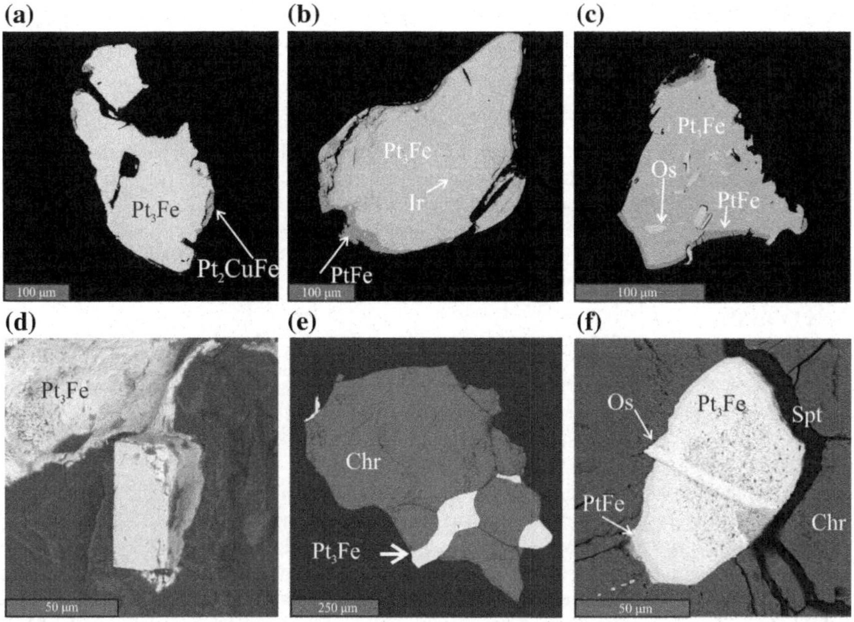

Fig. 1 Platinum minerals from Matysken massif (**a–c**) and associated placer (**d–f**): **a** isoferroplatinum Pt_3Fe grain with tulameenite Pt_2CuFe rim, **b** isoferroplatinum with inclusions of native Ir and tetraferroplatinum PtFe rim, **c** isoferroplatinum with native Os inclusions and tetraferroplatinum rim, **d** isoferroplatinum crystal in chromite-serpentine aggregate, **e** isoferroplatinum-chromite intergrowth, **f** isoferroplatinum grain with native Os inclusion and tetraferroplatinum rim in chromite. BSE-images

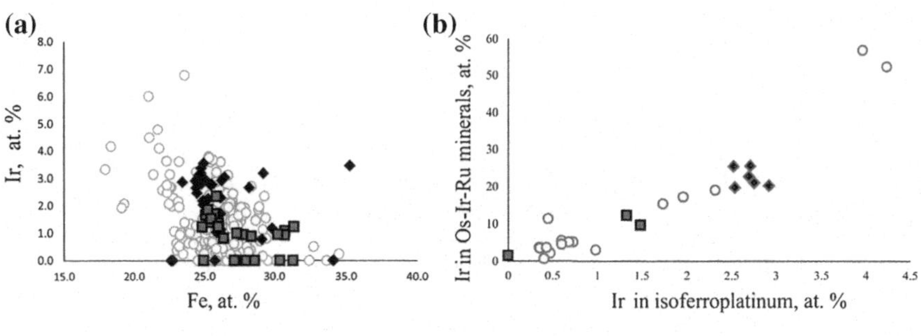

Fig. 2 Binary diagrams for Pt–Fe and Os–Ir–Ru alloys of Matysken massif and placer: **a** isoferroplatinum analysis plotted in coordinates Fe–Ir, **b** diagram in coordinates Ir in Pt–Fe alloys—Ir in Os–Ir–Ru alloys; 1—placer, 2—central part of dunite core, 3—peripheral part of dunite core. Note, that at (**b**) all points except two outermost ones represent hexagonal native osmium, while two other represent cubic native iridium

3. Pt–Fe and Os–Ir–Ru alloys from the central parts of the massif have higher Ir content compared with the same minerals from the outer parts, which corresponds to Nekrasov's model.

Acknowledgements This study was financially supported by the Russian Foundation of Basic Research, project no. 18-35-00126. We express our sincere gratitude to all people who took part in this work, especially—D. L. Korobov and R. N. Novakov.

References

Augé T, Genna A, Legendre O, Ivanov KS, Volchenko YuA. Primary Platinum Mineralization in the Nizhny Tagil and Kachkanar Ultramafic Complexes, Urals, Russia: A Genetic Model for PGE Concentration in Chromite-Rich Zones. Econ Geol. 2005; 100(4):707–732. https://doi.org/10.2113/gsecongeo.100.4.707.

Johan Z. Platinum-Group Minerals from Placers Related to the Nizhni Tagil (Middle Urals, Russia) Uralian-Alaskan-Type Ultramafic Complex: Ore-Mineralogy and Study of Silicate Inclusions in (Pt, Fe) Alloys. Min Pet. 2006;87(1):1–30. https://doi.org/10.1007/s00710-005-0117-1.

Kozlov AP, Chanturiya VA, Sidorov EG, Tolstykh ND, Telegin YuM. Large-Volume Platinum Ore Deposits in Zonal Mafic-Ultramafic Complexes of the Ural-Alaskan Type and the Outlook for Their Development. Geol Ore Depos. 2011. 53(5): 374. https://doi.org/10.1134/s1075701511050059.

Malitch KN, Stepanov SYu, Badanina IYu, Khiller VV. Bedrock Platinum-Group Element Mineralization of Zonal Clinopyroxenite–dunite Massifs of the Middle Urals. Dokl Earth Sci. 2017;476(2): 1147–1151. https://doi.org/10.1134/s1028334x17100051.

Nekrasov IYa, Ivanov VV, Lennikov AM i dr. Sostav Pt-Fe tverdykh rastvorov kak pokazatel' glubiny erozionnogo sreza platinonosnykh shchelochno-ul'traosnovnykh intruzivov. Dokl. AN SSSR. 1991;321(5):1049–1053. [Nekrasov IYa, Ivanov VV, Lennikov AM et al. Composition of Pt–Fe solid solution, as indicator of the level of erosion of the platinum-bearing ring alkaline–ultrabasic intrusions. Proc USSR Acad Sci. 1991;321(5):1049–1053 (in Russ.)].

Palamarchuk RS, Stepanov SYu, Khanin DA, Antonov AV. PGE Mineralization of Massive Chromitites of the Iov Dunite Body (Northern Urals). Moscow Univ Geol Bull. 2017;72 (6):445–454. https://doi.org/10.3103/s0145875217060096.

Razin LV. Geologic and Genetic Features of Forsterite Dunites and Their Platinum-Group Mineralization. Econ Geol. 1976;71:1371–1376. doi:https://doi.org/10.2113/gsecongeo.71.7.1371.

Sidorov EG, Kozlov AP, Tolstykh ND. Gal'moenanskii bazitgiperbazitovyi massiv i ego platinonosnost'. M.: Nauchnyi mir; 2012, 288 s. [Sidorov EG., Kozlov AP., Tolstykh ND. Galmoenan basic-ultrabasic massif and its potential for platinum. M.: Nauchnyi mir; 2012. (in Russ.)].

Stepanov SYu., Malitch KN, Kozlov AV, Badanina IYu, Antonov AV. Platinum Group Element Mineralization of the Svetly Bor and Veresovy Bor Clinopyroxenite–dunite Massifs, Middle Urals, Russia. Geol Ore Depos. 2017;59(3): 244–255. https://doi.org/10.1134/s1075701517030060.

Tolstykh ND, Sidorov EG, Laajoki KVO, Krivenko AP, Podlipsky M. The Association of Platinum-Group Minerals in Placers of the Pustaya River, Kamchatka, Russia. Can Min. 2000;38(5):1251–1264. https://doi.org/10.2113/gscanmin.38.5.1251.

Zhmodik SM, Nesterenko GV, Airiyants EV, Belyanin DK, Kolpakov VV, Podlipsky MYu, and Karmanov NS. "Alluvial Platinum-Group Minerals as Indicators of Primary PGE Mineralization (Placers of Southern Siberia)". Russian Geology and Geophysics, 2016;57 (10). Elsevier B.V.: 1437–1464. https://doi.org/10.1016/j.rgg.2016.09.002.

Experimental Study of the Reaction Pyrope + Enstatite = Phlogopite in the Presence of H_2O–KCl Fluid at 5 GPa in Application to the Problem of Modal Mantle Metasomatism

Evgeny V. Limanov, Valentina G. Butvina and Oleg G. Safonov

Abstract In order to model phlogopite formation in the upper-mantle garnet peridotites, it was performed an experimental study of the reaction En + 1/3Prp + [2/3KCl + 1/3H_2O] = 1/3Phl + 1/3Cl-Phl in the presence of a H_2O–KCl fluid with $X_{KCl} = 0.05–0.4$ at pressure of 5 GPa and temperatures of 900–1250 °C. Experiments showed that the decrease of the Al content in orthopyroxene and the increase of the Cl content in phlogopite are good indicators of the KCl activity in the fluid. These relationships can be used to quantify the KCl activity during the metasomatic processes in the upper mantle.

Keywords Modal metasomatism · Mantle · Phlogopite · Fluid · Experiment · Chloride

Introduction

Mantle metasomatism is a process of transformation of mantle rocks due to their interaction with various deep-seated fluids. The concept of mantle metasomatism has been developed in the 1970–1980s on the basis of studies of upper mantle xenoliths in kimberlites and alkali basalts (Dawson 1980; Harte 1983; O'Reilly and Griffin 2013; Sobolev 1977). According to this concept, the modifications of the mantle rocks via their interaction with external fluids and melts (regardless of their composition and origin), which leads to the formation of minerals atypical for mantle peridotites (amphiboles, phlogopite, apatite, carbonates, sulfides, titanite, ilmenite etc.), was collectively classified as the modal mantle metasomatism (O'Reilly and Griffin 2013). Modal mantle metasomatism is usually accompanied by partial melting. The H_2O and/or CO_2 activities are considered as the most

E. V. Limanov (✉) · V. G. Butvina · O. G. Safonov
D.S. Korzhinskii Institute of Experimental Mineralogy, Russian Academy of Sciences, Academica Osypyana ul., 4, Chernogolovka, Moscow Region 142432, Russia
e-mail: limanov.ev@iem.ac.ru

S. Votyakov et al. (eds.), *Minerals: Structure, Properties, Methods of Investigation*, Springer Proceedings in Earth and Environmental Sciences,
https://doi.org/10.1007/978-3-030-00925-0_18

important factors of the modal mantle metasomatism (O'Reilly and Griffin 2013 and references therein). However, mineralogical and geochemical data also indicate an active role of K and Na in this process. These components are dissolved in the mantle fluids as various salts, among which chlorides are of special interest (Safonov and Butvina 2013, 2016).

A characteristic mineral indicator of the modal mantle metasomatism is phlogopite (Dawson 1980; Harte 1983; O'Reilly and Griffin 2013; Sobolev 1977). Appearance of this mineral in the mantle rocks as a result of their interaction with potassium-bearing aqueous fluids is explained by the reaction (Aoki 1975):

$$\begin{aligned} &5MgSiO_3 + CaMg_2Al_2Si_3O_{12} + [K_2O + 2H_2O] \\ &= K_2Mg_6Al_2Si_6O_{20}(OH)_4 + CaMgSi_2O_6 \end{aligned} \quad (1)$$

or $5En + Grt + [K_2O + 2H_2O] = Phl + Di$, which combines two end-member reactions (Safonov and Butvina 2013, 2016):

$$1/2Prp + 3/2En + [1/2K_2O + H_2O] = Phl \quad (2)$$

$$1/2Grs + 9/2En + [1/2K_2O + H_2O] = Phl + 3/2Di \quad (3)$$

In mantle peridotites, these reactions result in gradual transformations of upper-mantle garnet lherzolites and harzburgites through garnet-free phlogopite-bearing peridotites to phlogopite wherlites. Series of such xenolith are well known in kimberlites (e.g. van Achterbergh et al. 2001).

In the above reactions, the alkaline component is taken as K_2O for clarity. However, in real fluids, K is present as chlorides, carbonates, etc. Present study shows preliminary results of the experiments at 5 GPa and 900–1250 °C of the reaction (2) in presence of the H_2O–KCl fluid.

Materials and Methods

As a starting material, there was used a mixture of pyrope gel, brucite and quartz (to produce enstatite and H_2O) with the following bulk composition (wt%): 47.82 SiO_2, 32.18 MgO, 11.76 Al_2O_3, 8.23 H_2O. KCl was added to produce the required starting X_{KCl} = KCl/(KCl + H_2O) from 0.05 to 0.4. The experiments were performed using a toroidal "anvil-with-hole" high-pressure apparatus (Litvin 1991) in the D.S. Korzhinskii Institute of Experimental Mineralogy RAS (IEM RAS) at a pressure of 5 GPa and temperatures of 900, 1000 and 1250 °C. Platinum capsules of 0.2 mm-wall thickness were used for the experiments. Run duration varied from 24 to 48 h (depending on run temperature). After experiments, samples were sealed into epoxy mounts and polished.

Run products were analyzed using CamScan MV2300 (VEGA TS 5130MM) electron microscope equipped with EDS INCA Energy 350 and Tescan VEGA-II

XMU microscope equipped with EDS INCA Energy 450 and WDS Oxford INCA Wave 700.

Results and Discussion

The results of the experiments are presented in Table 1. The products of the KCl-free experiments contain pyrope and Al-bearing enstatite. In the presence of KCl in the starting fluid, phlogopite is absent at 1250 °C, but appears in the products of the experiments at 1000 and 900 °C. All run samples contain fine-grained aggregates interpreted as quench products of a melt (Fig. 1a, b). An increase in KCl is accompanied by a decrease of the amount of pyrope and enstatite in the products of the experiments, so that in experiments with fluids with $X_{KCl} = 0.4$ they are rare or absent, and phlogopite is the dominant phase (Fig. 1a).

Experiments at 1000 °C show that pyrope is the first phase to disappear as the concentration of KCl in the fluid increases (Fig. 1b). Subsequently Al-bearing enstatite disappears. These relations indicate a series of reactions:

$$\mathrm{En} + 1/3\mathrm{Prp} + [2/3\mathrm{KCl} + 1/3\mathrm{H_2O}] = 1/3\mathrm{Phl} + 1/3\mathrm{Cl} - \mathrm{Phl} \qquad (4)$$

$$\mathrm{En} + 1/5\mathrm{MgTs} + [2/5\mathrm{KCl} + 1/5\mathrm{H_2O}] = 1/5\mathrm{Phl} + 1/5\mathrm{Cl} - \mathrm{Phl}, \qquad (5)$$

where MgTs is the Mg-Tschermack molecule ($MgAl_2SiO_6$) in the orthopyroxene solid solution and Cl-Phl is the $KMg_3AlSi_3O_{10}Cl_2$ end-member in the phlogopite solid solution.

Reaction (5) is manifested in a decrease of the Al content in orthopyroxene with the increase in X_{KCl} in the starting fluid (Fig. 2). With the increase of X_{KCl} in the starting fluid, the Cl content in mica increases as well, reaching 0.6–0.8 wt% at $X_{KCl} = 0.2$–0.4 (Fig. 3). Such Cl content in the synthetic phlogopite is consistent with the Cl content in phlogopites of metasomatized peridotites and inclusions in diamonds (see Fig. 3 in Safonov and Butvina (2013)). At $X_{KCl} = 0.4$ T = 900 °C phlogopite contains 3–4 wt.% of chlorine. This may be due to the fact that the system does not contain a water-salt fluid, but a KCl melt with water dissolved in it.

In the products of some experiments at $X_{KCl} > 0.1$, kyanite (Ky) appears. Its formation is also related to the decomposition of pyrope in the presence of the H_2O–KCl fluid according to reaction

$$\mathrm{Prp} + [\mathrm{KCl} + 1/2\mathrm{H_2O} + 1/2\mathrm{SiO_2}] = 1/2\mathrm{Phl} + 1/2\mathrm{Cl} - \mathrm{Phl} + 1/2\mathrm{Ky}. \qquad (6)$$

The presence of SiO_2 in the fluid can be due to the initial slight excess of this component in the starting mixtures or the sufficiently high solubility of SiO_2 in the fluid in equilibrium with the Phl + En assemblage (e.g. Schneider and Eggler 1986). High activity of SiO_2 in the saline fluid is indirectly indicated by the absence

Table 1 Run conditions and products of experiments on the reaction Prp + En + H_2O + KCl = Cl − Phl at 5 GPa

Run no.	T, °C	Time, h	X_{KCl}	Run products
PEH-1	1250	24	0	Opx + Grt + Q[a]
PEH-2	1250	24	0.05	Ol + Grt + Q
PEH-3	1250	24	0.1	Opx + Ky + Q
PEH-4	1250	24	0.2	Opx + Grt + Ky + Q
PEH-6	1000	48	0	Opx + Grt + Q
PEH-8	1000	48	0.05	Opx + Grt + Phl + Q
PEH-7,14	1000	48	0.1	Opx + Phl + Q
PEH-9	1000	48	0.2	Opx + Phl + Ky + Q
PEH-11	1000	48	0.4	Phl + Q
PEH-12	900	48	0.05	Opx + Grt + Phl + Q
PEH-16	900	48	0.1	Opx + Grt + Phl + Q
PEH-13	900	48	0.2	Opx + Grt + Ky + Phl + Q
PEH-15	900	48	0.4	Opx + Grt + Phl+Q

[a]Q—products of melt quenching

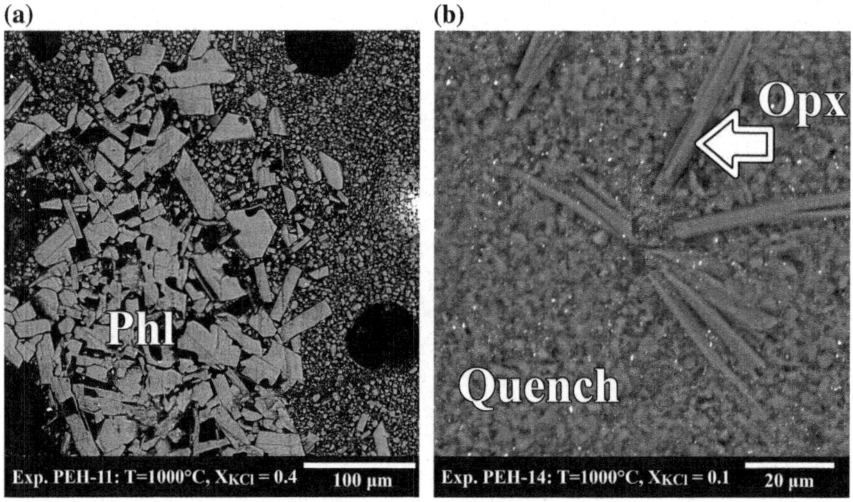

Fig. 1 Phases relations in the run products: **a** abundant phlogopite (Phl) coexisting with quench products in the run PEH-11 at T = 1000 °C, X_{KCl} = 0.4; **b** orthopyroxene (Opx) coexisting with quench products in the run PEH-14 at T = 1000 °C, X_{KCl} = 0.1

of forsterite in the run products (it was observed only in the PEH-2 experiment, Table 1), which is a typical product of phlogopite dissolution in aqueous fluids (e.g. Schneider and Eggler 1986).

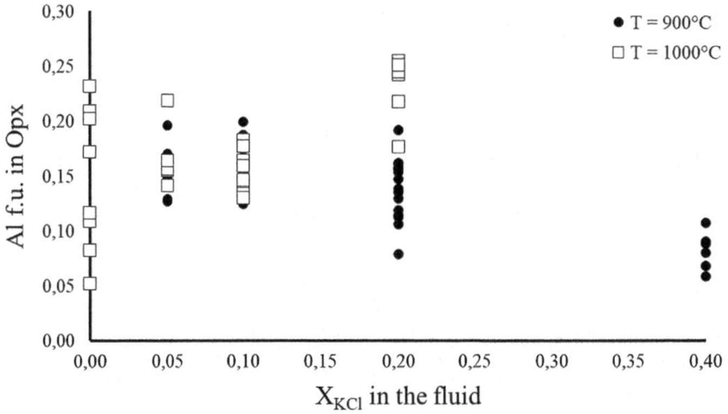

Fig. 2 Dependence of the Al content (a.p.f.u.) in orthopyroxene on the KCl content in the starting fluid (X_{KCl} = KCl/(KCl + H_2O) at 900 and 1000 °C

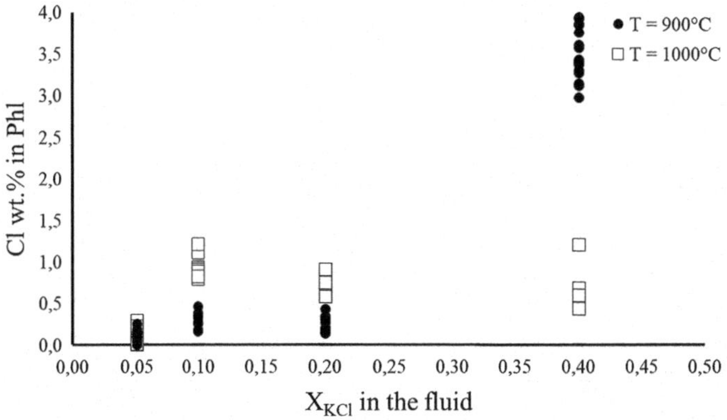

Fig. 3 Dependence of the Cl content (wt%) in phlogopite on the KCl content in the starting fluid (X_{KCl}= KCl/(KCl + H_2O) at 900 and 1000 °C

Conclusion

The experiments reproduce the major reaction occurring during transformation of upper-mantle garnet peridotites in the course of mantle metasomatism under the influence of the H_2O–KCl fluids, leading to the formation of phlogopite-bearing assemblages. Al-bearing orthopyroxene and garnet are not stable in the presence of H_2O–KCl fluid even at $X_{KCl} = 0.05$. The Al content in orthopyroxene and Cl in phlogopite according to reaction (5) are the direct indicators of the KCl activity in the fluid. These effects can be used to quantify this activity as well as KCl concentration in saline fluids in the processes of the modal mantle metasomatism.

Acknowledgement The study was supported by the Russian Foundation for Basic Research (project 16-05-00266).

References

Aoki K. Origin of phlogopite and potassic richterite bearing peridotite xenoliths from South Africa. Contrib. Mineral. Petrol. 1975;53:145–156. https://doi.org/10.1007/bf00372601.

Dawson JB. Kimberlites and their xenoliths. Berlin: Springer; 1980.

Harte B. Mantle peridotites and processes—the kimberlite sample. In: Hawkesworth CJ and Norry MJ, editors. Continental Basalts and Mantle Xenoliths. Cheshire: Shiva; 1983. p. 46–91.

Litvin YuA. Physicochemical studies of the melting of the deep matter of the Earth. Moscow: Nauka; 1991.

O'Reilly SY, Griffin WL. Mantle metasomatism. In: Harlov DE and Austerheim H, editors. Metasomatism and the chemical transformation of rock. Berlin, Heidelberg: Springer; 2013. p. 471–533.

Safonov OG, Butvina VG. Interaction of the model peridotite with the H_2O-KCl fluid: an experiment at a pressure of 1.9 GPa and its application to the processes of upper mantle metasomatism. Petrology. 2013;21(6):599–615. https://doi.org/10.1134/s0869591113060076.

Schneider ME, Eggler DH. Fluids in equilibrium with peridotite minerals: implications for mantle metasomatism. Geochimica et Cosmochimica Acta. 1986;*50*(5):711–724. https://doi.org/10.1016/0016-7037(86)90347-9.

Sobolev NV. The Deep-Seated Inclusions in Kimberlites and the Problem of the Composition of the Upper Mantle. Washington: American Geophysical Union; 1977.

Safonov OG, Butvina VG. Indicator reactions of K and Na activities in the upper mantle: natural mineral assemblages, experimental data, and thermodynamic modeling. Geochemistry. 2016;54(10):858–872. https://doi.org/10.1134/S0016702916100098.

van Achterbergh E, Griffin WL, Stiefenhofer J. Metasomatism in mantle xenoliths from the Letlhakane kimberlites: estimation of element fluxes. Contrib. Mineral. Petrol. 2001;141:397–414. https://doi.org/10.1007/s004100000236.

Application of Mössbauer, ESR, and FTIR Spectroscopy for Mineralogical and Technological Research of Titanium Ores

Vladimir P. Lyutoev, Alexander B. Makeev and Andrey Yu. Lysyuk

Abstract The study of the mineral composition of titanium ores from Timan titaniferous sandstones, titanium-magnetite deposits of Transbaikalia, the Kola Peninsula (Russia), and Panzhihua massif (China) was carried out using spectroscopic methods including Mössbauer ^{57}Fe, electron spin resonance, and Fourier-transform infrared spectroscopy. The spectroscopic data allowed the quantitative phase analysis of ores to be carried out as well as objective information on the ratio of ores minerals to be obtained. The results of investigation can be useful to determine the improvement direction of the ore enrichment technology.

Keywords Spectroscopic methods · Titanium ore · Iron-containing minerals · Ilmenite · Titanium-magnetite · Rutile · Mineral composition

Introduction

The industrial sources of titanium ore are as follows: ancient and modern ilmenite, rutile, leucoxene, complex ilmenite-rutile-zirconium coastal marine and alluvial deposits (over 60% of the world reserves); original magmatogenic ilmenite-magnetite, ilmenite-hematite, and ilmenite-rutile deposits in gabbro-anorthosites (more than 20% of the world reserves); anatase-perovskite-apatite deposits in the carbonatite weathering crusts. According to the titanium ore reserves, Russia occupies the second place in the world after China (ilmenite-titanomagnetite deposits of the Panzhihua massif). The explored reserves of such ores are

V. P. Lyutoev (✉) · A. Yu.Lysyuk
Institute of Geology of the Komi Science Centre, Ural Branch of the Russian Academy of Sciences (IG Komi SC UB RAS), Pervomaiskaya 54, 167982 Syktyvkar, Russia
e-mail: vlutoev@geo.komisc.ru

A. B. Makeev
Institute of Geology of Ore Deposits, Petrography, Mineralogy and Geochemistry, Russian Academy of Sciences (IGEM RAS), Staromonetny 35, 119017 Moscow, Russia
e-mail: abmakeev@igem.ru

S. Votyakov et al. (eds.), *Minerals: Structure, Properties, Methods of Investigation*, Springer Proceedings in Earth and Environmental Sciences,
https://doi.org/10.1007/978-3-030-00925-0_19

approximately evenly distributed among the original and placer deposits (Bykhovsky et al. 2007). Large titanomagnetite and apatite-ilmenite magmatogenic deposits are located in Transbaikalia, the Kola Peninsula, Karelia, and within the gabbroid belt of the western slope of the Southern Urals. In Russia, more than 80% of the explored titanium dioxide reserves of alluvial type are concentrated in the Yarega oil-titanium paleoplacer of the Southern Timan, mainly in the form of leucoxene. According to the geological structure and mineralogical features, the titaniferous sandstones of the Pizhma deposit in the Middle Timan are close to it. It should be noted that there are geological, geochemical and mineralogical reasons to interpret the genesis of Pizhma and possibly Yarega titaniferous sandstones as volcanogenic by lamprophyres followed by metamorphic ore strata transformation (Makeev et al. 2012). Those deposits possess huge reserves, but the technology of ore enrichment has not been fully worked out yet.

Russia is the largest producer of metal titanium and its alloys, but its own titanium deposits are not exploited. Production of pigment titanium dioxide—the main product for the world titanium industry—is almost completely absent in the country, which is substituted by supplies from the US, China, and Ukraine. The aim of the work was to conduct a comparative phase analysis of titanium minerals from Russia deposits using modern spectroscopic methods such as Mössbauer ^{57}Fe, electron spin resonance (ESR), and Fourier-transform infrared (FTIR) spectroscopy for determining the direction of improvement of the ore enrichment technology.

Materials and Methods

The study of the chemical and mineral composition was carried out on the example of the original ilmenite-titanomagnetite ore deposits: Chineiskoe, Slyudinskoe, Bystrinskoe, Kruchininskoe in Transbaikalia (Aldan Shield), Tsaginskoe and Achinskoe of Kevsk ore region in the Kola Peninsula (Baltic Shield), and Pizhma titaniferous sandstones in the Middle Timan. The schematic location of these deposits is given in Fig. 1. For comparison, samples of titanomagnetite ores from Chinese Panzhihua deposit were also studied by spectroscopic methods. The samples were represented as concentrates and ore lumps.

The chemical composition of ores and rocks, their morphological features were previously studied by the methods of X-ray fluorescence, scanning electron microscope, and microprobe using the equipment of IGEM RAS and MSU. Spectroscopic studies were conducted in the IG Komi SC UB RAS. Infrared spectroscopy and X-ray diffraction were used to analyze the mineral phase composition of the ore samples. Absorption infrared spectra were obtained by a Lumex FT-02 FTIR spectrometer in the range 400–4000 cm^{-1} at 256 scans and with an instrumental resolution 2 cm^{-1}. The specimens were prepared as pressed pellets of 800 mg KBr and 1.5 mg of a powdered specimen. Powder X-ray patterns of the samples were obtained using Shimadzu XRD-6000 (CuK$_{\alpha}$) diffractometer.

Fig. 1 Layout of the investigated titanium ores of titanium-magnetite-ilmenite deposits: 1—Panzhihua, 2—Chineiskoe; 3—Slyudinskoe; 4—Kruchininskoe; 5—Bystrinskoe; 6—Tsaginskoe; 7—Achinskoe; and Timan titanium sandstones: 8—Pizhemskoe, 9—Yaregskoe

Electron spin resonance (ESR) spectra of the samples were collected by a SE/X-2547 radiospectrometer ("RadioPAN", Poland) in X-frequency range with HF modulation of 100 kHz at room temperature. g-Factors were calibrated according to the Li^0:LiF reference material (g_0 = 2.00229). The samples from 50 to 150 mg were powdered in a jasper mortar. EasySpin software package for MathLab was used for simulation and analysis of experimental ESR spectra (Stoll and Schweiger 2006).

The Mössbauer ^{57}Fe spectra were recorded using a MS-1104Em spectrometer in the thin absorber mode in the range of −11 to +11 and −4 to +4 mm/s at room temperature. Powdered samples (10–20 mg) were used for spectra accumulation. The isomeric shift was determined relatively to α-Fe. Univem standard software was used for the processing of spectra. Relative spectral areas of the Mössbauer subspectra associated with the respective phases and values of their recoilless fractions were used for quantitative phase analyses (Eeckhout and De Grave 2003).

Results and Discussion

The technological scheme for obtaining titanium mineral concentrates from Pizhma titanium sandstones was studied, including: crushing, grinding, sludge removal, gravity enrichment, dimensional classification, and magnetic separation (Makeev and Lyutoev 2015). The results of experimental spectroscopic studies of concentrates of Pizhma ores confirmed their multi-phase composition and reflected the quantitative ratios of minerals and phases. Typical FTIR spectra of ore concentrates from Pizhemskoe deposit are shown in Fig. 2c. According to FTIR and X-ray diffraction data, the main mineral components of the concentrates are rutile and ferruginous oxides of titanium, quartz, siderite, and clay minerals. Unlike Yaregskoe deposit, a significant part of the ore component from the titaniferous sandstones of Pizhemskoe deposit is represented not only by leucoxene (quartz-rutile), but also by relic ilmenite, Fe-rutile, and pseudorutile.

The polymineral composition of ore components of low degree of crystallinity, the presence of high contents of ferruginous carbonates, clay silicates, and quartz in the concentrates greatly impede the direct quantitation of the ratio of minerals and phases, on the base of X-ray diffraction and chemical analysis data. The distribution of ore minerals by size and magnetic classes of concentrate was refined by means of Mössbauer spectroscopy and ESR. Their typical spectra and interpretation are presented in Fig. 2a, b. The content of quartz and rutile (the main mineral phases of

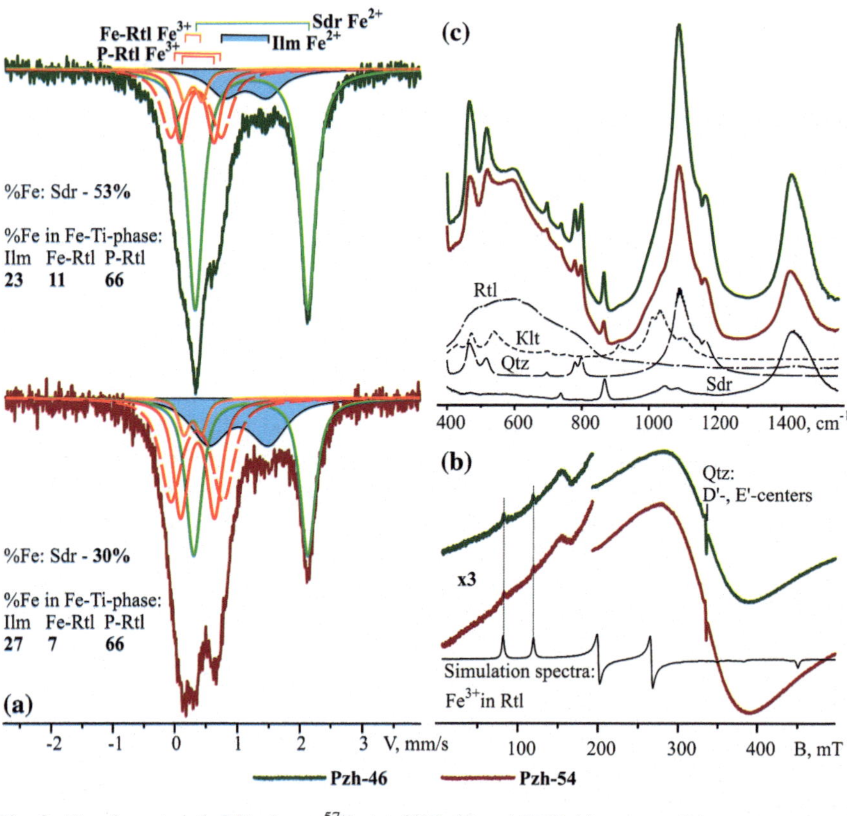

Fig. 2 The characteristic Mössbauer ^{57}Fe (**a**), ESR (**b**) and FTIR (**c**) spectra of the concentrates of Pizhma titanium ores and their interpretation: %Fe—area of the spectral contour of the component; Rtl—rutile, Fe-Rtl—Fe-rutile, P-Rtl—pseudorutile, Ilm—ilmenite, Sdr—siderite, Qtz—quartz, Klt—kaolinite. Phz-46 and Phz-54—samples of large (0.125–0.5 mm) and thin (0.04–0.125 mm) magnetic fraction of the bulk concentrate

leucoxene) was established to decrease in the magnetic fractions of concentrates. The ratio of leucoxene and magnetic titanium minerals (approximately 5:3) in bulk concentrate was calculated, as well as the ratio of pseudorutile, ilmenite, and Fe-rutile—approximately 20:4:6. The latter in the coarse-grained magnetic fraction of the concentrate is basically preserved. In the fine-grained magnetic fraction, the proportion of ilmenite increases, and the proportion of Fe-rutile decreases. The bulk magnetic fraction of the concentrate is enriched in siderite. An additional crushing of concentrates up to −0.3 mm size fraction was recommended in order to remove intergrowths with quartz. In order to clean samples from siderite, it is necessary to apply concentrate heating at the temperature of 600 °C.

In comparison with Chinese Panzhihua massif, investigations of the chemical and mineral composition of ilmenite-titanomagnetite ores from four Eastern Siberian deposits (Chineiskoe, Slyudinskoe, Bystrinskoe, and Kruchininskoe), from the Kola Peninsula (Tsaginskoe, Achinskoe in the Kevsk ore district) were conducted (Lyutoev et al. 2017). Microprobe and X-ray analysis, infrared and Mössbauer spectroscopy were used. First, for the typical ore samples Mössbauer spectra were obtained and deciphered, which allowed the ratio of iron-containing minerals of oxides and silicates to be quantified and the ratio of ilmenite and titanomagnetite to be determined for the main industrial types of ores. Typical Mössbauer and FTIR spectra of samples from some deposits of the ilmenite-titanomagnetite ores and their interpretation are presented in Fig. 3.

Since ilmenite is the main industrial titanium mineral in the original deposits of similar magmatic genesis (the extraction and processing of which is carried out in standard technological processes), the determination of its relationships with titanomagnetite, magnetite, and non-ore mass of silicate minerals is the main mineralogical-technological problem.

Three mineralogical-technological types of titanium ores were identified. It was established that there are no appreciable ilmenite amounts in ores of Bystrinskoe deposit, where the only ore mineral found was titanomagnetite. The low ilmenite content in the ore component at a level of about 30 wt% and lower (micrograined texture, thin intergrowth of ore Ti-Fe minerals) of Chineiskoe and Tsaginskoe deposits requires the application of contemporary technologies which are successfully used for the foreign deposits of Panzhihua (China) and Bushveld (South Africa). Ores with the ilmenite content of more than 50 wt% of the Eastern Siberian Kruchininskoe and Slyudinskoe deposits as well as Achinskoe deposit on the Kola Peninsula meet the industry requirements and are suitable for the standard technological processing. At the same time, the increased content of magnetic pyrrhotite ($\sim$10 mass%) in the ores of the Kruchininskoe deposit does not reduce the quality of titanium ores, only requires additional operations such as magnetic separation and flotation.

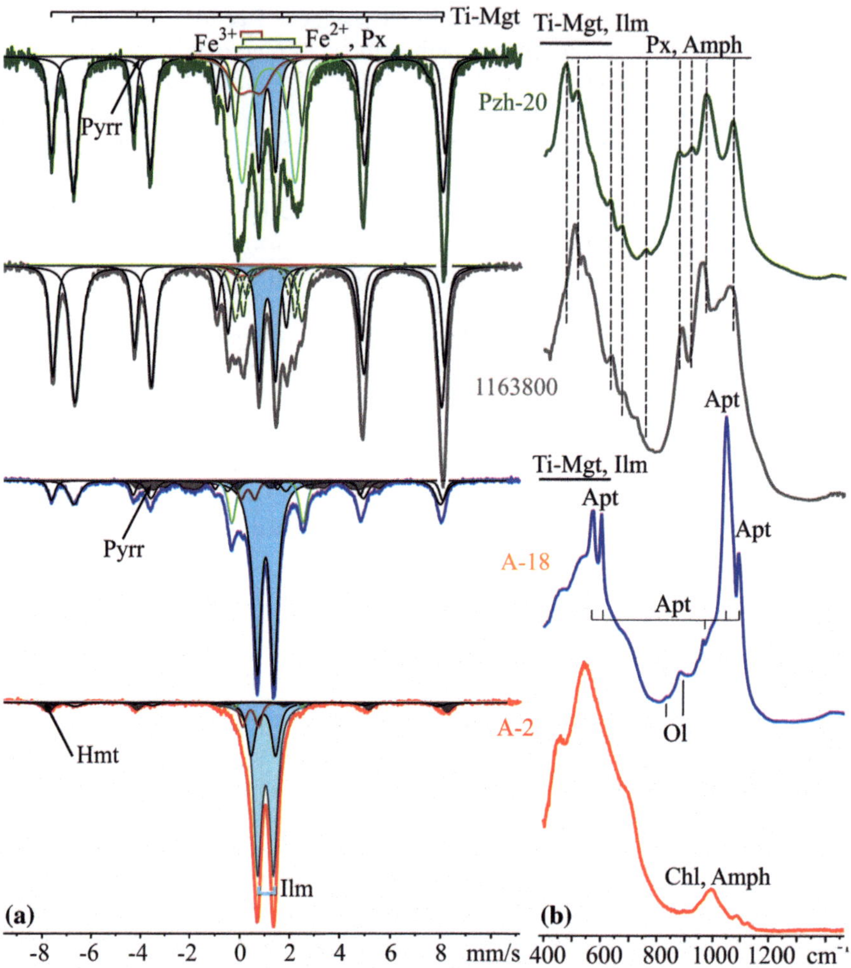

Fig. 3 Examples of interpretation of the Mössbauer (**a**) and IR absorption (**b**) spectra of the titanomagnetite ores of Panzhihua (Phz-20), Chineiskoe (1163800), Kruchininskoe (A-18) and Achinskoe (A-2) deposits: Ti-Mgt—titanomagnetite, Ilm—Ilmenite, Hmt—hematite, Pyrr—pyrrhotite, Px—pyroxene, Amph—amphibole, Apt—apatite, Ol—olivine, Chl—chlorite

Conclusions

Despite the long history of research of placer and original objects of titanium raw materials in the country, a low degree of its mineralogical and technological study is recognized. To a large extent, this is explained by the variable chemical and phase ore composition, the complex interrelationship of ore-forming minerals, resulting from decomposition of solid solutions and their subsequent transformations, and the low degree of ordering of ore minerals. Standard chemical and X-ray phase analyses do not provide reliable information about their quantitative mineral composition, which is extremely necessary for the development of effective technology for their processing (Pirogov et al. 2013).

Qualitative classification of ores and determination of the basic mineral components can be carried out by the FTIR method. The concentration ratio between Fe-Ti-oxide ore phases and non-metallic minerals can be estimated by the ratio of absorption values of the vibration Fe-O, Ti-O band at 400–600 cm^{-1} and in the high-frequency bands of stretching vibrations of silicates, carbonates, phosphates, and other non-metallic minerals in the spectral range of 1000–1400 cm^{-1}. The data of Mössbauer and ESR spectroscopy on the distribution of iron in minerals from titanium ores allowed the mineral composition of the ore component to be classified properly and direct quantitative phase analysis of iron-containing minerals to be carried out, obtaining objective information on the mineral ratio in ores.

Acknowledgements The authors are grateful to B. I. Gongalsky (IGEM RAS, Moscow, Russia), A. V. Mokrushin and Yu. L. Voitekhovsky (GI Kola SC RAS, Apatity, Russia) for the specimens provided for the investigations. The research was carried out within the framework of the topics with the state registration numbers AAAA-A17-117121270036-7 (IG Komi SC RAS) and 0136-2018-0020 (IGEM RAS).

References

Bykhovsky LZ, Kalish EA, Pakhomov FP, Tigunov LP. Aktual'nye problemy povysheniya investitsionnoi privlekatel'nosti mestorozhdenii titana v Rossii. Prospect and protection of mineral resources. 2007: 11; 14–18. [Bykhovsky LZ, Kalish EA, Pakhomov FP, Tigunov LP. Actual problems of increasing investment attractiveness of titanium deposits in Russia. Prospect and protection of mineral resources. 2007: 11; 14–18. (in Russ.)].

Eeckhout SG, De Grave E. Evaluation of ferrous and ferric Mossbauer fractions. Part II. Phys. Chem. Minerals. 2003: 30; 142–146. http://doi.org/10.1007/s00269-003-0300-z.

Makeev AB, Dudar VA, Samarova GS, Bykhovskiy LZ, Tigunov LP. Pizhemskoye titanovoe mestorozhdenie (Srednii Tiaman): aspecty geologicheskogo stroeniya i osvoeniya. Rudnik Budushchego. 2012: 1(9): 16–24. [Makeev AB, Dudar VA, Samarova GS, Bykhovskiy LZ, Tigunov LP. Pizhemsk titanium deposit (Middle Tiaman): aspects of a geological feature and development. Rudnik Budushchego. 2012: 1(9): 16–24. (in Russ.)].

Makeev AB, Lyutoev VP. Spectroscopy in process mineralogy. The Pizhemskoye deposit titanium ores concentrates mineral composition. Obogashchenie Rud. 2015: 5; 33–41. http://doi.org/10.17580/or.2015.05.06.

Lyutoev VP, Makeev AB, Lysyuk AYu. Exploring a possibility to determine titanium magnetite ores mineral composition by spectroscopy data. Obogashchenie Rud. 2017: 5; 28–36. http://doi.org/10.17580/or.2017.05.05.

Pirogov BI, Bronitskaja ES, Astakhova JuM, Volkov ES. Osobennosti veschestvennogo sostava titanomagnetitovykh rud magmaticheskogo genesisa, opredelayyuschie ikh obogatimost'. Prospect and protection of mineral resources. 2013: 2; 47–51. [Pirogov BI, Bronitskaja ES, Astakhova JuM, Volkov ES. Features of material composition of titanium magnetite ores of the magmatic genesis, defining their dressability. Prospect and protection of mineral resources. 2013: 2; 47–51. (in Russ.)].

Stoll S, Schweiger A. Easy Spin, a comprehensive software package for spectral simulation and analysis in EPR. J. Magn. Reson. 2006; 178(1): 42–55.

Investigation of the Mineral and Chemical Composition of the Cave Deposits in Emine-Bayir-Khosar, Crimea

Gleb S. Maksimov, Igor A. Nauhatsky, Elizaveta I. Timohina and Elena M. Maksimova

Abstract X-ray diffractometric and spectroscopic study of the structural and material composition of water-chemogenic deposits in Emine-Bayir-Hasar cave, Crimea, have shown that they are composed of several different minerals. These include calcite, pigeonite, hedenbergite, tridimite and cristobalite.

Keywords Cave deposits · Mineral composition · X-ray diffractometry

Introduction

Although the Crimea belongs to the classic karst areas with more than 1000 natural karst cavities, the mineralogy of water-chemogenic deposits, which are the most spectacular and well-known formations in the caves, has been poorly studied. The following minerals have been found in them: gypsum, aragonite, calcite, illite, quartz, kaolinite, goethite, pyrolusite, (Tischenko 2008).

Emine-Bayir-Hasar is one of the most beautiful caves in the Crimea, located on the mountain Chatyr-Dah, which is the part of the Main Ridge of the Crimean Mountains and occupies the central position in it (Fig. 1). Chatyr-Dah is a territory, in which there are almost all the variety of karst forms, both surface and subsurface, with more than 140 caverns of genetic classes of corrosion and erosion-corrosion. The Emine-Bayir-Hasar cave has a length of 1630 m, a depth of 125 m, and a volume of 160.5 thousand m^3. The general understanding of its morphometry can be obtained from (Lukyanenko 2001).

Emine-Bayir-Hasar is a multilevel system of passages which are separated from each other by landslides and water dripping in the halls: the Main hall, the Lake hall, the hall of Idols and others. In the Emine-Bayir-Hasar there are various genetic types of cave deposits: residual, landslip, water-chemogenic and mechanical

G. S. Maksimov · I. A. Nauhatsky · E. I. Timohina · E. M. Maksimova (✉)
V.I. Vernadsky Crimean Federal University, Ave. Vernadsky 4, Simferopol, Crimea, Russian Federation
e-mail: lenamax112@rambler.ru

S. Votyakov et al. (eds.), *Minerals: Structure, Properties, Methods of Investigation*, Springer Proceedings in Earth and Environmental Sciences,
https://doi.org/10.1007/978-3-030-00925-0_20

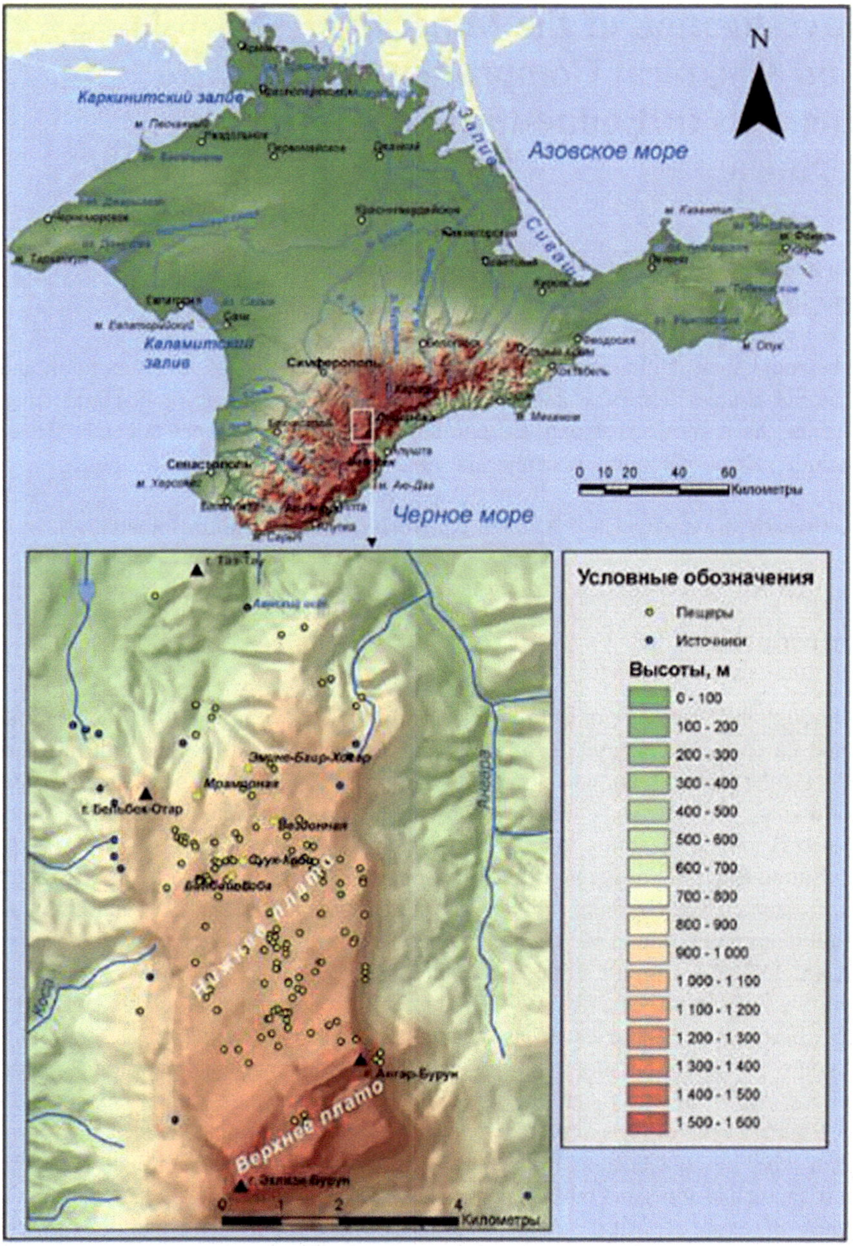

Fig. 1 The location of the Emine-Bayir-Hasar cave

Fig. 2 The cave deposits in the Emine-Bayir-Hasar

chemogenic. The Emine-Bayir-Hasar cave is characterized by an exclusive richness of speleothem decoration (Fig. 2). This is varied in the form of gravitational (tubular, cone-shaped, plate, etc.) and anomalous stalactites. There are also conical, pagoda-shaped stalagmites etc.

Materials and Methods

The material composition of two samples of water-chemogenic formations of the Emine-Bayir-Hasar cave was studied. The samples included the fragment of the lower part of stalactite with a hollow channel inside from the Dublyansky hall (Fig. 3).

The investigation of mineral and chemical structure of stalactite and crust was carried out by X-ray diffractometry and X-ray spectroscopy.

Results and Discussion

The structural analysis was carried out using a X-ray diffractometer with copper monochromatic radiation $\lambda(K\overline{\alpha}) = 1.54178$ Å. Registration of diffraction patterns of samples was carried out in 2θ angular range 10°–110° (Fig. 4).

The crystal structure of samples was determined according to Bragg's law (Gerasimov et al. 1975):

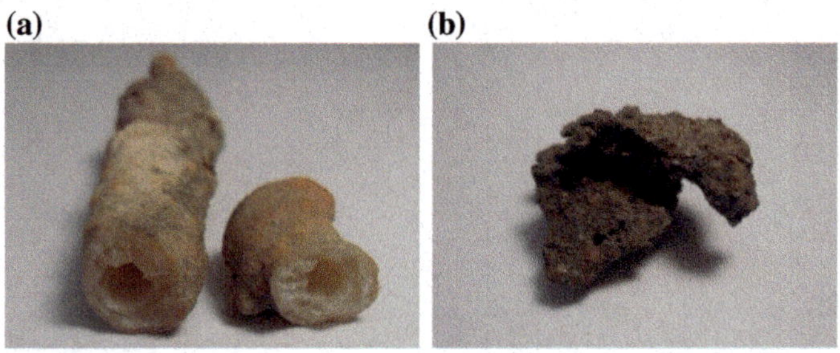

Fig. 3 The samples of water-chemogenic formations in the Emine-Bayir-Hasar cave: **a** the stalactite; **b** the crust

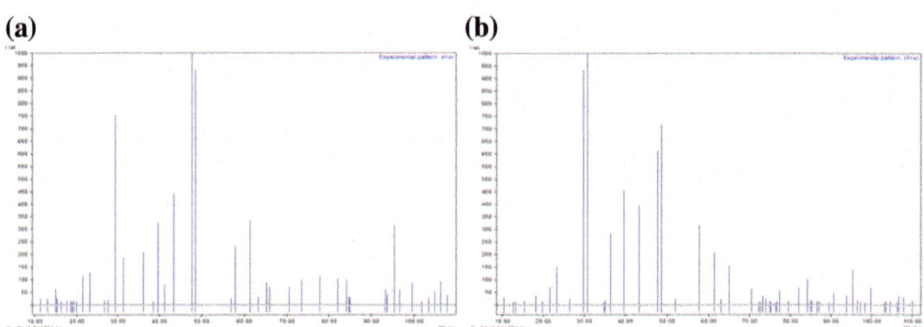

Fig. 4 The diffractograms of samples: **a** stalactite; **b** crust

Table 1 The phase composition of the samples (θ—Bragg's angle, I—intensity)

Stalactite			Crust		
2θ	I	Mineral	2θ	I	Mineral
14.99	63.3	SiO_2	20.15	6.5	Tridimite
19.02	9.7	SiO_2	20.38	15.9	Tridimite, pigeonite
26.86	17.9	SiO_2	23.15	96.8	Tridimite
29.52	751.6	Calcite	27.59	11.2	Tridimite
36.21	206.2	Pigeonite	29.78	1000.0	Hedenbergite
39.71	326.3	Pigeonite	34.95	7.9	Tridimite
43.41	443.2	SiO_2	35.26	5.1	Hedenbergite
47.69	1000.0	Calcite	47.73	430.8	Hedenbergite
48.61	931.8	Calcite	48.63	560.1	Pigeonite
57.8	232.1	Pigeonite	53.96	6.1	Pigeonite
61.38	332.8	SiO_2	57.70	236.5	Pigeonite
63.34	29.2	Pigeonite	63.19	53.8	Pigeonite
65.16	87.7	SiO_2	64.90	18.4	Hedenbergite
65.88	64.6	Calcite	65.17	117.7	Tridimite
82.04	102.3	Pigeonite	65.73	19.5	Tridimite
84.71	30.8	SiO_2	74.42	17.7	Hedenbergite
95.31	40.0	Calcite, SiO_2	81.27	7.6	Hedenbergite, pigeonite
96.63	60.1	Pigeonite	86.64	22.8	Hedenbergite, pigeonite
99.61	87.7	Calcite	95.22	140.5	Hedenbergite

$$n\lambda = 2d\sin\theta, \tag{1}$$

where

n is the order of diffraction;
λ is the X-ray wavelength, Å;
d is the interplane distance, Å;
θ is the Bragg's angle.

The d experimental values were compared with tabulated from the database (Crystallography Open Database, http://database.iem.ac.ru/mincryst/rus). We have identified the following phase composition of the samples (Table 1).

The elemental composition of the test samples was determined by the X-ray fluorescence analysis using Rigaku ZSX Primus (Table 2).

Table 2 Chemical composition of the samples

No.	Stalactite		Crust	
	Component	wt%	Component	wt%
1	MgO	0.5	MgO	0.2
2	Al_2O_3	0.8	Al_2O_3	2.6
3	SiO_2	1.1	SiO_2	4.2
4	P_2O_5	0.07	P_2O_5	0.2
5	S	0.006	S	0.004
6	K_2O	0.05	K_2O	0.6
7	CaO	97.1	CaO	84.7
8	TiO_2	0.3	TiO_2	0.4
9	Cr_2O_3	0.005	V_2O_5	0.02
10	MnO	0.01	Cr_2O_3	0.009
11	Fe_2O_3	0.09	MnO	3.9
12	Co_2O_3	0.004	Fe_2O_3	2.9
13	NiO	0.009	Co_2O_3	0.007
14	CuO	0.006	NiO	0.01
15	SrO	0.007	CuO	0.02
16	ZrO	0.001	ZrO	0.09
17			As_2O_3	0.003
18			Rb_2O	0.005
19			SrO	0.006
20			Y_2O_3	<0.0002
21			ZrO_2	0.01
22			BaO	0.06

Conclusion

The qualitative and quantitative X-ray phase analysis of the mineral composition showed that the stalactite was composed of calcite $CaCO_3$ (70.3%), pigeonite $(Ca,Fe,Al,Mg)Si_2O_6$ (22.6%) and phase of SiO_2 (high-temperature modification of cristobalite SiO_2), and the crust was composed of pigeonite $(Ca,Fe,Al,Mg)Si_2O_6$ (69.6%), hedenbergite $(Ca,Fe)Si_2O_6$ (22,7%), and tridimite SiO_2 (7.8%).

Calcite is the most common mineral of the Crimea, which is a rock-forming mineral of limestones and marls. It is interesting to observe the presence of cristobalite in the samples, which is very rare in the Crimea. Mineral tridymite was discovered for the first time in the Crimean cave deposits.

A detailed study of the water-chemogenic deposits is very important in practical terms. The density, size and shape of the deposits help determine the intensity of the water inflow. The position of different morphological groups of water-chemogenic deposits in relation to different elements of the cave microrelief and to other types of deposits records the levels of flooding (fringe) and the change of stages of sediment accumulation (crust, stalagmites). The color and mineralogical composition of deposits are indicators of paleogeographic conditions.

References

Crystallography Open Database. URL: http://database.iem.ac.ru/mincryst/rus.

Gerasimov V N, Dolivo-Dobrovol'skaya E M, Kamentsev I E, Kordat'eva V V, Kosoi A L, Lesyuk G I, Rozhdestvenskaya I V, Stroganov E V, Filatov S K, Frank-Kamenetskaya O V et al. A Manual on X-Ray Diffraction Investigations of Minerals ed. V A Frank-Kamenetskii. Leningrad: Nedra, 1975.

Lukyanenko EA. Peshchera Emine Bayır Hasar – novyj ehskursionyj ob"ekt v turistkoj infrastructure Kryma. Uchenye zapiski TNU. 2001; 14(1): 88–92.[Lukyanenko EA. Emine Bayır Hasar – new excursion object in the tourist infrastructure of the Crimea. Scientific notes of TNU. 2001. 14(1):88–92. (In Russ.)].

Tischenko AI. Mineralogicheskaya izuchenost' karstovykh polostej Kryma. Speleologiya I karstologiya. 2008;1: 81–84. [Tischenko AI. Mineralogical study of karst cavities of Crimea. Speleology and karstology. 2008;1:81–84. (In Russ.)].

Microminerals in Gallstones

Ekaterina V. Mashina and Vasily N. Filippov

Abstract Microminerals and inclusions were studied in cholesterol and pigment gallstones by electron microscopy. It has been established that the micromineral diversity in gallstones is due to the presence of a class of oxides, chlorides, sulfates, sulfides, silicates, carbonates, and native-metal phases of iron and copper. It has been noted that pigment gallstones, in contrast to cholesterol gallstones, are characterized by the accumulation of heavy metals such as bismuth, lead, niobium and gold.

Keywords Gallstones · Biominerals · Inclusions · Metals

Introduction

The scientific direction of biomineralogy is at the junction of a number of sciences, primarily medicine, zoology, botany, biochemistry, mineralogy, organic mineralogy, paleontology, geology of mineral deposits, etc. One of the topical issues of biomineralogy is the problem of pathogenic mineral formation in the human body. Along with kidney stones, dental calculi and others, gallstones are often formed in the human body. The process of gallstone formation has not been studied in detail because of the latent course of the initial stage of cholelithiasis. At present, a lot of studies have been undertaken to find out why and how gallstones are formed, but so far there is no generally accepted theory. In general, cholelithiasis is a multifactorial and multistage disease characterized by impaired cholesterol, bilirubin and mineral metabolism. The general factors of cholelithiasis development include: ecology, various metabolic disorders, food, unhealthy life style (alcoholism, smoking, etc.), and age (aging is associated with increased formation of gallstones) (Dadvani et al. 2009). According to a common classification, gallstones are divided into two main

E. V. Mashina (✉) · V. N. Filippov
IG Komi Scientific Center, Ural Branch of the Russian Academy of Sciences, Syktyvkar, Russia
e-mail: borovkova@geo.komisc.ru

S. Votyakov et al. (eds.), *Minerals: Structure, Properties, Methods of Investigation*, Springer Proceedings in Earth and Environmental Sciences,
https://doi.org/10.1007/978-3-030-00925-0_21

groups: pigment and cholesterol. Pigment gallstones occur in less than 20% of cases (Korago 1992). The main organic component of cholesterol stones is cholesterol, pigment is the salt of bilirubin (calcium bilirubinate, copper bilirubinate) (Li et al. 1995). The mineral component in the gallstone has not been studied enough. It is known that the main mineral phases in the gallstones are calcium carbonate (vaterite, aragonite and calcite) and calcium phosphate (carbonate-apatite and hydroxylapatite) (Efimova et al. 2005; Mashina 2014; Mashina et al. 2015). However, in addition to calcium and phosphorus, other chemical elements, which may be included in various minerals, are also present in gallstones. Based on this, the purpose of this study was to study microminerals and inclusions in cholesterol and pigment gallstones by electron microscopy.

Materials and Methods

The gallstones of the inhabitants of the Republic of Komi were investigated. Based on the structural studies (X-ray diffraction and IR spectroscopic analysis), the gallstones were divided into six groups: group 1—cholesterol (10 samples); group 2—cholesterol + calcium phosphate (4 samples); group 3—cholesterol + calcium carbonate (7 samples); group 4—pigmented (3 samples); group 5—pigment + calcium phosphate (2 samples); group 6—pigment + calcium carbonate (4 samples). The samples were studied by JSM 6400 analytical scanning electron microscope equipped with an energy-dispersive spectrometer "Link".

Results and Discussion

The results of our study showed that only halite and iron oxide, as well as various polymetallic compounds of titanium, iron, calcium, aluminum, magnesium, manganese and zinc, could be found in the cholesterol stones of group 1, apparently being organometallic compounds (mass. %): (a) Ti 21.38, Si 15.20, Mg 14.13, Ca 3.24, Pb 1.89, Cr 0.33; Ti 30.46, Pb 15.33, Si 6.72, Mg 5.38, Cr 3.72, Ca 1.89, Al 0.67; (b) Fe 56.74, Ca 6.49, Si 2.42, Al 1.03, Zn 0.82, Ti 0.60; Pb 41.01, Ti 11.13, Cr 10.28, Ca 4.75, Mg 3.39, Si 3.14; Ti 47.67, Si 3.60, Ca 3.52, Mg 2.79, Al 1.43; Si 23.43, Mg 20.77, Ti 5.86, Ca 2.51, Al 0.76; Si 18.77, Ti 10.63, Ca 9.01, Mg 7.85, Zn 2.72, Al 1.15; (c) Ca 28.48, Si 15.14, Al 7.18, Fe 6.07, Sr 1.33, Mg 1.21; Ca 22.93, Fe 17.07, Al 9.48, Si 9.34, Mg 1.42, Mn 0.42; Fe 55.13, Zn 20.84, Si 7.61, Mn 4.23, Ti 1.64, Cu 1.01; Fe 39.67, Mn 0.34 (Fig. 1a–c). Halite formed elongated dendritic crystals (Fig. 1d), and iron oxide had a spherical appearance. In the cholesterol samples of group 2, micromineralites were found: albite, barite, hematite and gypsum. Barite was traced in the form of joint clusters having a crystal-like appearance (Fig. 1e), and single grains of irregular shape. Albite had a

Fig. 1 Microminerals and inclusions in cholesterol gallstones: **a** (point 1)—Ti–Si–Mg–Ca–Pb–Cr; **a** (point 2)—Ti–Pb–Si–Mg–Cr–Ca–Al; **b** (point 1)—Fe–Ca–Si–Al–Zn–Ti; **b** (point 2)—Pb–Ti–Cr–Ca–Mg–Si; **b** (point 3)—Ti–Si–Ca–Mg–Al; **b** (point 4)—Si–Mg–Ti–Ca–Al; **b** (point 5)—Si–Ti–Ca–Mg–Zn–Al; **c** (point 1)—Ca–Si–Al–Fe–Sr–Mg; **c** (point 2)—Ca–Fe–Al–Si–Mg–Mn; **d**—halite; **e**—barite; **f**—albite, **g**—weddellite

plate-like shape (Fig. 1f). In the gallstones of the 3rd group, barite, dolomite and weddellite were found, and various inclusions of zinc, copper, iron and chromium were recorded. Weddellite formed enveloped crystals of about 20 μm (Fig. 1g); as a rule, this mineral was found in the composition of urinary stones.

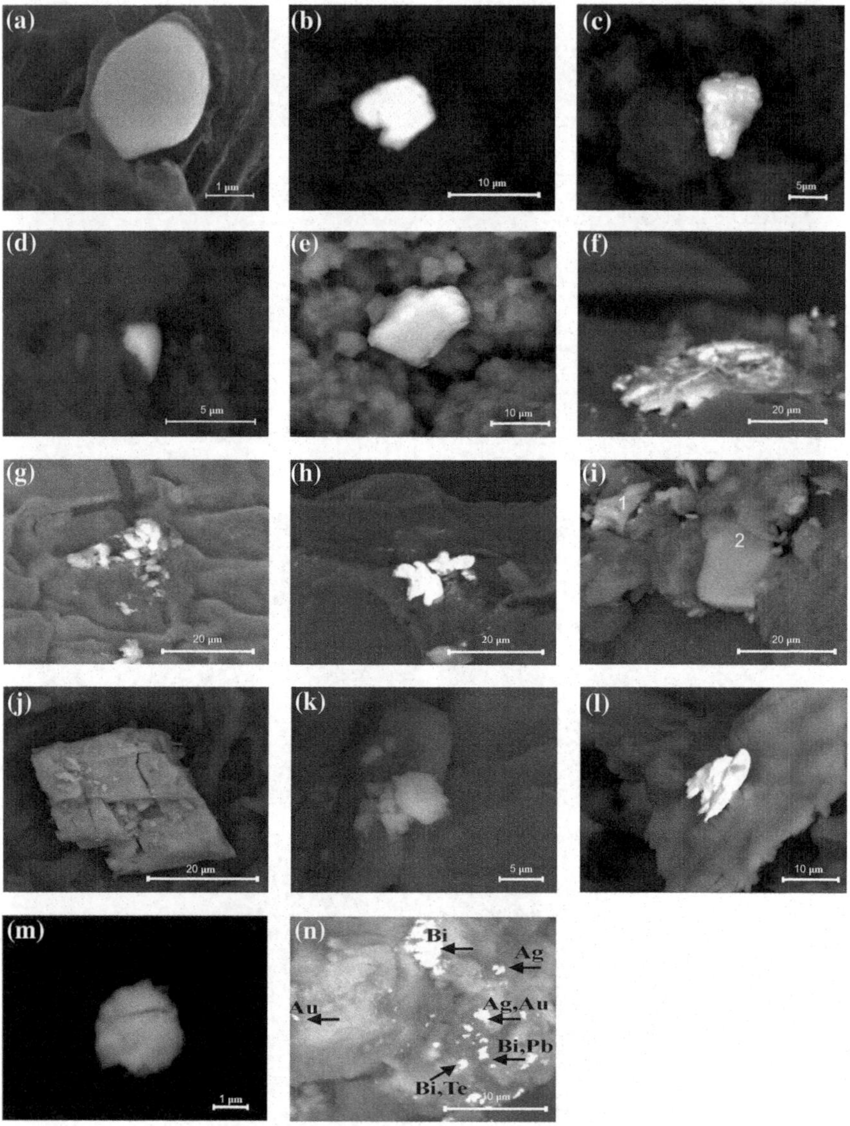

Fig. 2 Microminerals and inclusions in pigment gallstones: **a**—Fe–Ni; **b**—Fe–Ni–Mn; **c**—Fe–Ni–Si; **d**—Fe–Cu–Al–Zn; **e**—silicon oxide; **f**—Cu–Ni–Cr, **g**—Cu–Ni–Fe; **h**—Cu, **i**—1-albite, 2-microcline; **j**—dolomite; **k**—sylvite; **l**—zinc; **m**—Bi–Nb; **n**—Au, Ag

The electron microscopic study of the pigment gallstones showed that the following microminerals and inclusions were recorded in the samples of group 4: pyrite, silicon oxide, various native-metal compounds of iron (Fig. 2a–d). Silicon oxide was shown to be represented by prismatic grains (Fig. 2e). In the pigment gallstone of group 5, albite, microcline, as well as the inclusion of native copper and the native metal compounds of copper and nickel were found (Fig. 2f–i). Albite was found in the form of elongated angular grains up to 10 μm in size, microcline formed tabular crystals up to 15 μm (Fig. 2i). The samples of group 6 included: sylvite, dolomite, copper sulfate, copper oxide, weddellite, native-metal phases of iron (>70 mass. %) and copper (>60 mass. %): Fe–Ni–Si; Fe–Ni–Mn; Fe–Al; Fe–Ni; Cu–Zn–Pb. The dolomite had a rhombohedral appearance; the sylvite was represented by accumulations of individual grains (Fig. 2j,k). Crystalline grains were also found to be almost entirely (up to 98 wt%) composed of zinc (Fig. 2l) and microinclusions of bismuth (71 mass. %) and niobium (Fig. 2m), silver and gold (mass. %): Au 56; Ag 13; Au 12; Ag 16, Au 6 (Fig. 2n).

Macro- and microelements are not synthesized in the body, they are transferred primarily with food, water and air. Mineral substances entering the body form various compounds with high-molecular proteins. The authors (Lamanova et al. 2010) studying such pathogenic formations as the calcification of the blood system, believe that the disintegration of metal-binding proteins under the influence of various factors inevitably leads to the accumulation of free metal ions in the medium, which means that the formation of certain new minerals is possible according to specific oxidation-reduction and acid-base conditions in the source of mineralization. It should be noted that among the elements discovered by us there are toxic: aluminum, lead, barium, bismuth and potentially toxic: silver and gold.

Conclusion

Thus, the electron microscopy study of gallstones complements the information on the composition of the studied objects. It has been found that the micromineral diversity in gallstone is due to the presence of a class of oxides, chlorides, sulfates, sulfides, silicates, carbonates and iron-copper compounds that predominate among the native-metallic phases. It has been noted that pigment gallstones, in contrast to cholesterol gallstones, have a richer spectrum of microinclusions, and are also characterized by the accumulation of heavy metals, of which bismuth, lead, niobium and gold can be distinguished.

References

Dadvani SA., Vetshev PS., Shulutko AM., Prudkov MI. Zhelchnokamennaya bolezn'. M.: GEOTAR-Media; 2009. 176s [Dadvani SA., Vetshev PS., Shulutko AM., Prudkov MI. Gallstone disease. M: GEOTAR-Media; 2009. 176 p. (In Russ)].

Efimova YA., Kuz'micheva GM., Nikitina EA., Orlova SV. Rentgenografiya zhelchnykh kamnei. Voprosy biologicheskoi meditsinskoi i farmatsevticheskoi khimii. 2005; 2: 36–49 [Efimova YA., Kuzmicheva GM., Nikitina EA., Orlova SV. Radiography of gallstones. Questions of biological medical and pharmaceutical chemistry. 2005; 2: 36–49. (In Russ)].

Korago AA. Vvedenie v biomineralogiyu. Sankt-Peterburg: Nedra; 1992. 279s [Korago A. Introduction to biomineralogy. St. Petersburg: Nedra; 1992. 279 p. (In Russ)].

Lamanova LM., Boroznovskaya NN. Vnekletochnye mineral'nye zerna v tkanyakh serdechno-sosudistoi sistemy, metody ikh obnaruzheniya i diagnostiki. Vestnik Tomskogo gosudarstvennogo universiteta. 2010; 339: 193–200 [Lamanova LM., Boroznovskaya NN. Extracellular mineral grains in the tissues of the cardiovascular system, methods for their detection and diagnosis. Bulletin of Tomsk State University. 2010; 339: 193–200. (In Russ)].

Li WH., Shen GR., Soloway RD., Yang ZL., Tong XB., Wu E., Xu DF., Wu JG., Xu CX. Copper bilirubinate and black pigment gallstone. Biospectroscopy. 1995; 149–156. https://doi.org/10.1002/bspy.350010209.

Mashina EV. Fosfaty kal'tsiya v kholelitakh. Materialy mineralogicheskogo seminara s mezhdunarodnym uchastiem «Yushkinskie chteniya». Mai 19–22, 2014, Syktyvkar, Rossiya. s. 204 [Mashina EV. Calcium phosphates in gallstone. Materials of a mineralogical seminar with international participation of the "Yushkin readings". May 19–22, 2014, Syktyvkar, Russia, p. 204. (In Russ)].

Mashina EV., Makeev BA., Filippov VN. Karbonaty kal'tsiya v kholelitakh. Izvestiya Tomskogo politekhnicheskogo universiteta. Inzhiniring georesursov. 2015; 326(1): 34–39 [Mashina EV., Makeev BA., Filippov VN. Calcium carbonate in gallstone. Izvestiya Tomsk Polytechnic University. Engineering georesources. 2015; 326(1): 34–39. (In Russ)].

The Properties of Macrocyclic Complexes of Ruthenium

Mehriban Rahil Mikayilova and Rena Eldar Mustafayeva

Abstract A new method for the synthesis of rhenium and ruthenium phthalocyanines is proposed, which is distinguished by its availability and high yields. It is established that as a result of synthesis only one form of ruthenium phthalocyanine is formed, which can be attributed to the PcRuO structure on the basis of physical and chemical studies.

Keywords Ruthenium · Macrocyclic complexes · Phthalocyanine ligand

Introduction

Ruthenium has many valuable and interesting properties. For many mechanical, electrical and chemical characteristics, it can compete with many metals and even with platinum and gold. However, unlike these metals, ruthenium is very fragile, and therefore it is not possible to make any products from it yet. Apparently, the brittleness and intractability of ruthenium by mechanical treatment are due to the insufficient purity of the samples subjected to the tests. The physical properties of this metal depend very much on the method of production, and no one has yet been able to isolate ruthenium of high purity. Attempts to obtain pure ruthenium by sintering in briquettes, zone melting and other methods did not lead to positive results. For this reason, technically important characteristics such as tensile strength and elongation at break have not yet been accurately determined. Only recently, the melting point of ruthenium is accurately estimated at 2250 °C, and its boiling point lies somewhere around 4900 °C. Metal ruthenium very actively sorbs hydrogen. Usually the standard of the hydrogen sorbent is palladium, the cubic centimeter of which absorbs 940 cm^3 of hydrogen. The absorption capacity of ruthenium is higher. It sorbs 1500 volumes of hydrogen (Raevskaya 1979).

M. R. Mikayilova (✉) · R. E. Mustafayeva
Azerbaijan State Oil and Industry University, Azadlig avenue 20, AZ-1010,
Bakı, Azerbaijan
e-mail: mehriban-mikayilova@mail.ru

S. Votyakov et al. (eds.), *Minerals: Structure, Properties, Methods of Investigation*,
Springer Proceedings in Earth and Environmental Sciences,
https://doi.org/10.1007/978-3-030-00925-0_22

Another important property of ruthenium: at a temperature of −272.53 °C it becomes a superconductor. Compact metallic ruthenium is not soluble in alkalis, acids and even in boiling aqua regia, but partially soluble in nitric acid with the addition of strong oxidizing agents per chlorates or bromates.

Ruthenium can be dissolved in alkaline medium by hypochlorites or in acid medium by electrochemical method. When heated in air, ruthenium begins to partially oxidize. The maximum oxidation rate is observed at 800 °C. Up to a temperature of 1000 °C, ruthenium is always oxidized only to RuO_2, but if it is heated to 1200 °C and higher, it begins to convert to volatile RuO_4 tetra oxide, exhibiting a higher valence of 8.

RuO_4 is a very interesting compound. Under normal conditions, these are golden yellow needle-shaped crystals that already melt at 25 °C, turning into a brown-orange liquid with a specific odor similar to that of ozone. When touched with the slightest trace of most organic substances, ruthenium tetroxide momentarily explodes. At the same time, it dissolves well in chloroform and carbon tetrachloride. RuO_4 is poisonous: with prolonged inhalation of its vapors, the person begins to feel dizzy, there are attacks of vomiting and suffocation. Some chemists who worked with ruthenium tetroxide developed eczema.

The ability of ruthenium to form tetroxide played an important role in the chemistry of this element. By transferring to volatile RuO_4, it is possible to separate ruthenium from other noble and base metals and, after its recovery, obtain the purest ruthenium. In the same way, ruthenium impurities are removed from rhodium, iridium and platinum.

The struggle against radioactive ruthenium is paid much attention to by physicists, chemists, technologists and especially radio chemists of many countries. At the First and Second International Conferences on the Peaceful Uses of Atomic Energy in Geneva, several reports were devoted to this problem.

Ruthenium is the only platinum metal found in living organisms (according to some sources, it is also platinum). It concentrates mainly in the muscle tissue. The highest ruthenium oxide is extremely toxic and, being a strong oxidant, can cause fire of fire hazardous substances.

Ruthenium is a rare and very scattered element. In industry heterogeneous catalysis uses metallic ruthenium and its alloys. Ruthenium is used in the preparation of the most effective catalysts. The present study was carried out in the field of synthetic and structural chemistry of ruthenium complexes with organic nitrogen-containing compounds.

Materials and Methods

Using ruthenium $Ru_3(CO)_{12}$ as a initial compound and reacting with a phthalocyanine ligand in benzonitrile for 1.5 h followed by extraction with pyridine, it was possible to obtain in a small yield a ruthenium complex with phthalocyanine containing axially coordinated CO molecules and pyrolysine, Ru Pc (CO) (py).

Alkyl-substituted ruthenium phthalocyaninates were synthesized from free phthalocyanine with higher yields. $(C_5H_{11}O)$ SP_2H_2 and dissolved in boiling 2-ethoxyethanol, a solution of $RuCl_3*3H_2O$ in the same solvent (molar ratio of starting reactants 1:2) was added and the mixture was boiled for 24 h.

The cooled solution was poured into $MeOH/H_2O$ (3:1), filtered and dried. The reaction product was then purified by chromatography on neutral alumina Al_2O_3 (eluent-chloroform). However, this method failed to obtain a complex without axial ligands: the IR spectrum of the compound synthesized in this way contains a characteristic band at 1900 cm^{-1}, characterizing the valence vibrations of CO. In the ^{13}CNMR spectrum of this complex, a signal is observed in the region of 181.06 ppm, which also confirms the presence of a CO molecule in the complex. The yield$(C_5H_{11}O)_8$PcRu (CO) was 30%, for (2-Et-hexO)$_8$PcRu (CO) 25%.

Results

Thus, according to the methods of synthesis and the composition of substituted and unsubstituted ruthenium phthalocyaninates, it can be concluded that the final product of the reaction, regardless of the method of synthesis, almost always contains in its composition a carbonyl group of CO as an axial ligand.

It was found that the production of pure ruthenium phthalocyaninate, which does not contain any axially coordinated molecules, is possible with the thermal decomposition of complexes of the composition PcRu (CO) L or $PcRuL_2$. Therefore, many works are devoted to the synthesis and study of ruthenium complexes with phthalocyanines and various axial ligands.

Synthetic methods have been developed and Ru (IV) complex compounds with five types of donor ligands have been prepared: phenanthroline, nicotinic acid amide, imidazole, benzimidazole and their derivatives, and also phthalocyanine.

Conclusion

A new method for the synthesis of rhenium and ruthenium phthalocyanines is proposed, which is distinguished by its availability and high yields. It is established that as a result of synthesis only one form of ruthenium phthalocyanine is formed, which can be attributed to the PcRuO structure on the basis of physical and chemical studies.

Reference

Raevskaya MV. Physicochemistry of ruthenium and its alloys. Chemistry of ruthenium, rhodium, palladium, osmium, iridium, platinum. M.: 1979; 138–140.

Crystallochemical Analysis of Actinoid Diffusion Path in the Structures of Light Rare Earth Monazites

Nikita A. Muromtsev, Nikolai N. Eremin, Amina S. Ulanova and Ekaterina I. Marchenko

Abstract The interest in monazites is caused by assumptions about their use in disposal of nuclear waste. This process requires restraining actinide ions in the monazite structure. Our study has provided the results of computer modeling of light rare earth monazite structure, which give us useful information about diffusion paths of impurity ions.

Keywords Computer modeling · Migration path · Monazite · Rare earth elements · Plutonium · Lanthanum · Crystal structure

Introduction

As is known, natural samples of monazites $MePO_4$ are a complex isomorphic mixture of light lanthanides with significant admixtures of heavy rare earths and radioactive actinides. These compounds are necessary for the disposal of nuclear waste, for example, weapon-grade plutonium (Ewing 1995). In this work we have explored the possibility of actinide impurity ion diffusion through the crystal structure of various compositions of rare-earth monazites. The results can provide useful information about the characteristics of monazite performance.

N. A. Muromtsev · N. N. Eremin (✉) · E. I. Marchenko
Department of Geology, Lomonosov Moscow State University,
Leninskie Gory, 1, 119234 Moscow, Russia
e-mail: neremin@geol.msu.ru

A. S. Ulanova
Department of Natural Sciences, Lomonosov Moscow State University,
Dushanbe, Tajikistan

S. Votyakov et al. (eds.), *Minerals: Structure, Properties, Methods of Investigation*,
Springer Proceedings in Earth and Environmental Sciences,
https://doi.org/10.1007/978-3-030-00925-0_23

Modeling of Ion Migration Paths

Actinides can isomorphically replace a rare earth element in its structural position. In addition, actinides can be localized in interstitial insertion positions. The geometric analysis of the monazite structure type by the H-poisk software (Muromtsev 2018) showed four possible impurity atom insertion positions, which are nonequivalent and fixed in the structure with centers of inversion $\bar{1}: a(0.5\,0.5\,0.5), b(0.5\,0.5\,0.0),$ $c(0.5\,0.0\,0.5)\,and\,(0.0\,0.5\,0.5)$. These positions have different entourage (10–14 closest neighbors in first coordination sphere with radius 3.1 Å), which changes with different types of atom insertion. The conducted energy optimization of these areas in the insertion defect formation state has revealed the following patterns:

(1) For atoms of phosphorus and oxygen—*b* is the best position for insertion in all monazite structures;
(2) For atoms of rare-earth elements and plutonium, positions *b* and *d* are preferable (2–3 eV gain compared with *a* and *c* positions) for adding insertion atoms. For the larger cations (La, Ce, Pu, Pr, Nd), the *b* position is preferable, whereas for the smaller cations (Sm, Eu, Gd), the *d*-position is better.

The comparison of the initial and final interatomic distances in proprietary Me-defect area for $LaPO_4$ and $GdPO_4$ is given in Table 1.

As can be seen from the results of energy optimization, the defect area actually has significant geometric distortions. The maximum changes are obtained for Me–O_3 distances in the *d*-position (up to 0.9 Å) and Me–P in the *b*-position (up to 0.8 Å).

The possibility of insertion ions to move between positions of the crystal structure can be estimated by software package TOPOSpro (Blatov et al. 2014) using a certain atomic radii system. We have noted that in a series of experimental and theoretical papers by Gibbs and coauthors (Gibbs et al. 1992; Feth et al. 1993) the observed sizes of atoms in crystals are repeatedly shown to fluctuate between their ionic and covalent radii.

Taking into account this fact, in our analysis of possible ways of Pu^{3+} ion diffusion in monazites, we have tested 3 models: a model with ionic radii system R_I, another one with atomic radii R_A and a model with binding radii R_B. Values in R_B are evaluated in approximation of linear dependence atom radius and its charge. In Table 2 ionic radii used in crystallochemical analysis of plutonium diffusion pathways in monazites are presented.

Results and Discussion

The analysis of possible Pu migration paths in La- and Gd-monazites has shown that the migration of Pu^{3+} in these structures is extremely difficult both by the revealed void system and by isovalent substitution of Me-positions by Pu^{3+}.

Table 1 Comparison of the initial and final interatomic distances in the region of the insertion defect in the $LaPO_4$ and $GdPO_4$ structures, Å

$LaPO_4$			$GdPO_4$		
b-position of inserting Me			*b*-position of inserting Me		
Starting	*Optimized*	*Difference*	*Starting*	*Optimized*	*Difference*
La-O1 1.840	La-O1 2.368	(+0.528) (+0.528) (+0.347) (+0.347)	Gd-O1 1.834	Gd-O1 2.231	(+0.397) (+0.397) (+0.286) (+0.286)
La-O1 1.840	La-O1 2.368	(+0.772) (+0.772)	Gd-O1 1.834	Gd-O1 2.231	(+0.547) (+0.547)
La-O4 2.095	La-O4 2.442	(−0.341)	Gd-O4 1.983	Gd-O4 2.269	(−0.564)
La-O4 2.095	La-O4 2.442	(−0.341) (+0.297) (+0.297)	Gd-O4 1.983	Gd-O4 2.269	(−0.564) (+0.233) (+0.233)
La-P 2.341	La-P 3.113		Gd-P 2.209	Gd-P 2.756	
La-P 2.341	La-P 3.113		Gd-P 2.209	Gd-P 2.756	
La-O3 3.017	La-O3 2.676		Gd-O3 2.906	Gd-O3 2.342	
La-O3 3.017	La-O3 2.676		Gd-O3 2.906	Gd-O3 2.342	
La-La 3.029	La-La 3.326		Gd–Gd 2.964	Gd–Gd 3.197	
La-La 3.029	La-La 3.326		Gd–Gd 2.964	Gd–Gd 3.197	
d-position of inserting Me			*d*-position of inserting Me		
Starting	*Optimized*	*Difference*	*Starting*	*Optimized*	*Difference*
La-O3 1.472	La-O3 2.377	(+0.905) (+0.905) (+0.703) (+0.703)	Gd-O3 1.377	Gd-O3 2.230	(+0.853) (+0.853) (+0.661) (+0.661)
La-O3 1.472 La-La 2.103	La-O3 2.377	(−0.072)	Gd-O3 1.377	Gd-O3 2.230	(−0.146)
La-La 2.103	La-La 2.806	(−0.072) (+0.174) (+0.174)	Gd–Gd 2.018	Gd–Gd 2.679	(−0.146) (+0.194) (+0.194)
La-O4 2.508 La-O4 2.508	La-La 2.806	(−0.375)	Gd–Gd 2.018	Gd–Gd 2.679	(−0.299)
La-P 2.827	La-O4 2.436	(−0.375)	Gd-O4 2.421	Gd-O4 2.275	(−0.299)
La-P 2.827	La-O4 2.436		Gd-O4 2.421	Gd-O4 2.275	
La-O2 2.892 La-O2 2.892	La-P 3.001		Gd-P 2.731	Gd-P 2.925	
	La-P 3.001		Gd-P 2.731	Gd-P 2.925	
	La-O2 2.517		Gd-O2 2.74	Gd-O2 2.441	
	La-O2 2.517		Gd-O2 2.74	Gd-O2 2.441	

Table 2 Three models of ion radii used in crystallochemical analysis

Ion	R_I	R_A	R_B
O^{2-}	1.37	0.60	1.01
P^{5+}	0.17	0.93	0.52
La^{3+}	1.21	1.98	1.57
Pu^{3+}	1.18	1.95	1.54
Gd^{3+}	1.11	1.88	1.47

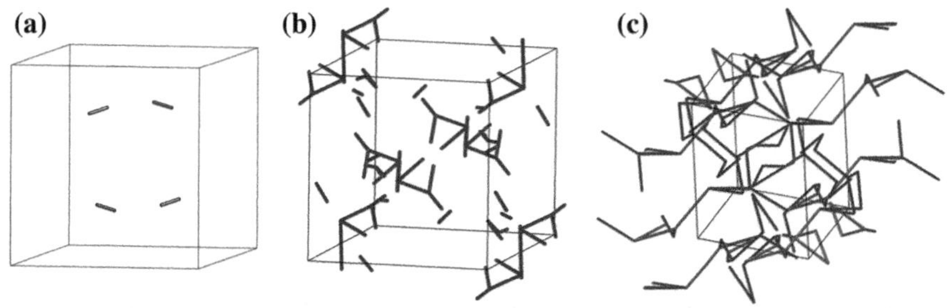

Fig. 1 Diffusion paths of impurity ions in the structure of $LaPO_4$: **a** – ions with a radius of 1.18 Å; **b** – ions with a radius of 0.70 Å; **c** – ions with a radius of 0.70 Å with the possibility of passage through LaO_9 polyhedra

Note that all three models of ion radii used have given us almost identical results. As can be seen from Fig. 1a, Pu ions are "sealed" in the crystal structure and have only a small freedom degree inside the LaO_9 polyhedron.

A similar situation is observed for smaller ions with a size more than 0.8 Å. Ions with radius of 0.7 Å can find two-dimensional diffusion channel system (Fig. 1b), which becomes three-dimensional if we take the Me position into account (Fig. 1c). In the case of Gd-monazite (smaller in its metric characteristics), these trends are more pronounced.

Conclusion

Thus, it can be concluded that all monazite structure type representatives have well preserved matrices limiting the solid-state diffusion of all actinides, because atoms smaller than the Pu atoms (Am, Cm, Bk and Cf) exceed the threshold radius value, equal to 0.7 Å.

References

Blatov VA., Shevchenko AP., Proserpio DM. Applied Topological Analysis of Crystal Structures with the Program Package ToposPro. Cryst. Growth Des. 2014; 14(7): 3576–3586.

Ewing RC., Weber WJ., Clinard FW. Radiation effects in nuclear waste forms for high-level radioactive waste. Prog. Nucl. Energy. 1995; 29(2): 63–127.

Feth S., Gibbs GV., Boisen MB., Myers RH. Promolecule radii for nitrides, oxides, and sulfides. A comparison with effective ionic and crystal radii. J. Phys. Chem. 1993; 97(44): 11445–11450.

Gibbs GV., Spackman MA., Boisen MB. Bonded and pro molecule radii for molecules and crystals. 1992; 77: 741–750.

Muromtsev NA., Kuzmina NA., Marchenko EI., Eremin NN. Teoreticheskii kristallohimicheskii analiz pustot v kristallicheskih strukturah polimorfnyh modifikacii Nb5Si3. Sbormik tezisov VIII Vserossiiskoi molodezhnoi nauchnoy konferencii “Mineraly: stroenie, svoistva, metody issledovaniya”, October 17–20, 2018. Yekaterinburg, Russia [Proceedings of the VIII All-Russian Youth Scientific Conference “Minerals: structure, properties, methods of research”, October 17–20, Yekaterinburg, Russia. (In Russ)].

U–Pb Age and Analysis of the Lu–Hf Isotope System of Zircon from Granitoids of the Final Phases of the Nepluyevsky Pluton (The Southern Urals)

Tatyana A. Osipova, Maria V. Zaitceva and Sergei Votyakov

Abstract The formation of granitoids of increased basicity and rocks of basitic composition on the Eastern slope of the Urals in the early Permian requires a reliable justification by geochronological data. U-Pb and Lu–Hf isotope systems in zircons separated from fine-grained granite of the Nepluyevsky pluton as well as Hf–Nd systematics of the rock were studied using LA-ICP-MS. Zircon LA-ICP-MS dating and the analysis of its Lu–Hf isotope system were performed using two mass spectrometers - quadrupole NexION 300S and multi-collector Neptune Plus and NWR 213 laser ablation system. The results of U-Pb dating for similar areas of NP-47 zircon grains by LA-ICP-MS and SHRIMP techniques are in satisfactory agreement. A difference in the age of zircons of different morphotypes was established: for the first type with a pronounced rhythmic zonation, a concordant U-Pb age was 360 ± 2 Ma (1σ, n = 19, SHRIMP- II) and 368 ± 6 Ma (1σ, n = 7, LA-ICP-MS); for the second type ("amoeboid-shaped" cores with "spotted" zoning), the concordant age was 430 ± 9 Ma (1σ, n = 1, SHRIMP-II). The comparison of the resulting isotopic Sm–Nd and Lu–Hf data suggests the xenogeneic, non-magmatic nature of the zircons contained in the rock, for which an older Silurian zircon of the second type could serve as a seed for growth. The age of the latter significantly exceeds the age of the Early Carboniferous rocks of the main phases of the pluton, broken by fine-grained granite.

Keywords Zircon U-Pb dating · Lu–Hf system · LA-ICP-MS · Neplyuevsky pluton

---

T. A. Osipova (✉) · M. V. Zaitceva · S. Votyakov
A.N. Zavaritsky Institute of Geology and Geochemistry, Ural Branch of the Russian Academy of Sciences, 15 Vonsovskogo Street, Ekaterinburg, Russia
e-mail: Osipova@igg.uran.ru

S. Votyakov et al. (eds.), *Minerals: Structure, Properties, Methods of Investigation*, Springer Proceedings in Earth and Environmental Sciences,
https://doi.org/10.1007/978-3-030-00925-0_24

Introduction

The formation of granitoids of increased basicity and rocks of basitic composition on the Eastern slope of the Urals in the early Permian requires a reliable justification by geochronological data (Osipova 2014).

The aim of the present work was to study U-Pb and Lu–Hf isotope systems in zircons separated from fine-grained granite of the Nepluyevsky pluton (L'vov 1964; Tevelev 2006; Popov 2003), and to analyze the Hf–Nd systematics of the rock.

Materials and Methods

Zircon grains (NP-47 sample), represented by prismatic idiomorphic transparent crystals (Fig. 1), can be divided into two morphotypes. The first type (characteristic for most crystals) exhibits a rough rhythmic zoning on CL images. The second type are irregular amoeboid-shaped cores, composed of zircon with a weakly pronounced "spotted" zoning and separated from the "rim" by curvilinear boundaries.

The data on the U-Pb isotope composition were obtained using LA-ICP-MS ("Geoanalyst" Common Use Center of Ural Branch of the Russian Academy of Sciences, Ekaterinburg) and SHRIMP-II (VSEGEI, St. Petersburg); the measurements of the Lu–Hf system parameters were carried out using LA-ICP-MS in both laboratories.

Zircon LA-ICP-MS dating and the analysis of its Lu–Hf isotope system were performed using two mass spectrometers - quadrupole NexION 300S and multi-collector Neptune Plus, and NWR 213 laser ablation system. The procedure for measuring Pb/U-isotopic ratios and the age calculation algorithm were described

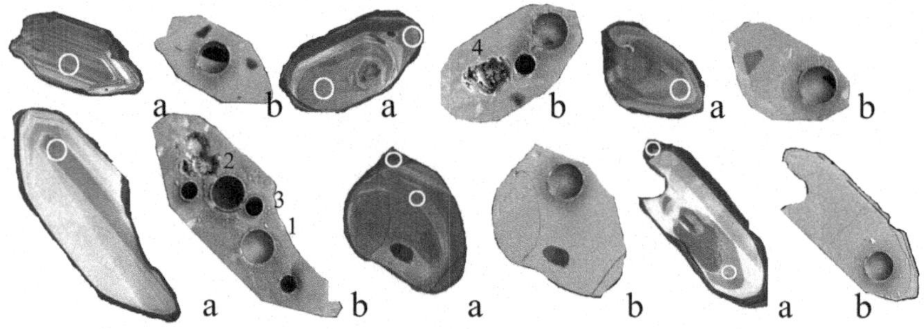

Fig. 1 CL and BSE images (**a**, **b**) of NP-47 zircon grains from granite of the Nepluyevsky pluton (the Southern Urals) after LA-ICP-MS: 1—U-Pb dating, 2, 3—Lu–Hf isotopic analysis (50 and 25 μm crater), 4—crater after SHRIMP-analysis and polishing of the top grain layer

by Zaitceva et al. (2016, 2017). The calculation of Lu/Hf and Hf/Hf isotope ratios as well as ε(Hf) values and model age performed using the Excel Hf-INATOR macros following Giovanardi et al. (2017). The optimization of the Neptune Plus operational parameters was carried out until the time and measurement error were minimized, while the maximum signals of the Hf isotopes were reached using the standard multi-element Neptune Solution (Hf concentration 200 ppb). The main criterion for the optimization (correctness) of the selected parameters was the consistency of the obtained values $^{176}Hf/^{177}Hf$ with the GeoReM database for Mud Tank, GJ-1, 91,500 and Plesovice zircon reference materials. The uncertainty of a single measurement of $^{206}Pb/^{238}U$ and $^{207}Pb/^{235}U$ isotopic ratios in the form of 1σ for the above standards was in the range of 1.5–5 and 3–18%, respectively; the uncertainty of a single measurement of $^{176}Hf/^{177}Hf$ ratio in the form of 2SD was in the range of 0.004–0.007% (all data were obtained from the ablation crater diameter of 50 microns; this uncertainty was increased 1.5–2 times with the crater decreasing down to 25 microns). The uncertainty of the single determination of ε(Hf) expressed in the form of 2SD varied in the range of 5–15% for the above reference zircons. The obtained U-Pb age values for the listed standards were in satisfactory agreement with the earlier data (Black 1978; Jackson 2004; Wiedenbeck 1995; Slama 2008). In the case of LA-ICP-MS dating of NP-47 zircon and the analysis of its Lu–Hf isotope system, GJ-1 zircon was used as the external (primary) standard, Mud Tank, 91,500 and Plesovice zircons as secondary ones. The analysis of Nd isotope composition of NP-47 granite sample was carried out using the Triton Plus mass spectrometer according to Fershtater et al. (2015).

Results and Discussion

The results of U-Pb dating for similar areas of NP-47 zircon grains, obtained by LA-ICP-MS and SHRIMP techniques are in satisfactory agreement (Table 1, Fig. 2). A larger number of analytical points obtained by SHRIMP technique helped to differentiate the age of zircons of different morphotypes: for the first type with a pronounced rhythmic zonation, the concordant U-Pb age is 360 ± 2 Ma (1σ, n = 19, SHRIMP- II) and 368 ± 6 Ma (1σ, n = 7, LA-ICP-MS); for the second type ("amoeboid-shaped" core with "spotted" zoning), the concordant age is 430 ± 9 Ma (1σ, n = 1, SHRIMP-II).

In order to confirm zircon authenticity, i.e. the presence (absence) of its genetic connection with the host granite, the approach proposed by Lokhov et al. (2009) based on the degree of coherence in the behavior of Sm–Nd and Lu–Hf systems during magmatic differentiation was used. The comparison of the resulting isotopic Sm–Nd and Lu–Hf data (Table 2, Fig. 3) suggests the xenogeneic, non-magmatic

Table 1 U-Pb dating of NP-47 zircon

Point	LA-ICP-MS-data			SHRIMP data		
	Concordia, Ma (±1σ)	$^{206}Pb/^{238}U^a$, Ma (±1σ)	$^{207}Pb/^{235}U^a$, Ma (±1σ)	Concordia, Ma (±1σ)	$^{206}Pb/^{238}U^a$, Ma (±1σ)	$^{207}Pb/^{235}U^a$, Ma (±1σ)
3_1	368 ± 6; MSWD = 0.0003	351 ± 16	288 ± 75	360 ± 2 MSWD = 0.041	362 ± 5	372 ± 30
4_2		378 ± 13	355 ± 63		363 ± 5	364 ± 12
8_1		371 ± 18	349 ± 89		366 ± 5	352 ± 24
6_1		368 ± 12	375 ± 58		353 ± 5	356 ± 11
14_1		388 ± 18	444 ± 65		355 ± 5	354 ± 13
16_1		346 ± 19	369 ± 64		363 ± 5	369 ± 15

[a]The age obtained from the equations of radioactive decay

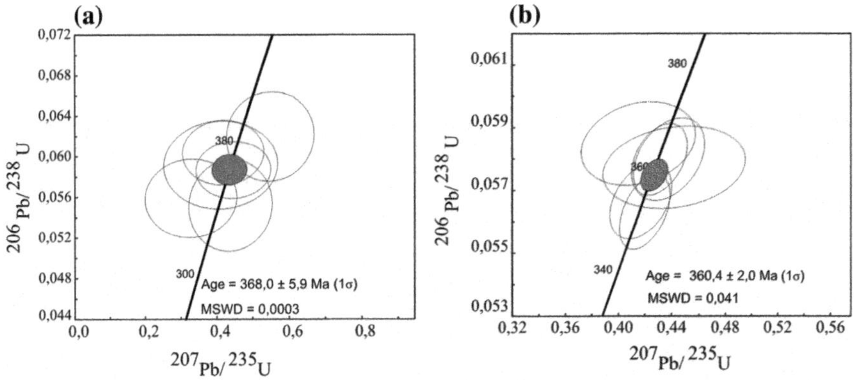

Fig. 2 $^{206}Pb/^{238}U$ vs. $^{207}Pb/^{235}U$ for the Np-47 zircon from the Nepluyevsky pluton: **a**—LA-ICP-MS, **b**—SHRIMP (VSEGEI)

nature of the zircons contained in the rock: the Upper Devonian zircons of the first type, whose points on the diagram εHf-εNd go beyond "terrestrial array" (TA) (Lokhov et al. 2009), could not crystallize from the melt according to the Nd isotope composition corresponding to the granite under study, and with a high degree of probability can be considered xenogeneic, i.e. metamorphic, for which an older Silurian zircon of the second type could serve as a seed for growth. The age of the latter significantly exceeds the age of the Early Carboniferous rocks of the main phases of the pluton, broken by fine-grained granite.

Table 2 Lu-Hf isotopic data for zircon NP-47

Point/ method of analysis		$^{176}Yb/^{177}Hf$ (±2SD)	$^{176}Lu/^{177}Hf$ (±2SD)	$^{176}Hf/^{177}Hf$ (±2SD)	$^{176}Hf/^{177}Hf_i^{(1)}$	ε (Hf)$^{(2)}$	$T_{DM}^{(3)}$ (Ma)	$T_{DM}^{c\,(4)}$ (Ma)
4_2	I	0.0236 ± 0.0010	0.00058 ± 0.00003	0.28270 ± 0.00003	0.28269	5.2	775	1039
6_1	I	0.0081 ± 0.0001	0.00016 ± 0.00002	0.28276 ± 0.00002	0.28276	7.6	676	876
14_1	I	0.0123 ± 0.0002	0.00029 ± 0.00002	0.28272 ± 0.00002	0.28271	5.9	745	989
16_1	I	0.0061 ± 0.0001	0.00011 ± 0.00003	0.28277 ± 0.00003	0.28277	7.8	670	866
Z7	I	0.0209 ± 0.0006	0.00041 ± 0.00003	0.28271 ± 0.00003	0.28270	5.5	763	1018
3_1	I	0.0128 ± 0.0003	0.00022 ± 0.00002	0.28275 ± 0.00002	0.28275	7.0	701	917
	II	0.00483 ± 0.00022	0.001039 ± 0.000017	0.28276 ± 0.00006	0.28276	7.4	685	891
4_1	I	0.0060 ± 0.0001	0.00013 ± 0.00005	0.28271 ± 0.00005	0.28271	5.6	754	1003
	II	0.00268 ± 0.00018	0.000127 ± 0.000001	0.28271 ± 0.00006	0.28271	5.7	751	998
13_2	I	0.0166 ± 0.0003	0.00043 ± 0.00002	0.28268 ± 0.00002	0.28267	4.5	802	1081
11_1	II	0.01324 ± 0.00054	0.000490 ± 0.000004	0.28268 ± 0.00003	0.28267	4.4	806	1088
13_1	II	0.00779 ± 0.00027	0.000288 ± 0.000001	0.28258 ± 0.00005	0.28257	2.5	939	1262

Note I, II – data of LA-ICP-MS, obtained at the "Geoanalyst" (IGG UB RAS) and VSEGEI, respectively; 1 – primary ratio of Hf isotopes, recalculated for an age of 360 million years (for point 13_1–430 million years) at a value of $1.867{\cdot}10^{-11}$ $years^{-1}$ of ^{176}Lu decay constant (Giovanardi and Lugli 2017); 2 – the value of ε(Hf) for the values of $^{176}Lu/^{177}Hf$ and $^{176}Hf/^{177}Hf$ - 0.0332 and 0.282772 (Giovanardi and Lugli 2017); 3 – model age of the source, calculated with the account of magma melting from depleted mantle at $^{176}Hf/^{177}Hf$ and $^{176}Lu/^{177}Hf$ - 0.28325 and 0.0384 (Black and Gulson 1978); 4 – the model age of the source, calculated from the two-stage model at $^{176}Lu/^{177}Hf = 0.015$, based on the magma melting from the middle continental crust formed from depleted mantle (Giovanardi and Lugli 2017)

Fig. 3 εHf (t) vs. εNd (t) for the NP-47 granite from the Nepluyevsky pluton: 1, 2 - LA-ICP-MS data from Geoanalytic CCU UB RAS and VSEGEI

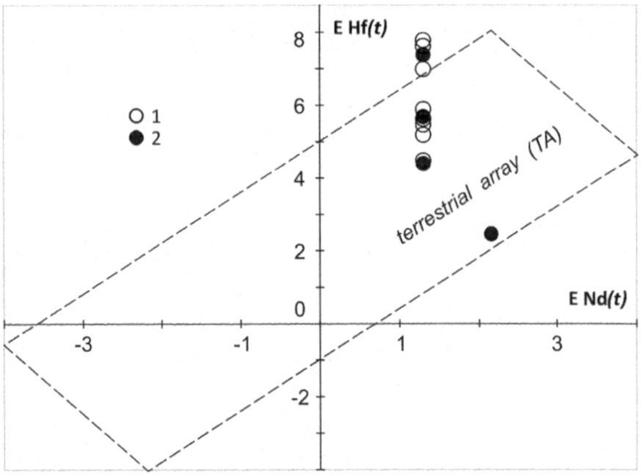

Conclusion

U-Pb and Lu–Hf isotope systems in zircons separated from fine-grained granite of the Nepluyevsky pluton as well as the analysis of the Hf–Nd systematics of the rock were studied using LA-ICP-MS. The results of U-Pb dating for similar areas of NP-47 zircon grains, obtained by LA-ICP-MS and SHRIMP techniques are in satisfactory agreement. A difference in the age of zircons of different morphotypes was established: for the first type with a pronounced rhythmic zonation, a concordant U-Pb age is 360 ± 2 Ma (1σ, n = 19, SHRIMP- II) and 368 ± 6 Ma (1σ, n = 7, LA-ICP-MS); for the second type ("amoeboid-shaped" nuclei with "spotted" zoning), the concordant age is 430 ± 9 Ma (1σ, n = 1, SHRIMP-II). The comparison of the resulting isotopic Sm–Nd and Lu–Hf data suggests the xenogeneic, non-magmatic nature of the zircons contained in the rock, for which an older Silurian zircon of the second type could serve as a seed for growth. The age of the latter significantly exceeds the age of the Early Carboniferous rocks of the main phases of the pluton, broken by fine-grained granite.

Acknowledgment The work was carried out at the UB RAS "Geoanalitik" Center for Collective Use within the state assignment of FASO of Russia (theme No. 0393-2016-020); the results of section "U-Pb-dating and analysis of the Lu–Hf-isotope system of zircons" were obtained within the RSF grant (project No. 16-17-10283); the results of section "Materials and methods" were obtained within the RFBR grant (project No. 17-05-00618).

References

Black LP., Gulson BL. The age of the Mud Tank carbonatite, Strangways Range, Northern Territory. J.Aust.Geol.Geophys. 1978; 3: 227–232.

Fershtater GB., Borodina NS., Soloshenko NG., Streletskaya MV. Novye dannye o prirode substrata yuzhnoural'skikh pozdnepaleozoiskikh granitov. Litosfera. 2015; 3: 5–16 [Fershtater

GB., Borodina NS., Soloshenko NG., Streletskaya MV. New data on the nature of substrate of Southern Ural late-paleozoic granites. Lithosphere. 2015; 3: 5–16. (In Russ]).

Giovanardi T., Lugli F. The Hf-INATOR: A free data reduction spreadsheet for Lu/Hf isotope analysis. Earth Science Informatics. 2017: 1–7.

Jackson SE., Norman JP., William LG., Belousova EA. The application of laser ablation-inductively coupled plasma-mass spectrometry to in situ U-Pb zircon geochronology. Chem. Geol. 2004; 211: 47–69.

Lokhov KI., Saltykova TE., Kapitonov IN., Bogomolov ES., Sergeev SA., Shevchenko SS. Korrektnaya interpretatsiya U-Pb vozrasta po tsirkonam na osnove izotopnoi geokhimii gafniya i neodima (na primere nekotorykh magmaticheskikh kompleksov fundamenta Vostochno-Evropeiskoi platformy). Regional'naya geologiya i metallogeniya. 2009; 38: 43–53 [Lokhov KI., Saltykova TE., Kapitonov IN., Bogomolov ES., Sergeev SA., Shevchenko SS. A correct interpretation of the U-Pb age for zircons based on the isotopic geochemistry of hafnium and neodymium (on the example of some magmatic complexes of the basement of the East European platform). Regional geology and metallogeny. 2009; 38: 43–53. (In Russ)].

L'vov BK., Vetrin VR., Ketris MP. Geologicheskoe polozhenie i petrograficheskie osobennosti granitoidov Dzhabyksko-Suundukskogo raiona (Yuzhnyi Ural). Voprosy magmatizma i metamorfizma. T.II. L.: Izd-vo LGU; 1964. s. 96–114 [Lvov BK., Vetrin VR., Ketris MP. Geological position and petrographic features granitoids of the Djabyk-Suunduk district (Southern Urals). Questions of magmatism and metamorphism. Vol. II. L.: Publishing House of Leningrad State University; 1964; 96–114. (In Russ)].

Osipova TA. Rb-Sr i U-Pb vozrast granitoidov zaklyuchitel'nykh faz Neplyuevskogo plutona (Yuzhnyi Ural): problemy datirovaniya. Mat. II mezhd. geol. konf. "Granity i evolyutsiya Zemli: granity i kontinental'naya kora", Avgust 17–20, 2014, Novosibirsk, Rossiya, ss. 161–162 [Osipova TA. Rb-Sr and U-Pb age of granitoids of the final phases of Neplyuevskiy pluton (Southern Urals): dating issues. Proceedings of the Second international geological conference "Granites and Earth's evolution: Granites and Continental Crust", August 17–20, 2014, Novosibirsk, Russia, 161–162. (In Russ)].

Popov VS., Tevelev AlV., Belyatskii BV., Bogatov VI., Osipova TA. The Rb-Sr isotopic Age of the Neplyuevka Pluton and Adjacent Intrusions, The Southern Urals. Doklady Earth Sciences. 2003; 391(5): 732–737.

Slama J., Kosler J., Condor DJ., Crowley JL., Gerdes A., Hanchar JM., Horstwood SA., Morris GA., Nasdala L., Norberg N., Schaltegger U., Schoene B., Tubrett MN., Whitehouse MJ. Plesovice zircon - A new natural reference material for U–Pb and Hf isotopic microanalysis. Chem. Geol. 2008; 249: 1–35.

Tevelev AV., Kosheleva IA., Popov VS., Kuznetsov IE., Osipova TA., Pravikova NV., Vostretsova ES., Gustova AS. Paleozoidy zony sochleneniya Vostochnogo Urala i Zaural'ya. Trudy laboratorii skladchatykh poyasov (vypusk 4). Pod red. prof. Nikishina A.M. M.: Geologicheskii f-t MGU; 2006 [Tevelev AV., Kosheleva IA., Popov VS., Kuznetsov IE., Osipova TA., Pravikova NV., Vostretsova ES., Gustova AS. Paleozoids of the joint zone of the Eastern Urals and the Trans-Urals. In: Nikishin AM., editor. Proceedings of the Laboratory of folded belts (issue 4). Moscow: Geological Faculty of Moscow State University; 2006. (In Russ)].

Wiedenbeck M., Alle P., Corfu F., Griffin WL., Meier M., Oberli F., Vonquadt A., Roddick JC., Speigel W. Three Natural Zircon Standards for U-Th-Pb, Lu-Hf, Trace-Element and REE Analyses. Geostandards Newsletter. 1995; 19(1): 1–23.

Zaitceva MV., Pupyshev AA., Shchapova YuV., Votyakov SL. U-Pb datirovanie tsirkonov s pomoshch'yu kvadrupol'nogo mass-spektrometra s induktivno-svyazannoi plazmoi NexION 300S i pristavki dlya lazernoi ablyatsii NWR 213. Analitika i kontrol'. 2016; 20(4): 294–306 [Zaitceva MV., Pupyshev AA., Shchapova YuV., Votyakov SL. U-Pb dating of zircons using a quadrupole mass spectrometer with inductively coupled plasma NexION 300S and NWR 213 laser ablation system. Analytics and control. 2016; 20(4): 294–306. (In Russ)].

Zaitceva MV., Votyakov SL. K metodike analiza Lu-Hf-izotopnogo sostava tsirkona na ICP-mass-spektrometre Neptune Plus s pristavkoi dlya lazernoi ablyatsii. Materialy III mezhdunarodnoi geologicheskoi konferentsii “Granity i evolyutsiya Zemli: granity i kontinental’naya kora”, Avgust 28–31, 2017, Ekaterinburg, Rossiya, ss. 102–104 [Zaitceva MV., Votyakov SL. To the procedure for analyzing the Lu-Hf isotope composition of zircon using the Neptune Plus ICP-mass spectrometer with laser ablation system. Proceedings of the III International Geological Conference “Granites and Earth’s evolution: the Mantle and the Crust in Granite Origin”, August 28–31, 2017, Ekaterinburg, Russia, 101–104. (In Russ)].

Crystal Chemistry of Carbonates and Clay Minerals from Bottom Sediments of Okhotskoe Sea as an Indicator of Climate Change

Nadezhda A. Palchik, Tatyana N. Moroz, Leonid V. Miroshnichenko and Vladimir P. Artamonov

Abstract The elemental and mineral composition of samples from bottom sediments of a deep-sea borehole in the Deryugin basin of the Ohotsk Sea have been investigated by diffraction, infrared spectroscopy, and X-ray fluorescence with synchrotron radiation methods. A clay samples were treated with ethylene glycol and subsequently analyzed. It was evident by this group of minerals that the main source of information about past climate change in the given region comes from illite-smectite and illite composition. The more vivid climate signal within the mineralogical historical records is the concentration of smectite layers within illite-smectite, the increase of which corresponds to warming and increased humidity as well as the opposite. This indicator reliably shows not only prolonged, but also short-term climatic events. The presence and composition of carbonates along the column was determined. It was shown that carbonates react more quickly to changes in environment than the transformation of the illite-smectite components. The gathered data may be compared to local and global environmental changes, including the paleoclimatic events in the system of Ohotsk Sea.

Keywords Carbonates · Clay minerals · X-ray diffraction · Fluorescence · Infrared spectra · Climate

Introduction

The detection of relationships between conditions of mineral formation and the features of their crystalline structures is one of the more important directions of research of crystal chemistry. The issue of mineral indicators and their associations

N. A. Palchik (✉) · T. N. Moroz · L. V. Miroshnichenko · V. P. Artamonov
V.S. Sobolev Institute of Geology and Mineralogy, Siberian Branch
of the Russian Academy of Sciences, Prosp. Koptyuga 3,
Novosibirsk 630090, Russia
e-mail: nadezhda@igm.nsc.ru

S. Votyakov et al. (eds.), *Minerals: Structure, Properties, Methods of Investigation*,
Springer Proceedings in Earth and Environmental Sciences,
https://doi.org/10.1007/978-3-030-00925-0_25

is relevant within studies of many different geological processes and phenomena - on the continental blocks as well as in seas and oceans (Chaudhri and Singh 2012; Parkash et al. 1980; Palchik et al. 2017; Zhdanova et al. 2017). Carbonates are the most variable when it comes to the conditions, within which they occur, than any other mineral groups. They are discovered in practically all types of species, including sources of ore, in caves, in hot springs, and in living organisms (Reeder 1983; Nechiporenko and Bondarenko 1988). Based on variations in the content of the isotopes of oxygen in oceanic carbonates, a clear connection is determined between ages of warming and cooling of the Earth corresponding to its location on the orbital path around the Sun. Clay minerals, due to the instability of their crystalline structures, are the "witnesses" of the most varying physical-chemical and thermodynamic processes that exist in the sedimentary layer of the Earth's crust. Their indicative abilities embrace the entire range of existing geological processes (Solotchina et al. 2017; Drits et al. 2004; Chamley 1998; Gorbarenko et al. 2002; Derkachev et al. 2009). The current interest of the scientific community is increasingly drawn to the study of sedimentary deposits of large bodies of water, as well as shallow lakes, because the power of a smaller volume of water lies in the fact that processes there happen faster and more intensely, and changes in climate or other conditions are recorded in a more distinct and reliable manner (Palchik et al. 2008; White and Schiebout 2008).

The goal of this research is to study the natural associations of minerals in the sedimentary deposits of the deep-sea trench of Ohotsk Sea, structural and crystal chemical features of individual phases and consequently their formations based on the changes in their environment, which is an important part of the complex approach to the reconstruction of quickly occurring changes in climate.

Materials and Methods

The objects of study were samples collected from the region of the Amur's estuary into the Sea of Okhotsk and from deep sea borehole SO 178-78 (thickness, 1890 cm) in the Deryugin basin in the Ohotsk Sea courteously provided by Dr. A.N. Derkachev of the Il'ichev Pacific Oceanological Institute of the Far Eastern Branch, Russian Academy of Sciences.

The phase composition of crystal substances and the structural state of the objects under study were determined via powder X-ray diffraction, which also allowed us to establish quantitative ratios of phases, characteristics of fine crystal structure, and crystallochemical features of the dispersed and mostly poorly crystallized minerals. Our X-ray studies were conducted using a Thermo Fisher Scientific (Ecublens) SARL ARL X'TRA powder diffractometer (Switzerland) with Cu$K\alpha$ radiation. The diffraction maps were scanned in an interval of 2θ from 2° to 55° with step of 0.05° with the scanning time of 5 s at each point. A qualitatively new level of these studies was provided by the implementation of effective mathematical modeling of complex X-ray diffraction profiles of multicomponent systems for the interpretation of the structural features of minerals in conjunction with the conditions of their formation. The research was conducted on the initial clay minerals and their structural transformations upon saturation with organic liquids.

The functional groups in the samples were identified using infrared spectroscopy. Infrared spectra were recorded at room temperature on a Bruker Vertex 70 Fourier transform infrared (FTIR) spectrometer with the OPUS software. The standard KBr (potassium bromide) pellet technique was used for measurements. The analysis was in absorption mode, scanning wavenumber region was from 400 to 4000 cm^{-1} with a resolution of 2 cm^{-1}, and using 30 scans. The background spectrum of each KBr pellet was subtracted from the sample spectra.

The elemental and microelemental analysis of the samples was performed following the conventional procedure at the Energy Dispersion X-ray Fluorescence Elemental Analysis Station of the VEPP_3 ring at the Institute of Nuclear Physics, Siberian Center for Synchrotron and Terahertz Radiation (Dar'in and Rakshun 2013). The emission spectra of the samples under study were excited by a beam of monochromatized radiation with energy of 25 keV. The samples were prepared by compressing the powder into tablets weighing 30 mg and with diameters of 6 mm. The emission spectra were processed using the specialized AXIL software (http://www-pub.iaea.org/MTCD/publications/PDF/IAEA-CMS-21_CD_web/PDF/CMS_21.pdf). External reference materials were used for quantitative calculations.

Results and Discussion

The elemental and microelemental analysis of sedimentary deposits of deep-sea borehole in the Deryugin basin in the Sea of Ohotsk and debris samples from the Amur river regions allowed us to determine that deposits of the borehole were not drift cone of sediments of the Amur river, as anticipated by the authors earlier (Kot 1998). The mineral composition of the samples from the Amur river, just like in the borehole included quartz, plagioclase, kaolinite, muscovite, disordered smectite and illite-smectite. The last two minerals in the trench were present in a substantially greater amount than in samples of debris from the river. Saturation of the samples with ethylene glycol led to the discovery of additional differences in the small angle regions, due to changes in phase relationships of phyllosilicates (Fig. 1). The content of such elements as Mn, Ba, Ni, Zn, Rb, etc. in the sediments of the trench was significantly higher than in the column. The presence of Th, U, and Zn of possibly anthropogenic nature was identified.

Recent sedimentation in the Deryugin basin possesses a series of special features, which sets this region apart from the deep-water part of Ohotsk and other marginal Eastern seas. Because the speed of sedimentation in this region is high, the lower age boundary of the 18-meter column is approximately 20,000 years old based on preliminary data. It is known that natural waters in this hollow region contain a minimal amount of oxygen, which creates specific conditions for

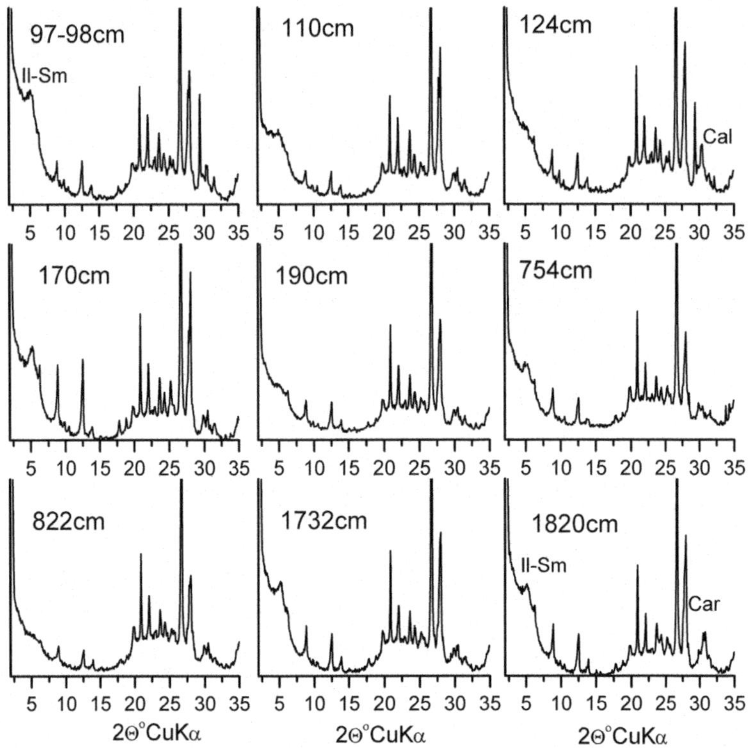

Fig. 1 X-ray diffraction patterns of the samples from different depth after treatment with ethylene glycol. Il-Sm—illite-smectite, Cal—calcite, Car—carbonate

sedimentation. Abnormally high concentrations of Mn, Fe, Ba, as well as Zn, V, and Sr have been revealed here. Additionally, the highest concentrations of Mn and Ba are observed in upper horizons (1–190 cm) and in lower ones (1332–1820 cm) (Palchik et al. 2015). Within the same intervals, we have discovered maximum carbonate content: in the upper portion of the column this is primarily calcite with a small admixture of dolomite, in the lower—Mn carbonates of the rhodochrosite-kutnohorite group, for which characteristic bands of CO_3 group lie around 876 и 866 cm^{-1}, respectively, in the IR spectrum (Fig. 2).

The upper (1–124 cm) and lower (1473–1820 cm) intervals displayed the maximum contents of carbonates (approximately 10% of the bulk composition), mainly calcite in the upper portion of the column and a discontinuous row of structurally disordered varieties of kutnohorite: Ca(Mn, Mg) $(CO_3)_2$, ankerite and rhodochrosite $MnCO_3$ in the lower portion. Trace quantities of carbonates were encountered in the middle portion of the section. The complex X-ray profiles of carbonates in angles ranging from 28° to 32° (2θ°Cu*K*α) were decomposed into individual peaks using the Pearson VII function (Fig. 3a, b). The model approach

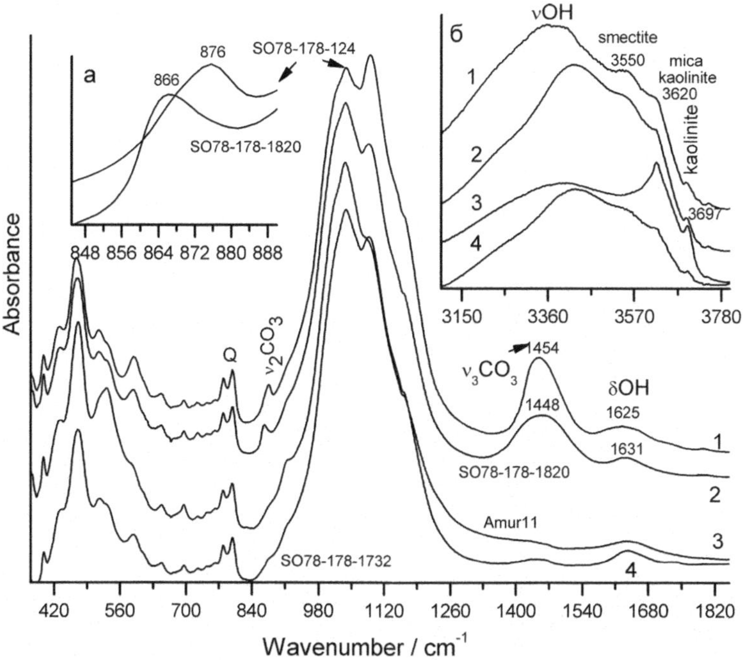

Fig. 2 Infrared spectra of samples from a deep sea borehole in the Deryugin basin: 1 – SO 78-178-1820, 2 – SO 78-178-1631, 4 – SO 78-178-1732, and 3 - from the region of the Amur River

allowed us to establish the position of the maximum and the integral intensity of the analytical peak of each carbonate phase, and to accurately assess their relations.

It was shown that the maximum content of carbonates in sea sediments was caused by existing global pulsations of melt water. The middle section of borehole contained trace amounts of carbonates. Based on the IR spectra, the mineral content of samples from the borehole of Deryugin and the debris from the Arum river differed insignificantly, recorded as carbonate and noncarbonate zones. The spectra reflected this by the presence of a band around 1435–1460 cm^{-1}. The clay components presented in research samples were kaollinite, smectite, mica, chlorite, which were absent in the Amur region sample (Fig. 2).

There is a wide variation in the content of clay along the column—from 10 to 60% (Fig. 1). Earlier on the example of the study of clay minerals from Holocene-pleistocene deposits of the Ohotsk sea, exposed by column LV 28-40-5 (Palchik et al. 2008), it was shown that the primary paleoclimatic load within the ensemble of clay minerals consisted of mixed-layer illite-smectite and illite. But the more vivid climatic signal in the mineralogical record appears to be the concentration of smectite layers in illite-smectite, the increase of which corresponds to an increase of temperature and humidity and vice versa. This indicator reliably indicates not only prolonged, but also short-term climatic events, which is reflected in

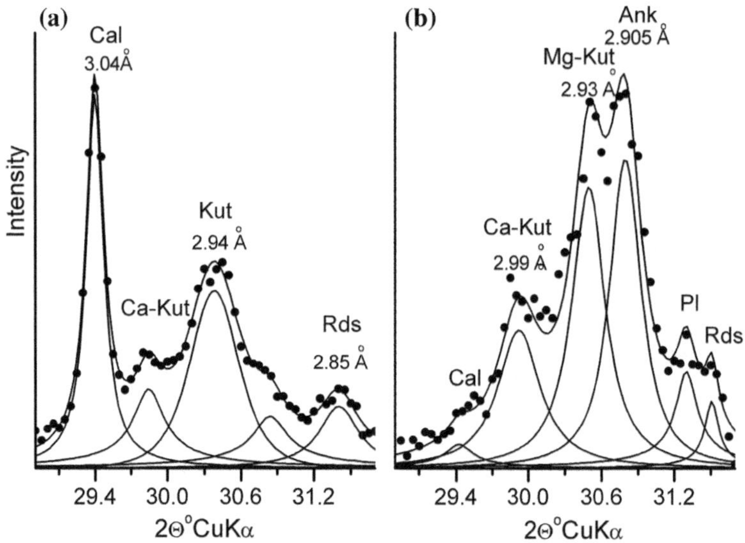

Fig. 3 Fitting of fragment of X-ray diffraction pattern of the samples: SO 78-178-124 (**a**), and SO 78-178-1820 (**b**). Cal—calcite, Ca-Kut—Ca-kutnohorite, Kut—kutnohorite, Mg-Kut—Mg-kutnohorite, Ank-ankerite, Pl—plagioclase, Rds—rhodochrosite

the diffraction patterns of samples treated with ethylene glycol (Fig. 1). As seen in Fig. 1, samples 98–97, 170, 1732, and 1820 contain noticeably larger quantities of smectite components, which is exposed after samples are treated with ethylene glycol in the region of reflection angles of 50 to 2θ, whereas samples 190 and 822 demonstrate an insignificant background increase in the given region, which indicates trace amounts of this mineral.

In accordance with the literature, warm and humid intervals in climate records are characterized by an increase of the content of Br, Ca, U, Ca/K, Sr/Rb, Sr/Ti, Y/Rb, while cold intervals are associated with maximum amounts of K, Ti, La, Ce, Th, Ni, Cr (Palchik et al. 2015). The upper 100 cm of the section contains large amounts of Group I elements and the illite-smectite component, which corresponds to a warm climate. The sample from 110 cm has a higher content of Group II elements and a slight decrease in illite-smectite, which provides evidence of short-term cooling during this period. The analysis of clay content after treatment with ethylene glycol (Fig. 1) and the elemental composition of research samples, as well as the presence of carbonates along the column, have shown that the latter reacts quicker to changes in environment, than the transformation of illite-smectite component. The study of the special features of the microelemental composition, crystal chemistry of carbonates and clay minerals using the method of mathematical modeling emission spectra of trench samples has shown that the collected data can be compared to global and local changes in environment, including paleoclimatic events in the system of the Ohotsk Sea.

Conclusion

Due to the analysis of clay content after treatment with ethylene glycol and the elemental composition of research samples, as well as the presence of carbonates along the column, we have discovered that the latter reacts quicker to changes in environment, than the transformation of illite-smectite component. The study of the special features of the microelemental composition, crystal chemistry of carbonates and clay minerals using the method of mathematical modeling emission spectra of samples from the deep-sea trench has shown that the collected data can be compared to global and local changes in environment, including paleoclimatic events in the system of the Ohotsk Sea. Further development of precise X-ray diffraction studies of fine organizational details in crystalline structures in combination with data from the mineralogical and geochemical analysis helps create hypotheses and form conclusions about the conditions of mineral formation and transformation in the Earth's crust.

It seems most interesting to study mineralogical and crystal-chemical characteristics of authigenic mineral phases, as the composition and structure of minerals is directly dependent upon the chemistry of sea waters, which in turn is controlled by the region's climate.

Acknowledgements The authors are grateful to Dr. A.N. Derkachev, of the Il'ichev Pacific Oceanological Institute, Far Eastern Branch of Russian Academy of Sciences, for the kindly presented samples of sedimentary deposits from the Ohotsk Sea, and to Dr. T.N. Grigorieva for the discussion of the results.

Instrumental research was conducted in the Analytical Center for multi-element and isotope analyses of SB RAS. The work was conducted within the governmental task, project # 0330-2016-0017.

References

Chamley H. Clay sedimentology. Springer Berlin Heidelberg; 1998. https://doi.org/10.1007/978-3-642-85916-8.

Chaudhri AR., Singh M. Clay minerals as climate change indicators—A case study. Am J Climate Change. 2012; 1: 231–239. https://doi.org/10.4236/ajcc.2012.14020.

Dar'in AV., Rakshun YaV. Determination procedure of the elemental composition of rock samples by x-ray fluorescence using synchrotron radiation from the VEPP-3 storage device. Nauchn. Vestn. Novosib. Gos. Tekhn. Univ. 2013; 2(51): 112–118.

Derkachev AN., Baranov BV., Karp BYa., Sukhoveev EN., Grigor'eva TN., Pal'chik NA., Moroz TN. Hydrothermal deposits: Evidence for Pliocene-Quaternary volcanism in the central part of the Sea of Okhotsk. Dokl. Earth Sc. 2009; 427(1): 728–732. https://doi.org/10.1134/S1028334X09050055.

Drits VA., Lindgreen H., Sakharov BA., Jakobsen HJ., Zviagina BB. The detail structure and origin of clay minerals at the Cretaceous/Tertiary boundary, Steevns Klint (Denmark). Clay Minerals. 2004; 39: 367–390. https://doi.org/10.1180/000985504394014.

Gorbarenko SA. Nürnberg D., Derkachev AN., Astakhov AS., Southon JR., Kaiser A. Magnetostratigraphy and tephrochronology of the Upper Quaternary sediments in the Okhotsk Sea: implication of ferrigenous, volcanogenic and biogenic matter supply. Marine Geology. 2002; 183: 107–129. https://doi.org/10.1016/s0025-3227(02)00164.

Kot FS. Rasseyannie metally v donnix otlozheniiyax reki Amur I zony cmesheniya v Ohotskom more. Geohimiya. 1998; 1: 102–107 [Kot FS. Trace Metals in the bottom sediments of the Amur River and of the Amur Mixing Zone in The Sea of Okhotsk. Geochemistry International. 1998; 36(1): 91–96].

Ntchiporenko GO., Bondarenko GP. Usloviya obrazovaniya morskih karbonatov (po eksperimental'nym dannym). M. Nauka, 1988. –136 c [Ntchiporenko GO., Bondarenko GP. Conditions for the formation of marine carbonates (according to experimental data). M: Science, 1988; 136 p.].

Palchik NA., Moroz TN., Grigireva TN., Dar'in AV., Mirishnichenko LV. Correlation between the mineral and microelement compositions of bottom sediments from the Sea of Okhotsk. Bulletin of the Russian Academy of Sciences. Physics. 2015; 79(1): 98–102. https://doi.org/10.3103/S1062873815010244.

Palchik NA., Moroz TN., Grigireva TN., Nikandrova NK., Mirishnichenko LV. Crystal-chemical characteristics of nontronites from bottom sediments of Pacific Ocean. Crystallography Reports. 2017; 62(1): 91–96. https://doi.org/10.1134/s1063774517010163. https://link.springer.com/article/10.1134/S1063774517010163.

Palchik NA., Solotchina EP., Goldberg EL., Stolpovskaya VN., Gorbarenko SA., Crystal chemistry of clay minerals in bottom sediments of the Okhotsk Sea as a Palioclimatic Indicator. Russian Journal of Inorganic Chemistry. 2008; 53: 867–874. https://doi.org/10.1134/S0036023608060090.

Parkash B., Sharma RP., Roy AK. The Siwalik Group Molasse, Sediments Shed by Collision of Continental Plates. Sedimentary Geology. 1980; 251: 127–159. https://doi.org/10.1016/0037-0738(80)90058-5.

Reeder RJ. Carbonates: Mineralogy and chemistry. Reviews in Mineralogy (T. 11). Mineralogical Society of America. 1983; 394 p.

Solotchina EP., Sklyarov EV., Solotchin PA., Zamana LV., Danilenko IV., Sklyarova OA., Tatkov PG. Authigenic carbonate sedimentation in Eravnoe group lakes (Western Transbaikalia): Response to Holocene climate change. Russian Geology and Geophysics. 2017; 58(11): 1390–1400. https://doi.org/10.1016/j.rgg.2017.11.005.

White PD., Schiebout J. Paleogene paleosols and changes in pedogenesis during the initial Eocene thermal maximum: Big Bend National Park, Texas, USA. Geological Society of America Bulletin. 2008; 120(11–12): 1347–1361 https://doi.org/10.1130/B25987.1.

Zhdanova A., Solotchina EP., Solotchin PA., Krivonogov S., Danilenko IV. Mineralogy of Yarkov pool (lake Chany) sediments as reflection of climate change in Holocene (south of Western Siberia). Russian Geology and Geophysics. 2017; 58(6): 692–701. https://doi.org/10.1016/j.rgg.2016.07.005.

http://www-pub.iaea.org/MTCD/publications/PDF/IAEA-CMS-21_CD_web/PDF/CMS_21.pdf

Quantitative Determination of Gas Phase Composition of Fluid Inclusions in Quartz from Krasnoye Gold Deposit (the Eastern Siberia) by Raman Microspectroscopy

Elizaveta A. Pankrushina, Sergei Votyakov, Natalya N. Ankusheva, Dmitry A. Zamyatin, Yulia Shchapova and Ekaterina E. Palenova

Abstract The analysis of the gas phase composition of fluid inclusions in quartz from Krasnoye gold deposit (Eastern Siberia) was carried out using Horiba LabRam HR800 Evolution Raman spectrometer. CO_2 was found to be the major component of fluid inclusions in quartz; it is formed due to fluid-rock interaction during the destruction of host black shales. CO_2 mol fraction (and CO_2 density) in inclusions in quartz from the gold-bearing and barren vein were estimated as 96.9–98.6 and 88.3–98.0% mol (1.23–1.29 and 1.25–1.30 g/cm^3), respectively. No differences in CO_2 and N_2 concentrations were found between the gold-bearing and barren quartz veins. The calculation results of the gas phase composition have been confirmed by gas chromatography. Based on our experience, study the composition of fluid inclusions which size is in the range 5–40 μm and located at the depth of about 100 μm from the surface of transparent minerals is promising when using HR800 Raman spectrometer.

Keywords Raman spectroscopy · Fluid inclusion · Gas phase · Composition · Quartz

E. A. Pankrushina (✉) · S. Votyakov · D. A. Zamyatin · Y. Shchapova
Zavaritsky Institute of Geology and Geochemistry, Ural Branch of the Russian Academy of Sciences, 15 Vonsovskogo Street, Ekaterinburg, Russia
e-mail: pankrushina@igg.uran.ru

D. A. Zamyatin · Y. Shchapova
Ural Federal University, 19 Mira Street, Ekaterinburg, Russia

N. N. Ankusheva · E. E. Palenova
Institute of Mineralogy, Ural Branch of RAS, Ilmeny Reserve, Miass, Russia

N. N. Ankusheva · E. E. Palenova
South Ural State University, Miass Branch, 16 Oktyabrya Avenue, Miass, Russia

S. Votyakov et al. (eds.), *Minerals: Structure, Properties, Methods of Investigation*, Springer Proceedings in Earth and Environmental Sciences,
https://doi.org/10.1007/978-3-030-00925-0_26

Introduction

Raman spectroscopy is a non-destructive technique used for the analysis of fluid inclusions with the lateral resolution above 1 μm; it is widely applied for qualitative assessment of gas phase composition of fluid inclusions (Bottrell and Miller 1990; Burke 2001; Frezzotti et al. 2012). The main advantage of Raman spectroscopy is the high locality of analysis and the non-destructive effect of laser radiation on fluid inclusion in minerals. However, the proper equipment selection and the analysis of metrological parameters for qualitative (quantitative) assessment of fluid inclusion composition are essential during the research method development. The aim of this work is to carry on the applicability test of Horiba LabRam HR800 Evolution Raman spectrometer for analysis of fluid inclusions on the example of quartz from Krasnoye gold deposit (the Bodaybo district, Eastern Siberia).

Materials and Methods

The investigated fluid inclusions in quartz from gold-bearing and barren quartz veins (Table 1) were collected from different depths in carbonaceous schists and aleurolites of the Krasnoye gold deposit (the Bodaybo district, Eastern Siberia), confined to the folded black shales of Vacha and Aunakit suites of Patomsky complex (Ankusheva et al. 2017). Raman spectra were measured using high resolution Horiba Jobin Yvon LabRam HR800 Evolution spectrometer privided with Olympus BX-FM microscope, 1800 g/mm grating. According to methodology, presented in (Pankrushina et al. 2018) 514 nm emission line of Ar laser (10–40 mW) was applied for excitation of Raman spectra; the 100 × objective (NA 0.9) was used; the acquisition time was 30 s, and the accumulation number was 10.

Table 1 Calculated values of CO_2, N_2 mol fractions and density of CO_2 in fluid inclusions in quartz from gold-bearing and barren quartz veins of Krasnoye gold deposit

No	Sample	Veins	$X_{CO_2} \pm \Delta X_{CO_2}$, mol%	$X_{N_2} \pm \Delta X_{N_2}$, mol%	$\rho_{CO_2} \pm \Delta\rho_{CO_2}$, g/cm^3
1	141425-135.6	Gold-bearing quartz veins	98.0 ± 0.4	2.0 ± 0.4	1.23 ± 0.07
2	141422-253		97.7 ± 1.0	2.3 ± 1.0	1.29 ± 0.03
3	141429-73.95	Barren quartz veins	97.6 ± 0.5	2.4 ± 0.5	1.26 ± 0.03
4	141424-86.9		97.2 ± 1.1	2.8 ± 1.1	1.27 ± 0.04
5	141414-130		96.4 ± 0.5	3.6 ± 0.5	1.30 ± 0.09
6	141422-222		97.8 ± 0.6	2.2 ± 0.6	1.25 ± 0.02
7	141419-244		87.1 ± 1.7	13.9 ± 1.7	–

Results and Discussion

Identification and Quantitative Determination of **Gas** ***Phase Composition of Fluid Inclusions*** Fluid inclusions of different shapes having the size from 5 to 40 μm have been found in each of the quartz samples (Fig. 1a, b). Their Raman spectra in gas bubble are determined as molecules of carbon dioxide and nitrogen (Fig. 1c, d).

Raman spectra of nitrogen consist of one band ν_0 at $\sim$2331 cm^{-1} (the vibrations of two nitrogen atoms) (Prieto et al. 2012). Two strong Raman modes at $\sim$1285 (ν_1) and $\sim$1388 ($2\nu_2$) cm^{-1} (so-called Fermi dyad) are observed in carbon dioxide Raman spectra (Hurai et al. 2015). Small satellite bands below 1285 and above 1388 cm^{-1} (so-called "hot bands") and peak with low intensity at $\sim$1370 cm^{-1} of $^{13}CO_2$ (Levine 1983; Hurai et al. 2015).

According to the methodology (Pankrushina et al. 2018), gas phase composition of fluid inclusion was estimated using the expression (Frezzotti et al. 2012):

$$X_a = [A_a/(\sigma_a\zeta_a)]/\sum[A_i/(\sigma_i\zeta_i)], \qquad (1)$$

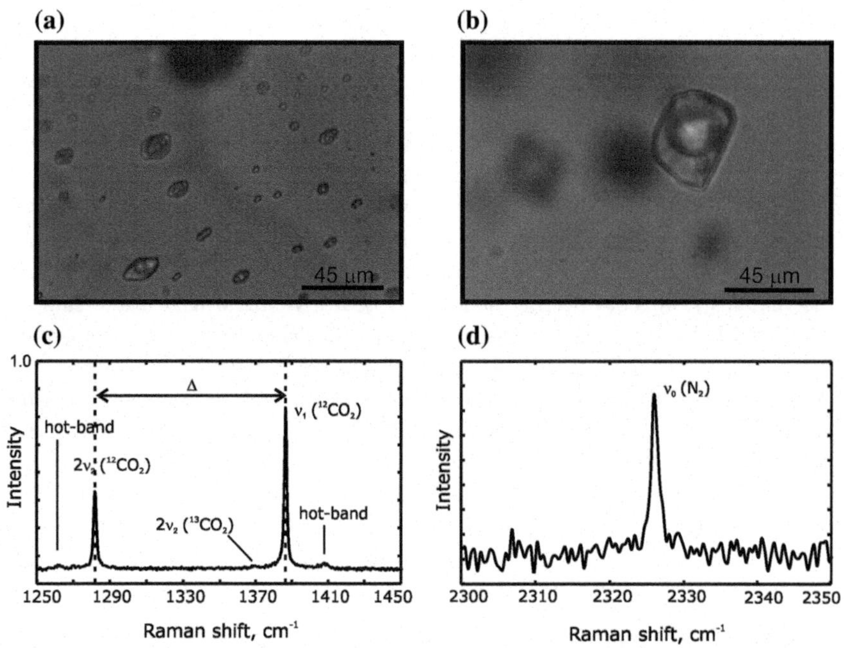

Fig. 1 Fluid inclusions in quartz (sample no. 141422-222) from Krasnoye gold deposit (**a**, **b**) and fragments of their Raman spectra (**c**, **d**). Excitation by 514 nm laser line, 30 mW

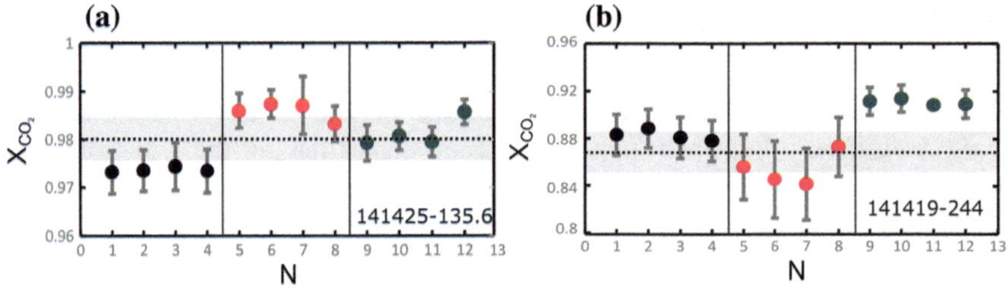

Fig. 2 Calculated values of X (CO_2) mole fraction in system CO_2 – N_2 in fluid inclusions in quartz (samples no. 141425-135.6 and 141419-244) from Krasnoye gold deposit. 1, 2, 3—serial numbers of fluids. The dashed line is the mean value. The gray field is a confidence interval

where X_a, A_a, σ_a and ζ_a—the mole fraction, peak area, Raman cross-section and instrumental efficiency for a-component, respectively. Index i presents the values for all components exist in the gas bubble in fluid inclusion.

Peak area A_i and its error ΔA_i can be evaluated using the peak fitting; it was shown in (Frezzotti et al. 2012) that σ_a values for CO_2 and N_2 Raman bands is taken as 1, 1.5 and 1, respectively; according to (Hurai et al. 2015) ζ_a-value is assumed to be equal to 1.

Using the expression (1), the mole fraction of CO_2 in fluid inclusions was found (Fig. 2). Measurement uncertainty has been calculated as indirect measurement error, including the contribution a number of factors: operational parameters of spectrometer, signal-to-noise ratio, the Raman cross-section and the instrumental efficiency for CO_2 and N_2.

It lies within the range from 96.9 to 98.6% mol in quartz from the gold-bearing vein and from 88.3 to 98.0% mol in quartz from the barren vein (Table 1). Significantly higher CO_2 concentration in fluid inclusions of quartz samples was found, which indicates the reactions with the organic matter of the host rocks (Xu et al. 2011). Not a big difference was found between the mole fraction of CO_2 phase in the samples from gold-bearing and barren quartz vein and also in the samples from different depths. Nevertheless, this observed difference does not allow the differences in the genesis conditions to be established.

According to the previous investigation of fluid inclusions in quartz from gold-bearing quartz-galena vein (No. 141425-135.6) by means of microthermometry and gas chromatography (Pankrushina et al. 2018), it was shown that the gas phase composition of inclusions is $X_{CO_2} = 98.15$, $X_{N_2} = 1.7$, $X_{CH_4} = 0.14\%$. The temperatures of fluid inclusion formation from Na-K-carbonate-hydrocarbonate fluids were 140–300 °C. The fluid pressure from three-phase inclusions with liquid CO_2 was estimated as equal to 1.2–1.6 kbar.

Estimation of CO_2 Fluid Density As an alternative to microthermometric measurements (Rosso and Bodnar 1995; Kawakami et al. 2003), CO_2 density (ρ, g/cm^3) in fluid inclusions can be calculated using the following expression with the value of splitting (Δ, cm^{-1}) between the bands in the Fermi dyad:

$$\rho = -0.03238697\,\Delta^3 + 10.08428\,\Delta^2 - 1046.189\,\Delta + 36163.67 \quad (2)$$

The Eq. (2) is valid for CO_2 density is in the range of 0.1–1.24 g/cm^3.

Using the expression (2), CO_2 density in fluid inclusions was calculated: 1.23–1.29 for quartz from gold-bearing veins, and 1.25–1.30 g/cm^3 for barren quartz (Table 1). Error of CO_2 density was calculated following (Pankrushina et al. 2018). The significant differences were not found between the CO_2 density in fluid inclusions of the samples neither from gold-bearing and barren quartz vein, nor from different depths.

Conclusion

Analysis of gas phase composition of fluid inclusions in quartz from Krasnoye gold deposit (Eastern Siberia) was carried out using Horiba LabRam HR800 Evolution Raman spectrometer. CO_2 was found to be the major component of fluid inclusions in quartz; it is formed due to fluid-rock interaction during the destruction of host black shales (Burke 2001). CO_2 mol fraction as well as CO_2 density in inclusions in quartz from the gold-bearing and barren veins were estimated to be 96.9–98.6 and 88.3–98.0% mol (1.23–1.29 and 1.25–1.30 g/cm^3), respectively. No differences were found between the fluid inclusion composition of the samples neither from gold-bearing and barren quartz vein, nor from different depths. The results of calculation of gas phase composition have been confirmed by gas chromatography. HR800 Raman spectrometer is promising for the study of the composition of fluid inclusions of about 10–40 μm in size located at the depth down to 100 μm from the surface of transparent minerals.

Acknowledgements The work was carried out at the UB RAS "Geoanalitik" Center for Collective Use; the results of sections "Methods. Identification and quantitative determination of gas phase composition of fluid inclusions. Estimation of CO_2 fluid density" were obtained within the RSF grant (project No. 16-17-10283); the results of sections "Materials. Geological interpretation" were obtained within the RFBR grant (project No. 16-05-00580a).

References

Ankusheva NN., Palenova EE., Shanina SN. Usloviya obrazovaniya kvarstevykh zhil zolotorudnykh mestorozhdeniy Kopylovskoye, Kavkaz, Krasnoe (Bodaybinsky rayon, V. Sidir) po dannym izucheniya fluidnakh vklucheniy. Metallorenia drevnikh i sovremennykh okeanov. 2017; 23: 183–186 [Ankusheva NN., Palenova EE., Shanina SN. The conditions of quartz veins formation of Kopylovskoye, Kavkaz and Krasnoye gold-bearing deposits, Bodaybo region, Eastern Siberia: fluid inclusion data. Metallogeny of ancient and modern oceans. 2017; 23: 183–186. (In Russ)].

Bottrell SH., Miller MF. The geochemical behavior of nitrogen compounds during the formation of black shale hosted quartz-vein gold deposits, north Wales. Applied Geochemistry. 1990; 5: 289–296. https://doi.org/10.1016/0883-2927(90)90004-O.

Burke EAJ. Raman microspectrometry of fluid inclusions. Lithos. 2001; 55: 139–158. https://doi.org/10.1016/S0024-4937(00)00043-8.

Frezzotti ML., Tecce F., Casagl A. Raman spectroscopy for fluid inclusion analysis. Journal of Geochemical Exploration. 2012; 112: 1–20. https://doi.org/10.1016/j.gexplo.2011.09.009.

Hurai V., Huraiová M., Slobodník M., Thomas R. Geofluids. Developments in Microthermometry, Spectroscopy, Thermodynamics and Stable Isotopes. Elsevier; 2015.

Kawakami Y., Yamamoto J., Kagi H. Micro-Raman densimeter for CO_2 inclusions in mantle-derived minerals. Applied Spectroscopy. 2003; 57(11): 1333–1339. https://doi.org/10.1366/000370203322554473.

Levine IN. Quantum Chemistry. Allyn & Bacon; 1983.

Pankrushina EA., Votyakov SL., Ankusheva NN., Pritchin ME., Kisin AY., Shchapova YV., Znamensky SE. Raman spectroscopy in the analysis of fluid inclusion composition in quartz from gold deposits of the Southern Urals: Methodological aspects. In AIP Conference Proceedings. American Institute of Physics. 2018; 2015: 20069. https://doi.org/10.1063/1.5055142.

Prieto AC., Guedes A., Dória A., Noronha F., Jiménez J. Quantitative determination of gaseous phase compositions in fluid inclusions by Raman microspectrometry. Spectroscopy Letters. 2012; 45(2): 156–160. https://doi.org/10.1080/00387010.2011.628737.

Rosso KM., Bodnar RJ. Microthermometric and Raman spectroscopic detection limits of CO_2 in fluid inclusions and the Raman spectroscopic characterization of CO_2. Geochimica et Cosmochimica Acta. 1995; 59(19): 3961–3975. https://doi.org/10.1016/0016-7037(95)94441-H.

Xu J., Hart CJ., Wang L., Chu H., Lin L., Wei H. Carbonic fluid overprints in volcanogenic massive sulfide deposits: examples from The Kelan volcanosedimentary basin, Altaides, China. Economic Geology. 2011; 106: 145–158. https://doi.org/10.2113/econgeo.106.1.145.

The Effect of Sitinakite Crystallinity Degree and Textural Characteristics on Its Sorption Properties

Igor A. Perovskiy

Abstract Titanosilicate with the structure of sitinakite of different degree of crystallinity was synthesized hydrothermally. Wastes of fluoride enrichment of leucoxene (titanium) ores of the Yaregskoye deposit (Russia, Komi Republic) were used as the initial precursor. The synthesized titanosilicate samples were characterized using X-ray diffractometry, IR spectroscopy, and low-temperature sorption-desorption of nitrogen. The regularities of sorption extraction of stable cations Sr^{2+}, Ba^{2+}, Cs^{+} on titanosilicate depending on the sorption temperature were studied. The highly crystalline titanosilicate was characterized by the highest sorption activity; its sorption capacity was 80 mg/g for Sr^{2+}, 113 mg/g for Ba^{2+} and 170 mg/g for Cs^{+}. High sorption properties with respect to stable cations show that the synthesized titanosilicate can be used as an inexpensive and effective sorbent for extracting radioactive isotopes of Sr, Cs, Ba.

Keywords Leucoxene ore · Titanosilicate · Sitinakite · Ion-exchange material

Introduction

Liquid radioactive waste containing radionuclides ^{137}Cs and ^{90}Sr represent the greatest danger for the biosphere. Selective extraction of ionic forms of these radionuclides from aqueous media is an important problem. To solve this extraction problem, anion and cation exchange, phytoremediation and vacuum evaporation are used. With a variety of proposed processes, the sorption methods with the use of inorganic sorbents remain effective and economically justified.

In recent years, natural titanosilicate, sitinakite, has gained wide popularity among inorganic sorbents. Sitinakite is a prototype of the synthetic material CST

I. A. Perovskiy (✉)
N.P. Yushkin Institute of Geology of the Komi Science Center of the Ural Branch of the Russian Academy of Science, Pervomayskaya st. 54, 167982 Syktyvkar, Russian Federation
e-mail: igor-perovskij@yandex.ru

S. Votyakov et al. (eds.), *Minerals: Structure, Properties, Methods of Investigation*, Springer Proceedings in Earth and Environmental Sciences,
https://doi.org/10.1007/978-3-030-00925-0_27

(IONSIV IE-910, IONSIV IE-911), which is used in selective extraction of ^{137}Cs and ^{90}Sr (Clearfield et al. 2000; Luca 2002; Dyer and Pillinger 1999; Tranter et al. 2005; Venkatesan et al. 2009). The high cost of titanium-containing precursors necessitates the search for new components for synthesis, particularly, inexpensive natural raw materials, such as leucoxene (titanium) ores of the Yaregskoye deposit of the Komi Republic, Russia. In this paper we continue to study the sorption properties of titanosilicate synthesized from such raw materials (Perovskiy and Burtsev 2013; Perovskiy et al. 2016).

The purpose of this work is to study the effect of crystallinity degree and texture characteristics of sitinakite, synthesized from leucoxene ore enrichment wastes, on sorption properties of Sr^{2+}, Ba^{2+}, Cs^{+} stable cations.

Materials and Methods

Reagents. NaOH (purity $\geq$ 98%, Fluka), $SrCO_3$ ($\geq$ 98%, Aldrich), $BaCl_2{\cdot}2H_2O$ (> 99%, Sigma-Aldrich), CsCl (purity $\geq$ 98%, Fluka), and HCl (puriss. spec., Sigma Tec) were used as received without additional purification. For the preparation of working solutions, deionized water with a specific resistance of 10 Mohm•cm was used. The hydrated precipitate used for the synthesis of titanosilicate was prepared by the original fluorammonium method of processing leucoxene concentrate of the Yaregskoye deposit (Perovskiy and Ignat'ev 2013). The chemical composition of the hydrated precipitate was determined using X-ray fluorescence analysis and presented in Table 1.

Synthesis. Titanosilicate was synthesized using the conventional hydrothermal method. The dried hydrated precipitate, 0.5 g, was treated with 37 mL of 1 M NaOH solution and dispersed for 20 min with a magnetic stirrer. The reaction suspension with a molar composition of $1TiO_2 - 1.2SiO_{2-5}.98Na_2O - 657.7H_2O$ was placed in 45 mL autoclave with a Teflon liner (Perovskiy et al. 2018). The synthesis was carried out at 210, 230 and 250 °C for 12 h. The reaction products were cooled down to room temperature, then collected by centrifugation, washed with distilled water and dried at 103 °C.

Characterization techniques. The synthesized samples were characterized using X-ray diffraction (Shimadzu XRD-6000 diffractometer with Cu Kα radiation in the range of reflection angles 2θ of 2 to 60°), infrared spectroscopy (Shimadzu IR-Prestige 21 spectrometer in the range of 400–4000 cm^{-1}). Adsorption and textural properties were evaluated by low-temperature (−196 °C) nitrogen adsorption-desorption as measured by the volume method using Quantachrome NOVA 1200e surface area and porosity analyzer. The pore surface area per unit mass of solid phase,

Table 1 Chemical composition of hydrated precipitate, %

SiO_2	TiO_2	Fe_2O_3	Al_2O_3	CaO	K_2O	NbO + ZrO_2
45.5	50.5	2.9	0.4	0.2	0.2	0.2

or specific surface, was determined by the BET method. The differential distribution of the mesopore volume by diameter was calculated by the Barrett-Joyner-Halenda (BJH) method. The relative error in determining the pore volume was ± 1%, surface area and pore size ± 10%. Prior to analysis, samples were degassed in vacuum for 2 h at 110 °C.

Sorption experiments. Sorption properties of synthesized titanosilicates were studied under static conditions on Sr^{2+}, Ba^{2+}, Cs^{+} stable cations. A titanosilicate sample with a weight of 30 mg was placed in a polypropylene tube, and 15 ml of the solution with a certain cation concentration was added. Sorption was carried out for 24 h with occasional shaking. At the end of the experiments, centrifugation was carried out and an aliquot of the solution was drawn. The initial (C_0) and equilibrium (C_p) cation concentrations were determined by ICP-AES spectrometer Vista MPX Rad.

Results and Discussion

Figure 1 represents the XRD patterns of the samples prepared by the hydrothermal method. According to powder x-ray diffraction data, the main identified phase was titanosilicate with the sitinakite structure (an ideal crystal-chemical formula is $KNa_2[Ti_4O_5(OH)(SiO_4)_2]{\cdot}4H_2O$), IMA No.1989-051). The decrease in the

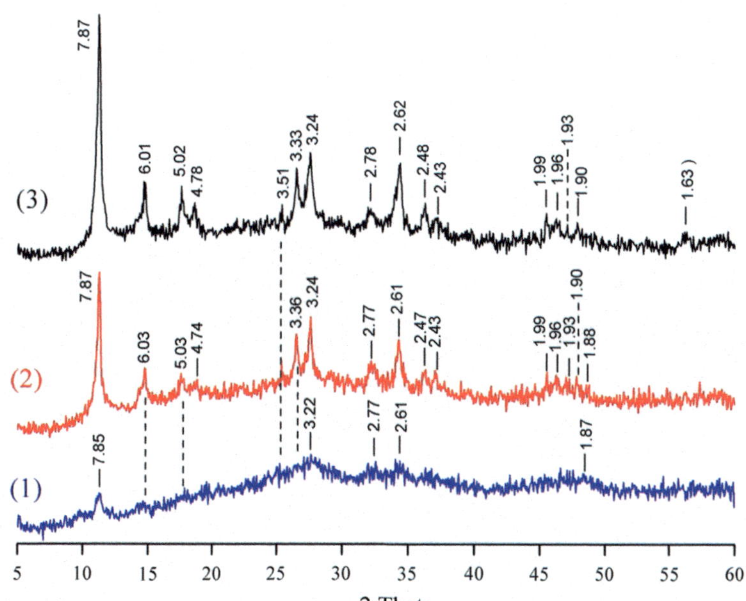

Fig. 1 XRD patterns of the synthesized sitinakite at a temperature of: 1—210 °C; 2—230 °C; 3—250 °C

temperature of hydrothermal synthesis does not affect the phase composition of the products but decreases the degree of crystallinity of titanosilicate.

The infrared spectra of the samples (Fig. 2) contain all the bands characteristic for natural titanosilicate (Chukanov 2014). The strong bands at 3230 and 3450 cm^{-1} belong to the stretching vibrations of the water molecule and the weak band at 1630 cm^{-1} to its bending vibrations. A band at 1400 cm^{-1} corresponds to the deformation vibrations of the NH^{4+} group. The presence of this band in the spectra of sitinakite is associated with the starting precursor, the precipitation of which was carried out with an ammonia solution. The described band in sitinakite can be explained by embedding the NH^{4+} group in its crystal lattice. A detailed study of the absorption bands in the characteristic region (400–1000 cm^{-1}) is given in (Perovskiy et al. 2018). The broadening of the characteristic bands associated with a decrease in the size of the crystallites of sitinakite is typical for sitinakit obtained at 230 and 210 °C.

The values of the specific surface area and pore volume of samples are shown in Table 2. As a result of the decrease in the synthesis temperature, an increase in the specific surface and pore volume of the sitinakite is observed, which is also associated with a decrease in the crystallite size.

The results of the sorption of metal cations on titanosilicate with different degrees of crystallinity are presented in Tables 3, 4 and 5. It has been found that a decrease in crystallinity slightly reduces sorption activity of sitinakite. The sorption capacity of titanosilicate synthesized at 230 and 250 °C increases with increasing

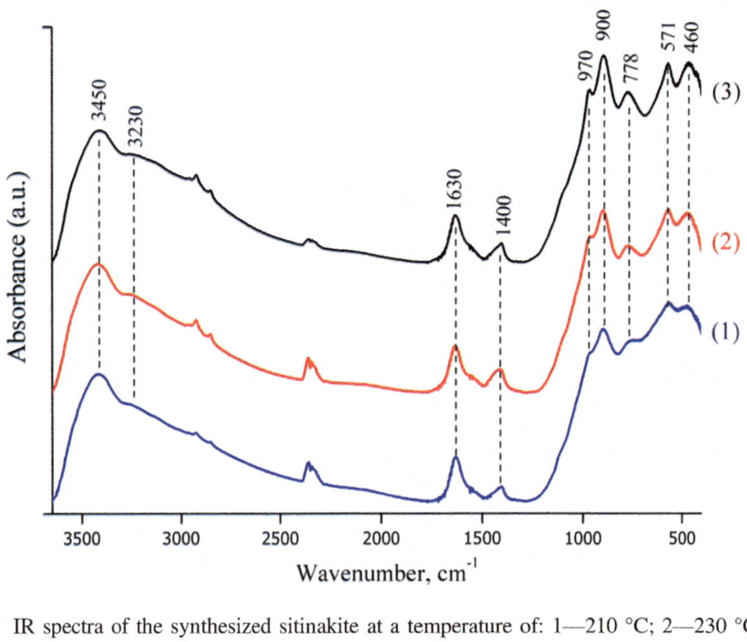

Fig. 2 IR spectra of the synthesized sitinakite at a temperature of: 1—210 °C; 2—230 °C; 3—250 °C

Table 2 Textural characteristics of the samples

Sample (°C)	Surface area (BET) (m^2/g)	Total pore volume (cm^3/g)	Micro pore volume (cm^3/g)
210	192.6	0.185	0.103
230	146.7	0.168	0.078
250	79.8	0.120	0.044

sorption temperature, which indicates the endothermic nature of the process. If titanosilicate is considered a weak electrolyte formed by a strong base and a very weak, insoluble titanium-silicic acid $H_{2(x+y)-z}Ti_xSi_yO_z \cdot nH_2O$, then with increasing solution temperature, the degree of dissociation of functional groups and, consequently, ion exchange capacity should increase (Perovskiy et al. 2018). This assumption is suitable for the sorption of Sr^{2+} and Ba^{2+} cations, where the concentration of Na^+ and K^+ cations desorbed into the solution increases. Such an effect is not observed with the sorption of Cs^+ cations, which is due to the incorporation of sorbed cations into the crystal structure of titanosilicate and the blocking of Na cations available for an exchange.

It should be noted that the titanosilicate obtained at 210 °C retains the selectivity for the Cs^+ cation at an elevated sorption temperature, while for high-crystalline samples the sorption capacity decreases. This fact can be associated with a large number of available active centers of sitinakite with a greater surface development.

Table 3 Sr^{2+} sorption capacity of titanosilicate (solution volume 0.015 L, sample mass 0.03 g, C_0 1000 mg/L)

Sample (°C)	Sorption temperature (°C)	C_p (mg/L)	Sorption capacity		Amount of desorbed Na (mmol/g)
			mg/g	mmol/g	
210	20	844.00	78.00	1.78	1.51
	60	844.33	77.83	1.78	1.53
	100	846.37	76.82	1.75	1.55
230	20	873.93	63.04	1.44	1.03
	60	847.01	76.50	1.75	1.54
	100	827.39	86.30	1.97	1.66
250	20	838.95	80.53	1.78	1.65
	60	824.50	87.75	2.00	1.74
	100	804.80	92.60	2.11	1.86

Table 4 Ba^{2+} sorption capacity of titanosilicate (solution volume 0.015 L. sample mass 0.03 g. C_0 1000 mg/L)

Sample (°C)	Sorption temperature (°C)	C_p (mg/L)	Sorption capacity		Amount of desorbed Na (mmol/g)
			mg/g	mmol/g	
210	20	780.80	109.60	1.60	1.60
	60	781.06	109.47	1.59	1.28
	100	781.42	109.29	1.59	1.32
230	20	838.80	80.60	1.17	1.28
	60	767.72	116.14	1.69	1.58
	100	760.74	119.63	1.74	1.57
250	20	773.40	113.30	1.65	1.70
	60	741.90	129.75	1.88	1.71
	100	721.70	139.10	2.03	1.81

Table 5 Cs^{+} sorption capacity of titanosilicate (solution volume 0.015 L. sample mass 0.03 g. C_0 1000 mg/L)

Sample (°C)	Sorption temperature (°C)	C_p (mg/L)	Sorption capacity		Amount of desorbed Na (mmol/g)
			mg/g	mmol/g	
210	20	676.50	161.75	1.22	1.08
	60	673.00	163.50	1.23	1.00
	100	684.00	158.00	1.19	0.91
230	20	699.50	150.25	1.13	1.06
	60	704.70	147.65	1.11	0.97
	100	717.00	141.50	1.06	0.87
250	20	659.50	170.25	1.28	1.12
	60	705.00	147.50	1.11	0.91
	100	753.50	123.25	0.93	0.62

Conclusions

In the process of hydrothermal synthesis at temperatures of 210, 230 and 250 °C, titanosilicates with the sitinakite structure were synthesized from the waste products formed during fluoride enrichment of leucoxene ores. IR spectroscopy and X-ray diffraction analysis have shown that a decrease in the synthesis temperature lowers the degree of crystallinity of titanosilicate, but does not affect the phase composition. A decrease in the synthesis temperature from 250 to 210 °C leads to an increase in the specific surface area of the titanosilicate from 79.8 to 192.6 m^2/g. The studies have shown that the sorption capacity of synthesized titanosilicate is slightly dependent on texture characteristics. Highly crystalline titanosilicate is characterized by the highest sorption capacity: 80 mg/g for Sr^{2+}, 113 mg/g for Ba^{2+} and 170 mg/g for Cs^{+}. At the same time, titanosilicate with low degree of crystallinity retains high sorption activity in Cs^{+} (sorption capacity is 160 mg/g) with increasing temperature due to the developed surface and the availability of functional groups, while the sorption capacity of the high-crystalline sample decreases to 123 mg/g. Thus, the use of waste enrichment of titanium ores and the reduction of the synthesis temperature to 210 °C significantly reduces the cost of titanosilicate production.

Acknowledgments *The work was carried out according to the research theme "Scientific basis for effective mining, the exploration and development of mineral resources base, development and implementation of innovative technologies, economic-geological zoning of Timan-Northern Ural region" (GR No. AAAA-A17-117121270037-4).*

References

Clearfield A., Bortun LN., Bortun AI. Alkali metal ion exchange by the framework titanium silicate $M_2Ti_2O_3SiO_4 \cdot nH_2O$ (M = H, Na). React Funct Polym. 2000;43 (1):85–95. https://doi.org/10.1016/S1381-5148(99)00005-X.

Chukanov NV. Infrared spectra of mineral species. Springer Geochemistry. Mineralogy. 2014. https://doi.org/10.1007/978-94-007-7128-4.

Dyer A., Pillinger M. Treatment of radioactive waste using a titanosilicate analog of the mineral zorite. Special Publication – Royal Soc Chem. 1999;239:261–269.

Luca V. Nb-substitution and Cs^+ ion-exchange in the titanosilicate sitinakite. Micropor Mesopor Mater. 2002;55, 1–13. https://doi.org/10.1016/S1387-1811(02)00353-0.

Perovskiy IA., Burtsev IN. Gidrotermalniy sintez sitinakita na osnove leikoksena Yaregskogo mestorozhdeniya. Vestnik instituta geologii Komi NTs UrO RAN, 2013;3:16–19. [Perovskiy IA., Burtsev IN. Hydrothermal synthesis of the sitinakite on the basis of leucoxene of the yaregskoe deposit. Vestnik of the Institute of Geology of Komi SC of UB of RAS. 2013; 3:16–19. (In Russ.)].

Perovskiy IA., Burtsev IN. Sorbtsionnye svojstva titanosilikata so strukturoj sitinakita, sintezirovannogo iz leikoksena. Perspektivnyie materialy. 2016;7:22 – 30. [Perovskiy IA., Burtsev IN. Sorption properties of titanosilicate with sitinakite structure synthesized from leucoxene. Advanced Materials, 2016;7: 22–30. (In Russ.)].

Perovskiy IA., Ignat'ev GV. Ftorammonijnyi sposob obeskremnivaniya leikoksenovogo kontsentrata Yaregskogo mestorozhdeniya. Prognoznaya otsenka tekhnologicheskikh svojstv poleznykh iskopaemykh metodami prikladnoi mineralogii. Sbornik statej po materialam dokladov VII rossiiskogo seminara po tekhnologicheskoi mineralogii. 2013,110–116. [Perovskiy IA., Ignat'ev GV. Ammonium fluoride method of desilication of leucoxene concentrate of Yaregskoye deposit. Forecast evaluation of the technological properties of minerals by applied mineralogy methods. Collection of articles following the VII Russian Seminar of the Technological Mineralogy. 2013;110–116. (In Russ.)].

Perovskiy I. A., Khramenkova E.V., Pidko E.A., Krivoshapkin P.V., Vinogradov A. V. Krivoshapkina E.F. Efficient extraction of multivalent cations from aqueous solutions into sitinakite-based sorbents // Chemical Engineering Journal. 2018; 354:727–739. https://doi.org/10.1016/j.cej.2018.08.030.

Tranter, T. J; Tillotson, R. D.; Todd T. A. Evaluation of ionsiv™ IE-911 as a cesium removal option for ineel acidic tank waste. Separ Sci Technol. 2005,40 (3):157–170. https://doi.org/10.1081/ss-20004190.

Venkatesan KA., Sukumaran V., Antony MP., Srinivasan TG. Studies on feasibility of using crystalline silicotitanates for the separation of cesium-137 from fast reactor high-level liquid waste. J Radioanal Nucl Chem. 2009;280(1):129–136. https://doi.org/10.1007/s10967-008-7422-1.

Mineral Salts from the Thermal Waters of the Kyndyg Deposit (The Republic of Abkhazia, the Western Caucasus)

Sergey S. Potapov, Olga Ya. Chervyatsova, Natalya V. Parshina and Sergey K. Vasil'yev

Abstract The chemical composition of the thermal waters of the Kyndyg field has been studied. Powerful deposits of mineral salts are formed from mineralized, slightly alkaline chloride-calcium-sodium waters in pipes and on the surface of the earth, where water flows. In the composition of salt deposits, minerals of the carbonate class, i.e. aragonite, calcite, magnesite have been found. The radioactivity of samples of mineral salts was studied using a low-background gamma spectrometer with a germanium detector. The samples have shown an increased concentration of radium-226 (^{226}Ra) to 7.6 ± 0.8 Bq/g. The healing effect of the therapeutic thermal waters of the Kyndyg deposit is due to the presence of radioactive radium-226 in the waters, the decay of which leads to the formation of a radioactive radon gas.

Keywords Thermal waters · Mineral salts · Carbonate minerals · Aragonite · Calcite · Magnesite · Radioactivity · Radium-226 · Radon gas · Kyndyg deposit

---

S. S. Potapov (✉) · N. V. Parshina
Institute of Mineralogy, Ural Branch of the Russian Academy of Sciences, Chelyabinsk Region, 456317 Miass, Russia
e-mail: s_almazov@74.ru

N. V. Parshina
e-mail: parschina@yandex.ru

O. Ya.Chervyatsova
FGBU State Reserve "Shul'gan-Tash", Burzyan district, v. Irgizly, Zapovednaya, 14, 453585 Bashkortostan, Russia
e-mail: kittary@yandex.ru

S. K. Vasil'yev
Emergency-Technical Center of the State Corporation "Rosatom", 3-rd Upper Lane, 2, lit. A, 194292 St. Petersburg, Russia
e-mail: spb@nwatom.ru

S. Votyakov et al. (eds.), *Minerals: Structure, Properties, Methods of Investigation*, Springer Proceedings in Earth and Environmental Sciences,
https://doi.org/10.1007/978-3-030-00925-0_28

Introduction

A natural high-thermal mineral spring is located in Kyndyg village of the Ochamchyrsky district near the Sukhum, the capital of Abkhazia (Fig. 1). Its mineralized hot waters form carbonate salts on the surface. Since we have previously dealt with salting problems in oilfield (Votyakov et al. 1995; Galeev et al. 1998; Potapov 1993, 1994) and thermal power equipment (Potapov 1998), as well as mineral formation from boiling solutions (Potapov et al. 1998), this phenomenon in the thermal waters of the Kyndyg deposit was of interest to us, and we started our research in August 2017.

Kyndyg is an unusual natural complex consisting of wells with mineral drinking water and open mineral springs for bathing, hydro massage, and healing mud. This is a famous and beloved place not only for tourists, but also for local residents. The water temperature reaches 110 °C as it bursts out of the bowels of the earth. The water flows from wells into the basins through metal and plastic pipes and gutters, cooling down 40–45 °C (Fig. 2). All the flowing springs are equipped with sunbeds and benches, on which you can lie and sit directly under the stream of healing water. Wellness hydro massage and swimming in the pool have a positive effect on the human body. "Water of life" of the Kyndyg hydrothermal spring presents an unforgettable feeling of rejuvenation, a surge of strength, and disappearance of insomnia. Due to a wide range of microelements (silicon, iodine, bromine, etc.) (Table 1), despite the short-term hydrotherapeutic procedures, the spring quickly provides blood supply to the skin tissues of both face and body, rejuvenating the whole organism.

Fig. 1 Location of the thermal water deposit Kyndyg

Fig. 2 Thermal shower and basin

Materials and Methods

Hydrochemical studies were conducted in the laboratory of the Institute of Mineralogy, Ural Branch of the Russian Academy of Sciences, Miass. We studied water pH, Eh, and electrical conductivity by electrochemical method (Hanna HI 9125, pH-meter-millivoltmeter pH 121, conductometer HI933000). We determined HCO_3^-, Cl^-, SO_4^{2-} using the titrimetric, mercuric and turbidimetric methods, Ca^{2+}, Mg^{2+}, K^+, Na^+ using atomic absorption spectrometry (Perkin-Elmer 3110 device), and trace elements using mass spectrometry with inductively coupled plasma by Agilent 7700x.

A DRON-2.0 diffractometer with CuKα-emission (Institute of Mineralogy, Ural Branch of the Russian Academy of Sciences, Miass, analyst E.D. Zenovich) was used to examine the minerals composing the crust. For visualization and study of morphology and elemental composition of the mineral salts, a TESCAN Vega 3 scanning electron microscope with an X-ACT (Oxford Instruments) energy dispersive spectrometer was used (Institute of Metal Superplasticity Problems of the Russian Academy of Sciences, Ufa, analyst I.I. Musabirov).

The radioactivity of mineral salt samples was studied using a low-background gamma spectrometer with a germanium detector and a SRP-68-01 radiometer.

Table 1 The chemical composition of the thermal waters of the Kyndyg-1 deposit

Anionic-cationic part (mg/dm^3)												
pH	Eh (mV)	G (μSm/cm)	HCO_3^-	Cl^-	SO_4^{2-}	NO_2^-	NO_3^-	NH_4^+	Ca^{2+}	Mg^{2+}	K^+	Na^+
7.72	250	1972	79.3	693.8	25.5	n.d.	n.d.	n.d.	245.6	41.55	59.0	123.0
Microelements (mcg/dm^3)												
Li	Be	Al	Sc	Ti	V	Cr	Mn	Fe	Co	Ni	Cu	Zn
92.5	0.010	1.40	0.352	3.64	0.076	1.37	3.68	130	0.38	0.184	0.67	2.66
As	Rb	Sr	Y	Zr	Nb	Mo	Cd	Sb	Ba	La	Ce	Pr
1.90	62.7	8595	0.135	0.018	<0.001	0.31	<0.008	0.014	208	0.155	0.022	0.013
Nd	Sm	Eu	Gd	Tb	Dy	Ho	Er	Tm	Yb	Lu	Ta	W
0.037	0.007	0.046	0.011	<0.003	0.005	0.002	0.009	0.001	0.008	0.002	<0.001	0.028
Tl	Pb	Bi	Th	U								
0.108	1.64	<0.004	0.009	0.046								

Note n.d.—not determined

Results and Discussion

The ion-cation and microelement composition of the water samples of the Kyndyg deposit is presented in Table 1.

Powerful deposits of mineral salts are formed from mineralized (total mineralization is 1267.75 mg/dm^3), slightly alkaline chloride-calcium-sodium waters in pipes and on the surface of the earth (Fig. 3). The curst thickness is from 2 to 10 cm or more. The crusts vary from white to yellow and orange. The crusts have banded, columnar, and radial-pentagonal texture. White crusts are more dense, uniform, and fine-grained. Yellow and orange crusts are composed of less dense aggregates of splintered-columnar mineral superindividuals (Fig. 4). The columns are easy to separate from each other, especially when the crusts are in a damp state, i.e. in aquatic environment.

The established composition of the salt deposits contains minerals of the carbonate class: aragonite, calcite, magnesite (Tables 2 and 3); in some samples, quartz occurs (see Table 3).

One can see the results of an analysis of two visually similar salt samples of the Kyndyg-2-2 morphotype (see Fig. 4) using an electron microscope (see Fig. 5). The morphology and dimensions of individuals are similar for both samples (see Fig. 5a–c for the first sample, Fig. 5d–e for the second sample). Aggregates have shape of sheaf-like stitches of pseudo-hexagonal long-prismatic crystals with a spear-shaped peak, often splintered. This shape is typical for aragonite and reflects the morphology of penetration twins along the plane (110). At large magnifications (Fig. 5e), one can see that the crystals have wide furrows along the elongation and numerous mineral microspherolite overgrowths on the flat faces of the aragonite crystals.

In the waters of the Kyndyg-1 well we discovered an increased content of strontium (8595 μg/dm^3), which is 47 times higher than the clarke for underground waters of the hypergenesis zone (Shvartsev 1985). As is known, this element acts as a stabilizer of the aragonite crystal lattice and contributes to its crystallization. According to EMF spectrometry, aragonite contains admixtures of strontium (up to 1.24 wt%) and iron (up to 1.87 wt%).

Fig. 3 Native white and orange mineral crusts

Fig. 4 Macro photos of samples of mineral salts

Based on the results of the chemical analyses transferred to the oxide form, we calculated molar ratios between the main components of carbonate minerals ($CaO + CO_2$) and the metal-impurities FeO and SrO (Fig. 6) for carbonate deposits and water of the Kyndyg-1 spring. The value of iron in aragonite is greater than that in the spring water; the inverse is true for strontium. Such a distribution can appear because iron concentrates on oxygen-oxidation barrier and accumulates in the aragonite crystals in the form of mechanical microimpurities of ferric hydroxides. Strontium incorporates the aragonite crystal lattice in the form of isomorphic impurity in an insignificant amount under the conditions of calcium ion excess that is "energetically more favorable" for calcite structure, thus strontium ratio decreases in calcite in comparison to the nourishing solution.

According to the Abkhaz State Environmental Monitoring Center (Deputy Director of ASEMC, O. Pustovarov), the radiation background near the basins for hydrotherapeutic procedures filled from a high-temperature well in Kyndyg village (near the river Chasha) is 0.23–0.24 μSv/h, which slightly exceeds the natural

Table 2 Radiogram of K-2-2 sample of the salt deposits from the thermal waters of the Kyndyg deposit (1), reference aragonite $CaCO_3$ (2), calcite $CaCO_3$ (3), and magnesite $MgCO_3$ (4)

K-2-2 sample (1)		Aragonite (JCPDS*) (2)			Calcite (JCPDS*) (3)			Magnesite (JCPDS*) (4)		
d (Å)	I	d (Å)	I	hkl	d (Å)	I	hkl	d (Å)	I	hkl
3.844	2	–	–	–	3.85478	16.50	10-2	–	–	–
3.394	87	3.39558	100	111	–	–	–	–	–	–
3.271	57	3.27275	57.40	021	–	–	–	–	–	–
3.025	32	–	–	–	3.03555	100	104	–	–	–
2.731	11	–	–	–	–	–	–	2.74089	100	104
2.701	29	2.70032	53.60	012	–	–	–	–	–	–
2.482	47	2.48438	29.20	102	2.49480	13.60	110	–	–	–
2.41	14	2.41029	15.70	031	–	–	–	–	–	–
2.37	23	2.37175	42.30	112	–	–	–	–	–	–
2.34	53	2.34137	29.40	130	–	–	–	–	–	–
2.276	5	–	–	–	2.28463	28	113	–	–	–
2.19	21	2.18939	14.60	211	–	–	–	–	–	–
2.105	55	2.10568	22.40	220	2.09443	17	202	2.10212	51.80	113
1.977	100	1.97688	77.30	221	–	–	–	–	–	–
1.903	7	–	–	–	1.91240	21.90	10-8	–	–	–
1.88	44	1.88174	34.10	041	1.87532	29.70	116	–	–	–
1.815	24	1.81428	32.60	132	–	–	–	–	–	–
1.743	10	1.74208	34.60	113	–	–	–	–	–	–
1.728	15	1.72481	15.70	023	–	–	–	–	–	–
1.699	4	–	–	–	–	–	–	1.69982	25.50	116
1.597	4	–	–	–	1.60409	11.20	21-2	–	–	–

*Joint Committee on Powder Diffraction Standards

Table 3 Results of gamma spectrometric analysis of samples of salt deposits

№	Sample №	Sample location Well №	Sample type and its mineral composition (according to X-ray phase analysis data)	Specific activity ^{226}RA (Bq/g)
1	K-1-1	Kyndyg-1	White salt crust Calcite, aragonite	0.2 ± 0.06
2	K-1-2	Kyndyg-1	Orange salt crust Aragonite, calcite, quartz, magnesite	1.3 ± 0.2
3	K-2-1	Kyndyg-2	Light gray salt crust Aragonite, calcite, magnesite	1.3 ± 0.2
4	K-2-2	Kyndyg-2	Light yellow salt crust Aragonite, calcite, magnesite	1.3 ± 0.2
5	K-2-3	Kyndyg-2	Orange salt crust Aragonite, calcite	7.6 ± 0.8

Fig. 5 SEM photo of mineral salts

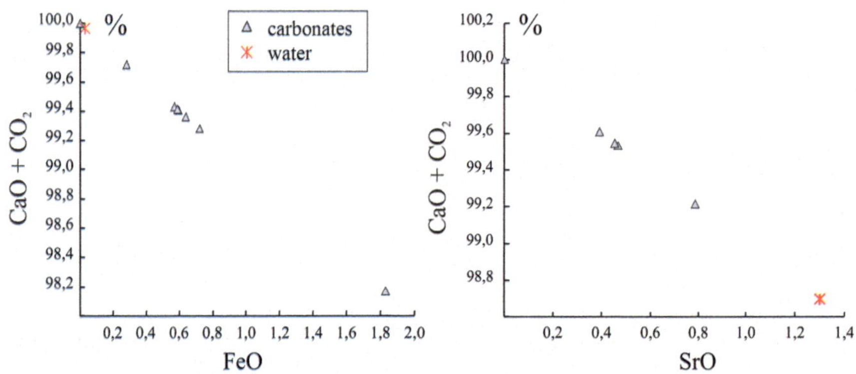

Fig. 6 Molar ratios between the main components of carbonate minerals ($CaO + CO_2$) and the metals-impurities FeO and SrO

background values. The indicator of specific total β-radioactivity does not meet regulatory requirements.

The radiation background about 3 m away from the high-thermal well on the territory of the boarding house "Evkaliptovaya roshcha" is 0.45–0.65 µSv/h, which exceeds the natural radiation background of the territory of the Republic of Abkhazia and does not meet the requirements of SanPiN 2.6.1.2800-10 "Hygienic requirements for limiting radiation exposure of population due to natural sources of ionizing radiation." The radiation background near the accumulative basin, filled from the same high-temperature well, is 0.23–0.25 μSv/h, which is only slightly higher than the natural background values.

The technique used to determine specific total beta-radioactivity consists in the cumulative determination of all the radionuclides contained in water including radon ^{222}Rn, for which the interference level in drinking water should not exceed 60 Bq/kg in accordance with SanPiN 2.6.1.2800-10 "Hygienic requirements for limiting radiation exposure of population due to natural sources of ionizing radiation."

We attempted to study radioactivity of the mineral salt samples using a low-background gamma spectrometer with a germanium detector. In the samples, we detected an increased concentration of radium-226 (^{226}Ra); one can see specific activities for the studied samples in Table 3. We did not find any artificial radionuclides. The obtained values are below the level related to radioactive waste (RW) (10 Bq/g), while the specific activity of the sample K-2-3 is close to this value (see Table 3, Fig. 4).

As is known, radium-226 is a radioactive nuclide of radium with atomic number 88 and mass number 226. The report on the discovery of a new radioactive element "radium" in pitchblende (later it turned out to be radium-226) was made on December 26, 1898 by P. Curie and M. Sklodowska-Curie in collaboration with G. Bemon (Vdovenko and Dubasov 1973). Radium-226 belongs to the radioactive family of uranium-238 (the so-called series of uranium-radium). The activity of one gram of this nuclide is approximately 36.577 GBq. The value of the non-system unit for measuring the curie activity (3.7×10^{10} Bq) was originally defined as the radioactivity of radium emanation (i.e., radon-222) in radioactive equilibrium with 1 g of ^{226}Ra (Mukhin 1993).

Radium-226 is directly formed as a result of α-decay of the nuclide ^{230}Th:

$$^{230}_{90}\mathrm{Th} \rightarrow ^{226}_{88}\mathrm{Ra} + ^{4}_{2}\mathrm{He}.$$

Radium-226 undergoes α-decay; its decomposition results in the formation of nuclide ^{226}Rn, known as radioactive radon gas or radium emanation:

$$^{226}_{88}\mathrm{Ra} \rightarrow ^{226}_{86}\mathrm{Rn} + ^{4}_{2}\mathrm{He}.$$

Consequently, the healing effect of the therapeutic thermal baths of the Kyndyg field is due to the presence of radioactive radium-226 in the waters (and in the mineral salts formed therefrom), the decay of which leads to the formation of radioactive gas radon that has a beneficial effect on human body.

The measurements of the γ radiation from the surface of salt samples that we carried out using the SRP-68-01 radiometer gave values at the natural background level (Table 4).

Table 4 The results of measuring the γ-radiation (μR/h) from the surface of samples of mineral salts deposited from the thermal waters of the Kyndyg deposit

№	Sample №	Sample location	Description	Measurement of gamma radiation			
				1	2	3	Average
1	K-1-1	Kyndyg-1	White salt crust	12	17	18	16
2	K-1-1a	Kyndyg-1	Large white salt crust	18	19	18	18
3	K-1-2	Kyndyg-1	Orange salt crust	18	19	17	18
4	K-2-1	Kyndyg-2	Light gray salt crust	17	17	18	17
5	K-2-2	Kyndyg-2	Light yellow salt crust	19	18	18	18
6	K-2-3	Kyndyg-2	Orange salt crust	19	18	18	18
7	K-2-4	Kyndyg-2	Columnar thick flat crust of gray calcite	19	18	15	17

Conclusions

Thus, the low values of γ-radiation of thermal waters and mineral salts formed from them are close to the background ones, and radioactive radium-226 present in the waters (and in the mineral salts formed therefrom), decays and leads to the formation of radioactive radon gas, which has a beneficial influence on human body and stipulates the healing effect of the therapeutic thermal waters of the Kyndyg deposit.

Acknowledgements The authors are grateful to K.A. Filippova, L.G. Udachina, G.F. Lonshchakova, M.N. Malyaryonok, N.I. Valizer for hydrochemical studies; E.D. Zenovich for the operative survey of radiographs; I.I. Musabirov for his work on a scanning electron microscope; Yu.S. Lyakhnitskiy and V.A. Muftakhov for contributing to the study of the radioactivity of the samples; and also to the employee of Novoafonskaya cave complex V.V. Marholiya for organizing the trip to the Kyndyg deposit and to the director of the Institute of Ecology of Academy of Sciences of Abkhazia R.S. Dbar for the information support.

References

Galeev R.G., Diyashev R.N., Sattarova F.M., Potapov S.S. 1998, Issledovanie mineral'nogo sostava i prichin otlozheniy soley v neftepromyslovom oborudovanii [Research of mineral composition and causes of salt deposits in oilfield equipment]. Neftyanoe khozyaystvo [Oil Industry], no. 5, pp. 41–45.

Mukhin K.N. 1993, Svoystva nuklonov, yader i radioaktivnykh izlucheniy [Properties of nucleons, nuclei and radioactive radiations]. Eksperimental'naya yadernaya fizika [Experimental Nuclear Physics], Moscow, vol. 1, pp. 173–174.

Potapov S.S. 1993, Mineralogiya solevykh otlozheniy v skvazhinakh i neftepromyslovom oborudovanii mestorozhdeniy Zapadnoy Sibiri. Avtoreferat dissertatsii na soiskanie uchenoy stepeni kandidata geologo-mineralogicheskikh nauk [Mineralogy of salt deposits in wells and oilfield equipment of West Siberia fields. The dissertation author's abstract on competition of a scientific degree of the candidate of geologo-mineralogical sciences], Ekaterinburg, 20 p.

Potapov S.S. 1994, Solevye otlozheniya v neftepromyslovom oborudovanii: obzor napravleniy issledovaniy, preimushchestva mineralogicheskogo podkhoda [Salt deposits in oilfield equipment: review of research areas, advantages of mineralogical approach]. Izvestiya vysshikh uchebnykh zavedeniy. Gornyy zhurnal [News of the Higher Institutions. Mining Journal], no. 9-10, pp. 71–77.

Potapov S.S., Votyakov S.L., Borisov D.R. 1998, Mineralogiya i spektroskopiya tekhnogennykh i antropogennykh (bytovykh) nakipey [Mineralogy and spectroscopy of technogenic and anthropogenic (household) incrustrations]. Ural'skiy mineralogicheskiy sbornik [Ural Mineralogical Collection], no. 8, pp. 151–170.

Potapov S.S. 1998, Mineral'nyy sostav kotel'nykh nakipey i otlozheniy v sisteme teplosnabzheniya doma otdykha "Turgoyak" [Mineral composition of boiler incrustations and sediments in the heat supply system of 'Turgoyak' holiday house]. Tez. dokl. III regional'nogo soveshchaniya "Mineralogiya Urala" [Abstract reports of the III regional meeting "Mineralogy of the Urals"], vol. 2, pp. 81–83.

Shvartsev S.L. 1985, Geokhimiya podzemnykh vod zony gipergeneza. Geokhimiya prirodnykh vod [Geochemistry of groundwater in the hypergenesis zone. Geochemistry of natural waters]. Trudy Vtorogo Mezhdunararodnogo simpoziuma [Proceedings of the Second International Symposium], Leningrad, pp. 108–113.

Vdovenko V.M., Dubasov Yu.V. 1973, Analiticheskaya khimiya radiya [Analytical chemistry of radium], Leningrad, 190 p.

Votyakov S.L., Potapov S.S., Borisov D.R., Krasnobaev A.A. 1995, Spektroskopicheskie svoystva tekhnogennykh karbonatov iz neftepromyslovogo oborudovaniya [Spectroscopic properties of technogenic carbonates from oilfield equipment]. Ural'skiy mineralogicheskiy sbornik [Ural Mineralogical Collection], no. 5, pp. 66–81.

Technology for Quality Assessment of Quartz Raw Materials

Vyacheslav M. Ryzhkov, Maria A. Igumentseva and Mikhail V. Shtenberg

Abstract Quartz glass due to its unique properties occupies a special position among all industrial glass. The issue of the production of quartz suitable for the melting of high-purity quartz glass has become very acute. In the article, a new technology for assessing the quality of quartz materials is proposed, which is used at the initial stage of development of quartz deposits, and includes the enrichment of the initial material, quartz glass melting and the subsequent analytical studies. The experimental data obtained using the developed technology are given, which makes it possible to conclude that the proposed scheme can be used for the attestation of quartz raw materials at the initial stage during the development of quartz deposits.

Keywords Quartz glass · Refining and enrichment · Melting

Introduction

Quartz glass due to its unique properties occupies a special position among all industrial glass. The development of the up-to-date branches of science and technology, such as optics and light engineering, aviation and cosmonautics, the chemistry of high purity substances and instrument engineering, fiber optic communication technologies, nuclear power engineering, medicine, etc., is largely determined by the level of the use of fused quartz materials.

Currently, the issue of the production of quartz suitable for the melting of high-purity quartz glass has become very acute (Götze and Möckel 2012). Multistage, labour-consuming and costly methods for the approbation of quartz raw materials are used in modern industry. In order to solve the problem of involving the quartz raw material deposits in production, it is necessary to develop an express method for assessing their quality, which will reduce labour costs and accelerate the obtaining of results on the suitability of raw materials.

V. M. Ryzhkov (✉) · M. A. Igumentseva · M. V. Shtenberg
Institute of Mineralogy, Ural Branch of RAS, Ilmeny Reserve, Miass, Russia
e-mail: Ryzhkov_v_m@mail.ru

S. Votyakov et al. (eds.), *Minerals: Structure, Properties, Methods of Investigation*, Springer Proceedings in Earth and Environmental Sciences,
https://doi.org/10.1007/978-3-030-00925-0_29

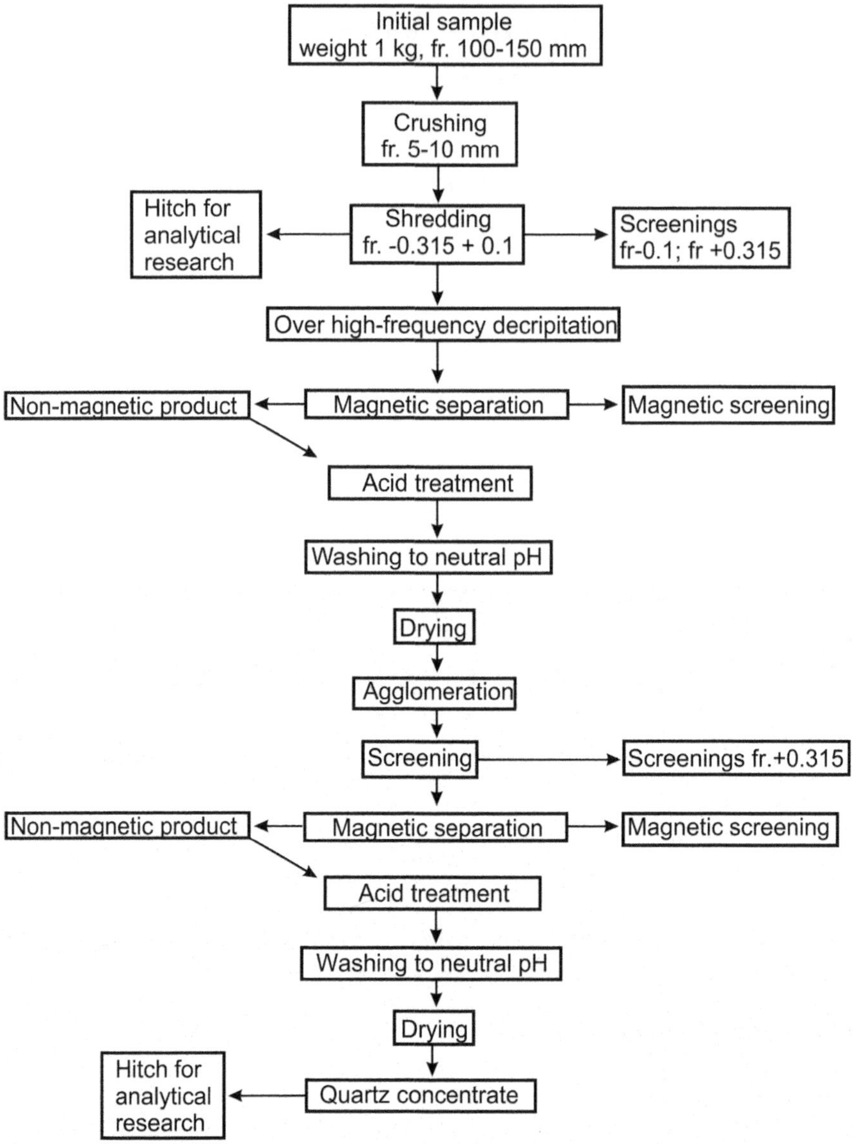

Fig. 1 Scheme of deep enrichment of quartz raw materials

Materials and Methods

Quartz samples from deposits differing in quality were selected while developing the technology. A quartz sample collected from the vein № 175 of the Kyshtym deposit was used as a reference.

The enrichment of quartz grains was carried out according to the technology developed at the Institute of Mineralogy UB RAS (Nasyrov 2011) (Fig. 1).

The melting of quartz glass was performed using the modernized melting plant "GRANAT-2M" in the argon protective atmosphere.

Melting mode:

1. Heating to 1300 °C under vacuum and material surface degassing for 0.5 h.
2. Argon supply of 0.1×10^5 Pa, heating to a temperature of 1850 °C and melting of the glass for 30 min.
3. Cooling to a temperature of 1350 °C and holding for 0.5 h to remove the thermal stress in the glass while maintaining the pressure of 0.1×10^5 Pa.
4. Cooling to room temperature over 3 h.

One of the ways of quartz raw material attestation as well as the technology of its enrichment is the smelting of quartz glass ingots and the subsequent analysis of their optical, macro-microstructural and other characteristics (Nasyrov et al. 2007).

In order to study the optical properties of quartz glasses, a 2-mm thick disk is cut from the glass block with $d \approx 30$ mm and polished.

The attestation criteria for the raw material purity, the efficiency of its enrichment, as well as the conditions for melting glass are, in particular, the characteristics of the optical transmission spectrum in a defined wavelength region (Nasyrov et al. 2007).

The spectra in the ultraviolet and visible regions are obtained using SF-56 scanning UV spectrophotometer with a double monochromator and an automated cuvette compartment. Shadow photos are obtained using Durable-Academy 400 overhead projector.

Results and Discussion

Optical properties of smelted quartz glasses. The transmission spectra in the ultraviolet, visible and infrared regions of smelted quartz glass samples are shown in Fig. 2.

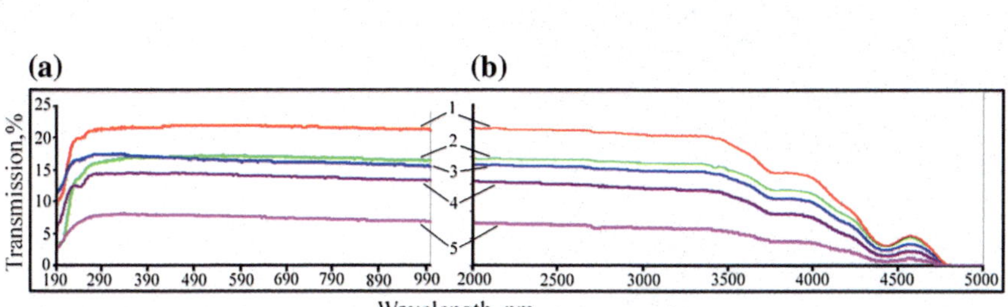

Fig. 2 Optical transmission spectra of samples in the short-wavelength (**a**) and long-wave (**b**) spectral regions. 1 - vein number 175; 2 - Larinskoe deposit; 3 - Kundravinskoye field; 4 - Argazinskoye field; 5 - Tolstikha vein

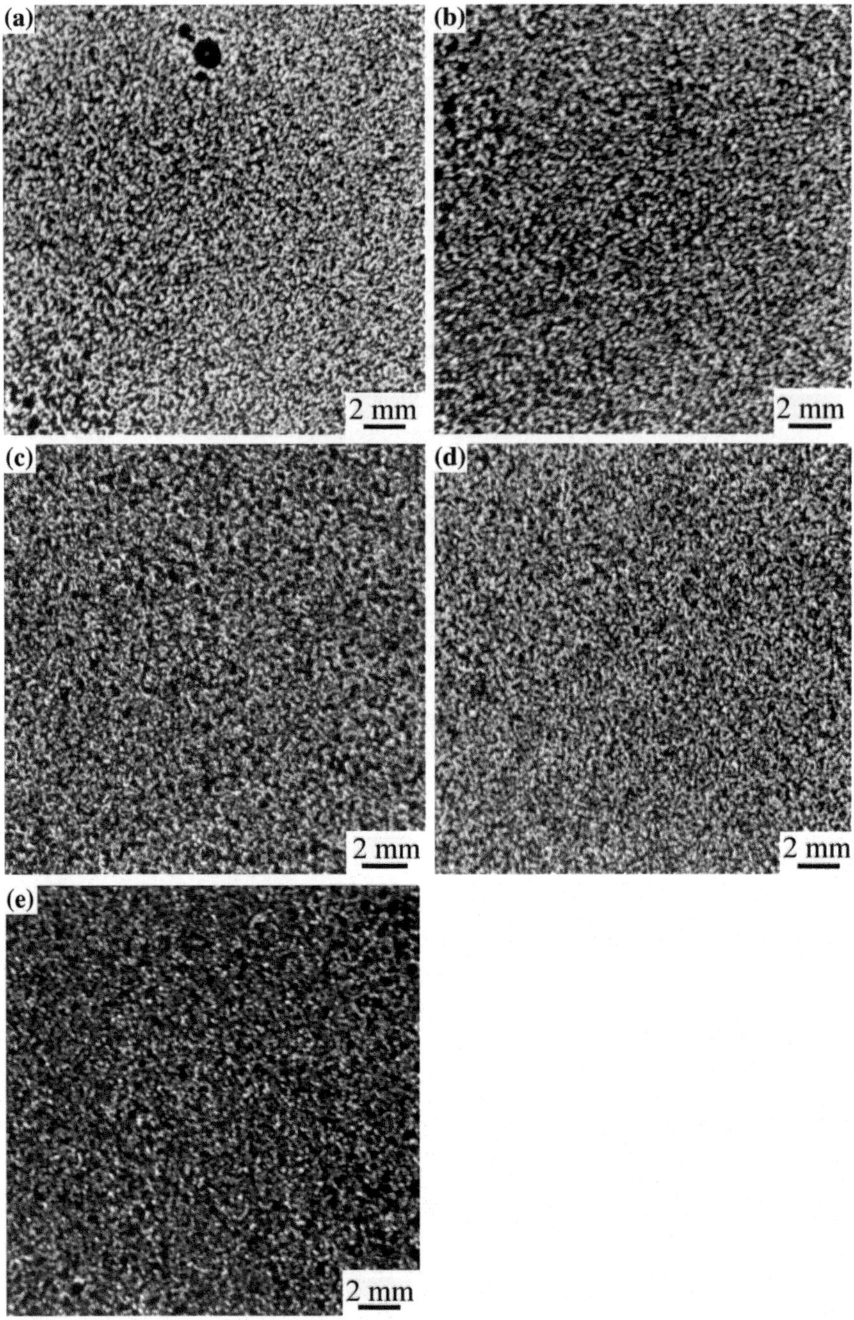

Fig. 3 Shadow photographs of quartz glasses (**a** - vein No. 175, **b** - Argazinskoye deposit, **c** - Larinskoye deposit, **d** - Kundravinskoye deposit, **d** - Tolstikha vein)

Shadow images of smelted quartz glasses. On shadow photographs taken with a large magnification using the optical installation, the smallest defects of the glass are clearly visible. The optical installation has a large depth of field, and a picture plane summarises all the defects occurring in the glass layer. Gas bubbles are displayed in black circles due to the strong reflection at the interface. The molten mineral inclusions ("svili") are clearly visible as well (Fig. 3).

Conclusion

The results obtained allow us to conclude that the proposed enrichment scheme is suitable for use in industry. The presented method of evaluation of quartz raw materials significantly reduces labour costs and accelerates the obtaining of results on the suitability of raw materials at the initial stage of quartz vein development.

References

Götze J., Möckel R. (editors). Quartz: Deposits, Mineralogy and Analytics. Springer Geology. Berlin Heidelberg: Springer; 2012. https://doi.org/10.1007/978-3-642-22161-3_2.

Nasyrov R.Sh. Sposob polucheniya vysokochistogo koncentrata iz prirodnogo kvarca. Patent RU 2431601. MPK C01B33/00. 2011 [Nasyrov R.Sh. Method for obtaining high purity concentrate from natural quartz. Patent RU 2431601. IPC C01B33 / 00. 2011. (In Russ)].

Nasyrov R.Sh., Bykov V.N., Korablev A.G., Shakirov A.R., Igumentseva M.A. Testovye naplavy kvarcevogo stekla kak metod ocenki kachestva kvarcevyh koncentratov. Razvedka i ohrana nedr. 2007; 10: 46–47 [Nasyrov R.Sh., Bykov V.N., Korablev A.G., Shakirov A.R., Igumentseva M.A. Test meltings of the quartz glass as a method of quality assessment of quartz concentrates. Exploration and Protection of Mineral Resources. 2007; 10: 46–47. (In Russ)].

The Composition of Sulfur Isotopes in Minerals from The Sheki-Hiech Cave, The Caucasus, Chechen Republic, Russia

Sergey A. Sadykov, Sergey S. Potapov, Olga Ya. Chervyatsova and Roman S. Dbar

Abstract The Sheki-Hiech cave is unique due to abundant sulfur and gypsum mineralization. The results of mineralogical and geochemical-isotopic studies of the samples from the Sheki-Hiech cave are presented in this paper. The isotopic composition of sulfur varies from −23.97 to −6.47‰ CDT. The cave is an example of an active of sulfuric acid speleogenesis.

Keywords Sulfur isotopes · Sulfate · Gypsum · Sheki-Hiech cave · Sulfuric acid speleogenesis

Introduction

Two expedition trips to study the karst regions of the Chechen Republic were undertaken in October 2016 and in April-May 2017. The first expedition was organized by the speleological Commission of the Russian Geographical Society in the Republic of Crimea (Chairman G. V. Samokhin) and GBU "Argun state historical and architectural Museum-reserve" (Director S.-E. M. Dzhabrailov). Teachers and students of the Crimean Federal University (Simferopol), the staff of

S. A. Sadykov (✉) · S. S. Potapov
Institute of Mineralogy, Ural Branch, RAS, Chelyabinsk Region,
456317 Miass, Russia
e-mail: sergei_sadykov@mail.ru

S. S. Potapov
e-mail: s_almazov@74.ru

O. Ya.Chervyatsova
FSBI State Nature Reserve "Shulgan-Tash", Burzyan District, Irgizly,
Zapovednaya, 14, 453585 Bashkortostan, Russia
e-mail: kittary@yandex.ru

R. S. Dbar
Institute Ecology, ASA, St. Krasnomayatskaya 67, Sukhum, Abkhazia
e-mail: romandbar@mail.ru

S. Votyakov et al. (eds.), *Minerals: Structure, Properties, Methods of Investigation*,
Springer Proceedings in Earth and Environmental Sciences,
https://doi.org/10.1007/978-3-030-00925-0_30

the Argun reserve (Grozny) and the Institute of Mineralogy, Ural branch, Russian Academy of Sciences (Miass) took part in the expedition. The main task of the first expedition was the inventory of known and the search for new caves in the territory of Galanchozh and Chateau districts of the Chechen Republic, the Russian Federation. The second expedition solved the search problems and provided a detailed mineralogical study of the Sheki-Hiech cave in the valley of the Sharo-Argun River between the villages of Ulus-Kert and Zonny in the region of the vanished village Buccazi of the Shatoy district.

Materials and Methods

Site location. The Sharo-Argun River is located at an altitude of 480 m above sea level in the area of research. The watersheds are located at 1000–1100 m, the relative depth of the division is 500–600 m. The river valley has a V-shaped cross-section. The steepness of the slopes is 45–60° in the lower part of the valley, in some areas the slopes become almost vertical with the formation of plains and canyons. The width of the river channel is 30–50 m, the floodplain is not wide, about 5–15 m and is composed of alluvial deposits. The Sharo-Argun river receives three small left and one right tributary at the investigated site. Feeder tributaries having rain and underground origin are due to unloading karst sources of cave systems of the upper stratigraphic level, which are probably the caves of the classic (meteoric) karst. The mineralization of these karst springs is 0.3–0.5 g/l and the temperature is 8 °C.

The Sharo-Argun River receives five large sulfur springs, in addition to the water flow from the ravine-beam hydrological system in the area. Four springs pour down the right river bank and one down the left bank, the flow is 1–2 m^3/s. The left-bank source is located within the floodplain alluvium. Perhaps part of the water is discharged directly into the riverbed. The average temperature of sulfur (hydrogen sulfide) sources is 15–18 °C, the total mineralization is 8–10 g/l. The water temperature of the left-bank source was 16 °C at the time of water and bacterial-algal mats sampling (May 4, 2017). The springs are located in the floodplain of the river at a distance not exceeding 1 km from each other.

The slopes of the valley of the Sharo-Argun River are composed of dense limestone of the Cretaceous age; they form steep, almost vertical, ledges up to 100 m above the river level. The caves are confined to two hypsometric levels—at the base of the cliffs and at an altitude of about 50–60 m. Our research group has studied and drawn the scheme of one of the largest to date cave of the Chechen Republic—Sheki-Hiech (Sulfur cave)—with length of 137 m according to the results of topographic survey. The cave is unique due to abundant mineralization of sulfur and gypsum. In the places of accumulation of sulfur deposits, threadlike colonies of sulfur bacteria (bacterial mats) in the form of gray and brown mucus are observed. Apparently, they actively participate in mineral formation.

In geological terms, the area is represented by the interlacing of heavily dislocated argillites, siltstones, sandstones and limestone. Apparent thickness of limestone reaches 150–200 m. Limestone is thick, the angle of incidence is 60–75°, the azimuth angle drops to 240°. On the site there are surface and underground karst forms. The surface forms are represented by small funnels, maps, niches and grottoes. The cave belongs to the underground form.

The entrance to the cave Sheki-Hiech has the form of a grotto with a height of about 2 m and a width of 4–7 m. In the middle part of the cave a stream with a pH of ≈ 2–4 flows.

Sheki-Hiech cave system is now living and producing minerals almost in front of researchers. Therefore, close attention has been paid to the study of mineralogy, geochemistry and genesis of this cave. The results of studies of mineralogy and geochemistry of isotopes in samples from the Sheki-Hiech cave are presented in this paper.

Analytical methods. Electron microscopic studies were carried out on a microscope JEOL JSM 6390LV with EDS attachment INCA Energy 450 X-max 80 at the Institute of Geology and Geochemistry of UB RAS, Yekaterinburg (analyst L. V. Leonova). The survey was carried out with carbon sputtering at an accelerating voltage of 20 kV in the mode of back-scattered (BSE) and reflected (SEI) electrons.

The isotopic composition of sulfur ($\delta^{34}S$ ‰, CDT) of mineral samples was determined using Thermo Finnigan Delta^{Plus} Advantage mass spectrometer coupled with the EA Flash1112 elemental analyzer via ConFlo III interface at the Institute of Mineralogy of the RAS, Miass. NBS-123 reference material was used for measurements. Measurement error in the series was 0.09‰, CDT. Isotope ratios were measured in a continuous gas flow.

Results and Discussion

Gypsum in the Sheki-Hiech cave is the most widespread mineral. Scanning electron micrographs show the morphological diversity of gypsum crystals. Planar elongated gypsum crystals of columnar and lamellar appearance are relatively rare. Among the elongated gypsum crystals there are isometric gall-type germination twins, as indicated by the characteristic shading on the faces. Planar well-faceted crystals of gypsum indicate relatively calm conditions of growth with a small saturation of the solution. More common split forms of gypsum were formed in conditions of significant supersaturation of the solution. Cleavage begins with the appearance of the block individuals, sometimes with full array structure of subindividuals. A strong cleavage formed radially-radiant (sterolithography, or spherolitic) units.

Different morphology of gypsum crystals (planar, double, block, split) in the unit indicates the ever-changing environment of crystal formation and growth. The crystal-forming environment, always supersaturated, shows no signs of dissolution and/or regeneration on the gypsum crystals.

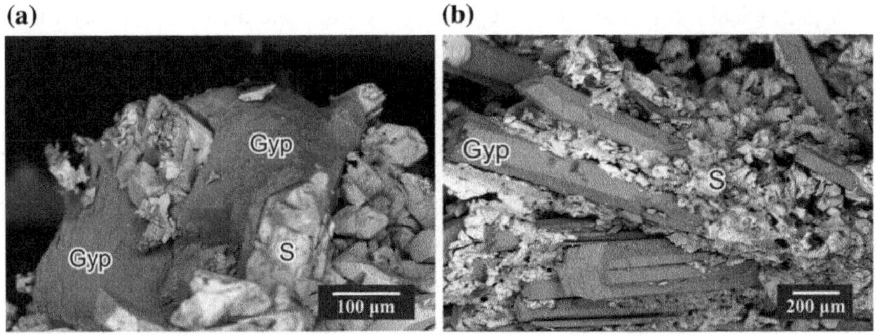

Fig. 1 The relative position of sulfur and gypsum in the unit. SEM image

The relationship between the crystals of gypsum and sulfur in aggregates is illustrated in Fig. 1. It is observed as the growth of the later crystals of sulfur on the crystals of gypsum (Fig. 1a), and almost simultaneous and joint growth of plaster and sulfur with the formation of induction (compromise) surfaces. Otherwise, sulfur crystallizes with a little delay with filling of space between gypsum crystals (see Fig. 1b).

This indicates the cycling (probably seasonal) crystallization of gypsum and sulfur. Some crystals of gypsum contain organic inclusions of preserved filamentous colonies of sulfur oxidizing bacteria (Fig. 2).

Sulfur is also widely distributed in the Sheki-Hiech cave, but the spatial distribution is clearly localized. Sulfur is not found in the far parts of the cave. Sulfur forms crystals in the area of underground hydrogen sulfide water flow into the cave cavity through the channel of subvertical ascending feed ("feeder") (Klimchouk 2013) and is deposited on the surface of walls, floor and foreign objects (for example, on the trunks of trees of temporary dwelling in the entrance part of the cave) up to the entrance to the cave, which is apparently limited to the air flow. Thus, the cave profile shows a certain mineralogical zoning in the distribution of the main mineral species—gypsum and sulfur.

Fig. 2 Filamentous colonies of ceramictile, canned in a crystal of gypsum. SEM image

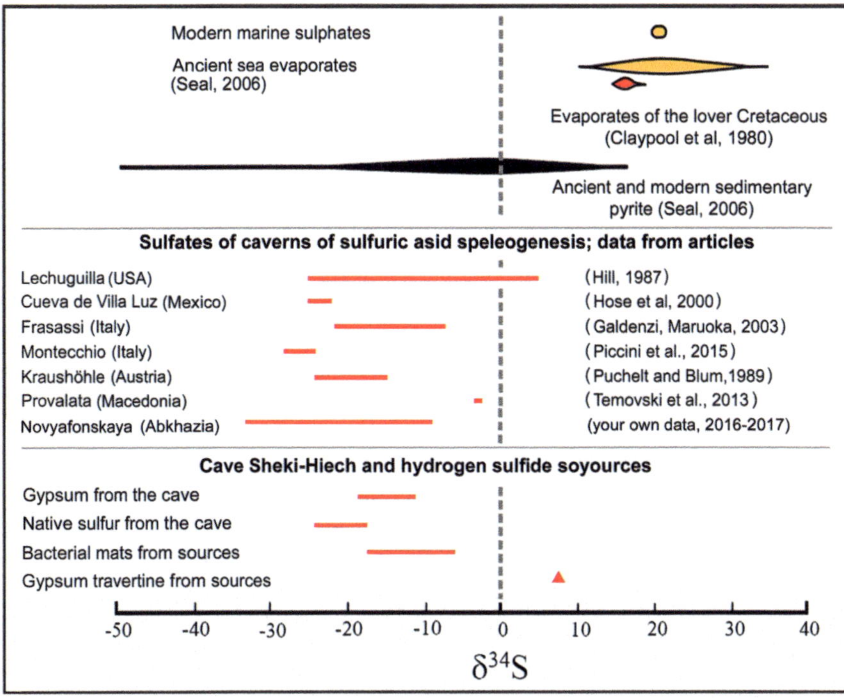

Fig. 3 Isotopic composition ($\delta^{34}S$ ‰, CDT) of native sulfur and gypsum of the Sheki-Hiech cave in comparison with that of sulphates of sulfuric acid speleogenesis caves, as well as sulfates and sulfides of other geological objects according to the literature data (Seal 2006; Claypool et al. 1980; Galdenzi and Maruoka 2003; Hill 1987; Piccini et al. 2015; Puchelt and Blum 1989; Temovski et al. 2013; Chervyiatsova et al. 2016)

The isotopic data obtained by us in the minerals of the Sheki-Hiech cave in comparison with sulfates of sulfuric acid speleogenesis caves in different regions of the world and sulfur-containing mineral formations of other geological objects according to the literature data are shown in Fig. 3.

As can be seen, crystalline native sulfur has the lightest isotopic composition (−17.27 ÷ −23.97‰, CDT). The average of five samples is −21.42‰, CDT. The isotope composition of gypsum has a wider value range from −10.09 to −18.22‰, CDT. However, in the sample where it was difficult to separate native sulfur and gypsum, isotope ratios have values close to the average value of native sulfur (−21.57‰, CDT). Isotope composition of bacterial mat samples is as follows: −18.32; −17.38 and −6.47‰, CDT. The heaviest isotope composition has a sample of gypsum travertine from the source at the exit of the Magomed-Hiech cave +8.47‰ CDT.

The light isotope composition in the sulfur studied is the result of the recovery of sulfate ions by anaerobic bacteria (Faure 1986).

Large variation of sulfur isotopic ratios in gypsum samples from different parts of the Sheki-Hiech cave can be explained by the kinetic regime of the formation of sulfates. Sulfuric acid speleogenesis takes place in the cave "hic et nunc", this process is dynamic and active, it will continue making adjustments to the morphology of the cave and the environment. Accordingly, isotopic ratios will change in newly formed sulfur-containing minerals.

It can be seen in Fig. 3 that the isotopic composition of minerals of the Sheki-Hiech cave (sulfur and gypsum), bacterial mats and associated sulfate mineral formations falls into the isotope composition range of classical sulfate caves that are well-studied and referred to the process of sulfuric acid speleogenesis. Only one sample of travertine plaster taken from a sulfuric acid source on the daytime surface is enriched with a heavy sulphur isotope. Probably, travertine plaster was formed without significant fractionation mainly due to evaporation of sulphate contained in water.

The study of the isotopic composition of sulfur confirms the hypothesis of the sulfuric acid process in the formation of the Sheki-Hiech cave. The process of the cave formation is not finished, as shown by field observations. Further studies of the Sheki-Hiech cave will help trace the evolution changes of the cave morphology as well as mineral and isotope composition.

Conclusions

According to the observation results, surveys and data provided in the article, the Sheki-Hiech cave is very unusual. The results of our studies clearly demonstrate a series of signs indicative for hypogene speleogenesis, moreover, it can be said that the Sheki-Hiech cave is an example of contemporary sulfuric acid speleogenesis. The source of H_2S is typically a process of anaerobic bacterial sulfate reduction, occurring in the lower hydrodynamic zone in the presence of sulfate and hydrocarbons (Forti et al. 2002). The peculiarity of this process is the subaerial implementation (less often in the oxygen shallow zone of hydrogen sulfide reservoirs) of the corrosion process.

Acknowledgements The authors are grateful to the Chairman of the Speleological Commission of the Russian Geographical Society in the Republic of Crimea G. V. Samokhin, Director of the state institution "Argun State Historical and Architectural Museum" S.-E. M. Dzhabrailov, private entrepreneur M. A. Daldaev for organizing the expedition, as well as employees of the Institute of Metals Superplasticity Problems of the Russian Academy of Sciences S. N. Sergeyev and I. I. Musabirov for providing SEM-photos.

The work is executed at financial support of RFBR under the project No. 17-55-40005 Abh-a on the theme: "The Manifestation of the sulfuric acid speleogenesis in Abkhazia: the morphological, mineralogical and isotope-geochemical aspects".

References

Chervyiatsova OYa., Potapov SS., Sadykov SA., Leonova LV., Dbar RS. Distribution, morphology, sulfur isotopic composition and genesis of gypsum deposits in Novoafonskaya cave (Abkhazia). Mineralogy. 2016, 3, 79–94.

Claypool GE, Holser WT, Kaplan IR, Sakai H and Zak I. The age curves of sulfur and oxygen isotopes in marine sulfate and their mutual interpretation. Chem. Geol. 1980; 28, 199–260.

Faure G. Principles of Isotope Geology. 2nd ed. New-York; John Wiley & Sons, 1986.

Forti P, Galdenzi S, Sarbu SM. The hypogenic caves: a powerful tool for the study of seeps and their environmental effects. Continental shelf research. 2002; 16 (22), 2373–2386.

Galdenzi S, Maruoka T. Gypsum deposits in the Frasassi Caves, central Italy. Journal of Cave and Karst Studies. 2003; 2 (65), 111–125.

Hill CA. Geology of Carlsbad cavern and other caves in the Guadalupe Mountains, New Mexico and Texas. New Mex. Bur. Min. Mineral Resour. Mem. 1987; 117, 1–150.

Klimchouk AB. Hypogene Speleogenesis, its hydrogeological significance and role in karst evolution. Simferopol: DIP; 2013. 180 p.

Piccini L, De Waele J, Galli E, Polyak VJ, Bernasconi SM, Asmerom Y. Sulphuric acid speleogenesis and landscape evolution: Montecchio cave, Albegna river valley (Southern Tuscany, Italy). Geomorphology. 2015; 229, 134–143.

Puchelt H, Blum N. Geochemische Aspekte der Bildung des Gipsvorkommens der Kraushuhle/ Steiermark. Oberrhein. Geol. Abh. 1989; 35, 87–99.

Seal RR. Sulfur isotope geochemistry of sulfide minerals. Reviews in mineralogy and geochemistry. Chantilly: G MSA. 2006. 61. P. 633–677.

Temovski M, Audra P, Mihevc A, Spangenberg J, Polyak V, McIntosh W, Bigot JY. Hypogenic origin of Cave Provalata, Republic of Macedonia: rare case of successive thermal carbonic and sulfuric acid speleogenesis. Int. J. Speleol. 2013; 42, 235–246.

Analysis of Solid Solutions with Spinel Structure

Elena N. Salata, Elena M. Maksimova and Igor A. Nauhatsky

Abstract The solid solutions with spinel structure along the junction Fe_3O_4 - $NiMn_2O_4$ and $MnFe_2O_4$ - Ni_2MnO_4 were studied by the powder X-ray diffraction (XRD). Measurements of lattice parameters and compositions for single-phase solution lattice demonstrate the validity of Vegard's law for some areas of plots.

Keywords Solid solutions · Spinel · X-ray diffraction

Introduction

Recent developments in crystal chemistry and the synthesis of new classes of ionic crystals have greatly emphasized the interest in solid solution properties and the distribution of ions in different coordination sites. One of the promising classes of inorganic materials is complex oxides with spinel structure. These compounds are widely used to create new functional materials (Harikishore et al. 2016; Golestani-Fard et al. 1987; Wicham 1964).

Intensive development of technology requires constant improvement of materials. The properties of the materials are determined, first of all, by their chemical composition. However, it is not always possible to select the necessary composition for obtaining a material with a set of specific properties, even when such a composition exists objectively, because of the complexity of the problem. In fact, with an empirical solution of the problem by scanning all sorts of combinations of chemical elements, a large number of experiments are required. Therefore, the development of methods for the scientific search for compounds with predetermined properties is an acute problem of materials science.

E. N. Salata (✉) · E. M. Maksimova · I. A. Nauhatsky
V.I. Vernadsky Crimean Federal University, Ave. Vernadsky 4,
Simferopol, Crimea, Russian Federation
e-mail: lenamax112@rambler.ru

S. Votyakov et al. (eds.), *Minerals: Structure, Properties, Methods of Investigation*,
Springer Proceedings in Earth and Environmental Sciences,
https://doi.org/10.1007/978-3-030-00925-0_31

Materials and Methods

Polycrystalline samples of spinels of complex compositions synthesized in the Fe-Ni-Mn-O system were studied. The sintering mixture was prepared according to the ceramic technology by mixing and grinding Fe_2O_3, NiO, MnO_2 oxides of the qualification "pure for analysis" and subsequent alternating grinding and pre-firing twice in air at 1070 K for 8 h. Tablets were pressed from the mixture. Samples intended for firing in oxygen in order to avoid undesirable transition $Mn^{4+} \rightarrow Mn^{3+}$ were formed from a starting mixture of oxides without pre-firing.

As the concentration of iron increased, the firing temperature was increased. The synthesis of the solutions was carried out in air. The duration of exposure was 20 h, which ensured the attainment of an equilibrium state. Samples synthesized in air were quenched by rapidly moving them into cold water (Zinovik et al. 2009). The synthesis conditions and the composition of the samples are shown in Table 1.

Results and Discussion

Powder X-ray diffraction patterns were obtained using a DRON-3 diffractometer with CuKa radiation (1,54178 Å), using NaCl as an internal standard. The lattice parameter refinement was performed by the UnitCell software (Holland and

Table 1 Synthesis conditions and the composition of the samples

№ sample	Solid solutions	Composition, x	Atmosphere	Firing and hardening temperature, K
1	$(NiMn_2O_4)_x(Fe_3O_4)_{1-x}$	0.15	Air	1670
2		0.25	Air	1620
3		0.33	Air	1520
4		0.50	Air	1470
5		0.60	Air	1370
6		0.80	Air	1200
7		1.0	Air	1085
8	$(Ni_2MnO_4)_x(MnFe_2O_4)_{1-x}$	0	Air	1520
9		0.25	Air	1470
10		0.33	Air	1420
11		0.50	Air	1270
12		0.60	Oxygen	1020, slow cooling
13		0.65	Air	
14		0.85	Air	
15		1.0	Air	

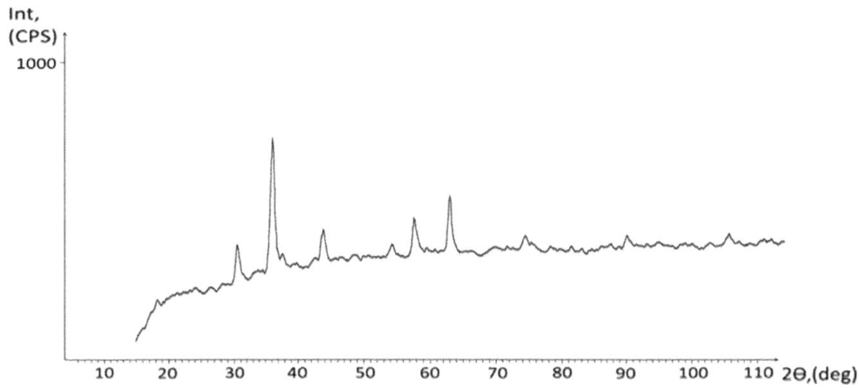

Fig. 1 The typical diffractogram of samples of solid solutions with spinel structure in the Fe-Ni-Mn-O system

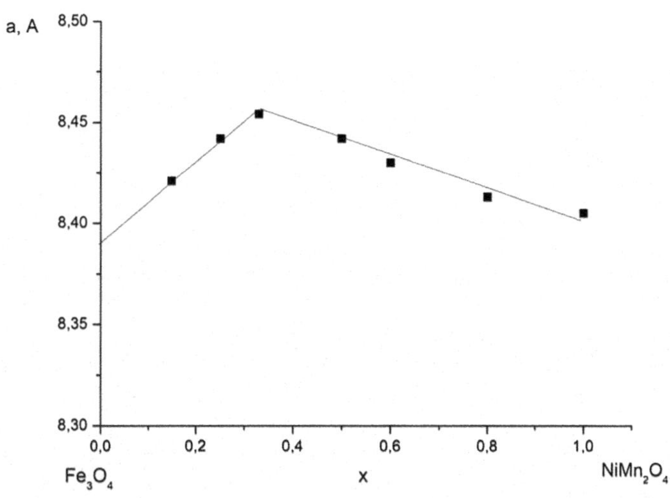

Fig. 2 The dependence of the spinel structure parameter "*a*" on the composition of solid solutions in the system $(NiMn_2O_4)_x\ (Fe_3O_4)_{1-x}$

Redfern 1997) and by graphical extrapolation (Frank-Kamenetskii et al. 1975) on the data collected in a scan range 15° < 2θ < 110° (Fig. 1).

The resulting parameters of crystal lattices "*a*" of samples of solid solutions with spinel structure in the Fe-Ni-Mn-O system are given in Figs. 2 and 3.

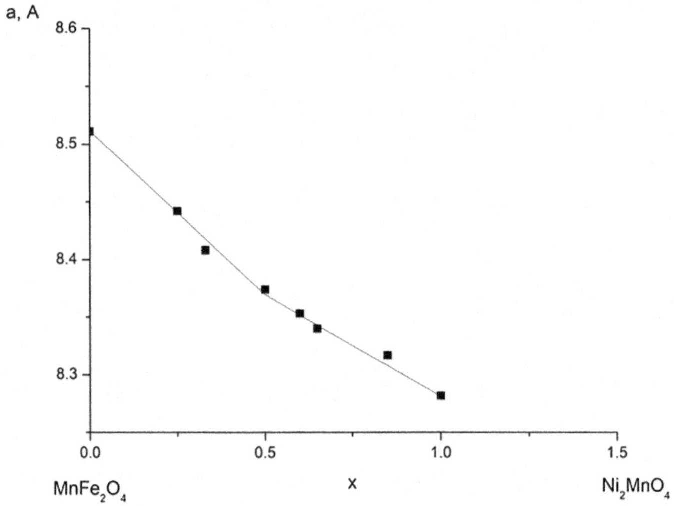

Fig. 3 The dependence of the spinel structure parameter "a" on the composition of solid solutions in the system $(Ni_2MnO_4)_x(MnFe_2O_4)_{1-x}$

Conclusion

The X-ray phase analysis has shown that all samples have a single-phase composition, except for samples № 14 and № 15 of the system $(Ni_2MnO_4)_x(MnFe_2O_4)_{1-x}$ at x = 0.85 and x = 1. In these samples, in addition to the peaks characteristic for the spinel structure, there are additional peaks of non-spinel phase. That may be due to insufficient amount of pressure during firing of the samples.

The dependence of the parameter of the elementary crystal cell "a" on the composition of single-phase solutions of both systems is generally complex, but in some areas can be described by Vegard's law. The change in the character of the "a" dependence on (x) in the system $(NiMn_2O_4)_x(Fe_3O_4)_{1-x}$ at x = 0.33 and in the system $(Ni_2MnO_4)_x(MnFe_2O_4)_{1-x}$ at x = 0.5 is more likely total, including the change of chemical compounds and ionic composition in the solid solution.

References

Frank-Kamenetskii VA, Gerasimov VN, Dolivo-Dobrovol'skaya EM, Kamentsev IE, Kordat'eva VV, Kosoi AL, Lesyuk GI, Rozhdestvenskaya IV, Stroganov EV, Filatov SK, Frank-Kamenetskaya OV. A Manual on X-Ray Diffraction Investigations of Mineralsed. Leningrad: Nedra, 1975

Golestani-Fard F, Azimi S, Mackenzie KJD. Oxygen evolution during the formation and sintering of nickel-manganese oxide spinels for thermistor applications. Journal of materials science. 1987. 22: 2847–2851

Harikishore DKR, Hiroo K, Yeoung-Sang Y. Spinel ferrite magnetic adsorbents: Alternative future materials for water purification? Coordinate Chemistry Reviews. 2016. 315: 90–111

Holland TJ, Redfern SA. Unite cell refinement from powder diffraction data: the use of regression diagnostics. Mineralogical Magazine. 1997

Wicham DG. Solid-phase equlibria in the System NiO-Mn2O3-O2. J. Inorg. Nuckl. m. 1964. 26 (8): 1369–1377

Zinovik EV, Prisedsky VV, Maksimova EM, Nauhatsky IA. Sintez i prognozirovanie svoystv so strukturoy shpineli v sisteme Ni-Mn-Fe-O. Nauchnye trudy Donetskogo natsional'nogo tehnicheskogo universiteta. 2009. 13(152): 75–84. [Zinovik EV, Prisedsky VV, Maksimova EM, Nauhatsky IA. Synthesis and prediction of properties with spinel structure in the system Ni-Mn-Fe-O. Scientific notes of Donetsk National Technical University. 2009. 13(152): 75-84. (In Russ.)]

Hf Isotopic Composition of Zircons from the Granodiorites of the Talovsky Intrusion as the Evidence for the Juvenile Paleoproterozoic Crust of the Vorontsovsky Terrane, Eastern Sarmatia

Konstantin A. Savko, Maria V. Zaitceva, Sergei Votyakov and Sergey V. Tsybulyaev

Abstract The Vorontsovsky terrane occupies the Eastern part of the Volga-Don orogen, which is located between the Archean Kursk block (Sarmatia megablock) and the Volgo-Uralia megablock (Eastern European platform). Along with geochemical and geochronological studies, the studies of the Hf isotopic composition of accessory zircons present great interest and can give important information about the melt sources of the numerous granitoid intrusions and the age of the continental crust of the Vorontsovsky terrane. Zircons from the granodiorites of the Talovsky massif, located in the Eastern part of the Vorontsovsky terrane, which intrudes into the Paleoproterozoic metaterrigenic sequences of the Vorontsovsky series, were studied using the Neptune Plus multi-collector mass spectrometer with NWR 213 laser ablation system. Obtained positive values of $\varepsilon_{Hf}(2050)$ ranging from +2.1 to +5.4 point out to a very brief crustal prehistory of the source of their parental melts. This fact excludes the presence of the Archaean crust in the Vorontsovsky terrane. Thus, Hf isotopic composition of zircons from the granodiorites of the Talovsky intrusion is another independent evidence for the juvenile Paleoproterozoic crust of the Vorontsovsky terrane separating the megablocks of Sarmatia and Volgo-Uralia.

Keywords Hf isotopic composition · Zircons · Granodiorites of the Talovsky intrusion · Vorontsovsky terrane · Laser ablation · MC-ICP-MS · Juvenile paleoproterozoic crust

---

K. A. Savko (✉) · S. V. Tsybulyaev
Voronezh State University, 1 Universitetskaya pl., Voronezh, Russia
e-mail: ksavko@geol.vsu.ru

M. V. Zaitceva · S. Votyakov
A.N. Zavaritsky Institute of Geology and Geochemistry of the Ural Branch of the Russian Academy of Sciences, 15 Vonsovskogo Str., Ekaterinburg, Russia
e-mail: zaitseva.mv1991@gmail.com

S. Votyakov et al. (eds.), *Minerals: Structure, Properties, Methods of Investigation*, Springer Proceedings in Earth and Environmental Sciences,
https://doi.org/10.1007/978-3-030-00925-0_32

Introduction

The Vorontsovsky terrane occupies the Eastern part of the Volga-Don orogen, which lies between the Archean Kursk block (Sarmatia megablock) and the Volgo-Uralia megablock (Eastern European platform). A rapid accumulation of terrigenous strata of the Vorontsovsky series occurred on the active continental margin of the Vorontsovsky terrane, which had ended 2.1 Ga ago. During this period, a collision event had taken place as a result of the Sarmatia and Volgo-Uralia collision which triggered the regional metamorphism with an age of about 2.07 Ga (Savko et al. 2015). During the period of about 2.05–2.07 Ga, the breakup of collisional orogen was going on in the Eastern Sarmatia, which had been formed after the collision of Sarmatia and the Volgo-Uralia. Its characteristic feature is synchronous (about 2.06 billion years) intrusion of post-collisional granitoids, magnesian diorites, basites and ultrabasites in the Vorontsovsky terrane (Savko et al. 2014; Terentiev and Savko 2016). Along with geochemical and geochronological studies, the study of Hf isotopic composition of accessory zircons is of great interest; it can give important information about the melt sources of the numerous granitoid intrusions and the age of the continental crust of the Vorontsovsky terrane.

The aim of the study is to determine Hf isotopic composition of zircons from granodiorites of the Talovsky intrusion of the Vorontsovsky terrane.

Materials and Methods

Samples. Zircons from granodiorites of the Talovsky massif, located in the Eastern part of the Vorontsovsky terrane, which breaks the Paleoproterozoic metaterrigenic deposits of the Vorontsovsky series, were studied.

The isometric-shaped massif has an area of about 45 km^2; it is composed of gabbrodiorites, diorites, quartz diorites, granodiorites and granites. According to the isotope dating of zircons (SHRIMP II), the age of the massif granodiorites (sample 8034/287) is 2050 $\pm$ 7 Ma and that of biotite granites (sample 8051/225.8) is 2063 $\pm$ 6 Ma (Savko et al. 2014; Savko and Skryabin 2012). The rocks are enriched in Fe, Ti, P, high field strength elements (Zr, Hf, Y) and REE with strong fractionation and large negative Eu anomalies; they refer to the post-collisional granitoids of the A2 type according to the geochemical characteristics (Eby1992). These granitoids are characterized by a positive $\varepsilon_{Nd}(T)$ ($\varepsilon_{Nd}(2050) = +2.3$) value according to Sm-Nd isotope-geochemical data (Savko et al. 2014).

The accessory zircons are represented by idiomorphic, often spear-shaped crystals (length 0.295–0.441, width 0.074–0.185 mm, prismatic habit with sharp pyramidal vertices, coefficient of elongation 1.71–5.84), varying from colorless to brownish. The grains are often fractured and contain the inclusions of other mineral phases as well as fluid inclusions. In CL-images a weak concentric zoning is

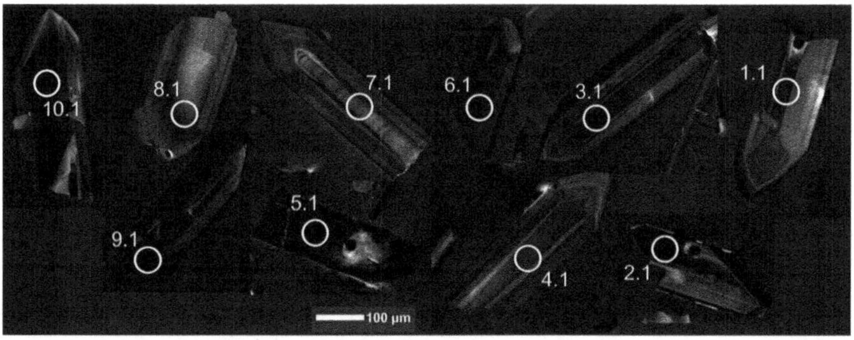

Fig. 1 CL-images of zircons from granodiorites of the Talovsky massif with the points of Hf isotopic analysis

observed, which in the long-prismatic crystals creates the illusion of their laminar structure (Fig. 1). In individual grains, the sectoral structure is clearly manifested, which most often is fixed at the tops of the crystals. The presence of relict nuclei has not been found.

Analytical techniques. The study of Lu-Hf isotope composition of zircons was performed using the Neptune Plus multi-collector mass spectrometer with NWR 213 laser ablation system. The equipment used is located in class ISO7 cleanroom.

The optimization of the mass spectrometer and laser ablation system operational parameters as well as the selection of reference materials was carried out. The procedure for correcting the effects of fractionation and the ion mass-discrimination as well as the isobaric interference of ^{176}Yb and ^{176}Lu on ^{176}Hf was developed. The experimental data processing was optimized using Mud Tank and GJ-1 (Black and Gulson 1978; Jackson et al. 2004) zircon reference samples.

The uncertainty of the single ^{176}Hf/^{177}Hf ratio measurement expressed in the form of 2SD was in the range of 0.005–0.008%. The uncertainty of the single determination of the ε(Hf) value expressed in the form of 2SD varied in the range of 5–9% for the above reference zircons. The obtained Hf/Hf isotopic ratios for the Mud Tank and GJ-1 reference materials were consistent with the GeoReM database data.

The operating parameters of the laser ablation system were as follows: the energy density of the laser radiation was 14 J/cm^2, the pulse repetition frequency was 20 Hz, and the spot diameter was 25 μm. The Hf-INATOR Excel macros (Giovanardi and Lugli 2017) was used for the processing of Lu-Hf data.

The measurements of the U-Pb zircon isotope age were performed using a SHRIMP II ion microprobe at the Center for Isotope Investigations of the VSEGEI (St. Petersburg) (Savko and Skryabin 2012). The studies of the Lu-Hf isotope composition were carried out at the same points (Fig. 1), where the U-Pb isotope age was determined by SHRIMP II.

Table 1 Hf isotopic composition of zircon grains from granodiorites of the Talovsky massif

Point number	$^{176}Yb/^{177}Hf$	± 2σ	$^{176}Lu/^{177}Hf$	± 2σ	$^{176}Hf/^{177}Hf$	± 2σ	$^{178}Hf/^{177}Hf$	± 2σ	$\varepsilon_{Hf}(t)$	± 2σ	T_{DM}	T^{c}_{DM}
1_1	0.027724	0.000783	0.000867	0.000017	0.281619	0.000017	1.467290	0.000042	3.8	0.6	2278	2423
2_1	0.038137	0.003586	0.001097	0.000032	0.281678	0.000032	1.467208	0.000085	5.4	1.1	2218	2324
3_1	0.037397	0.001423	0.001153	0.000022	0.281623	0.000022	1.467218	0.000074	3.6	0.8	2289	2439
4_1	0.043862	0.001524	0.001315	0.000028	0.281637	0.000028	1.467271	0.000067	3.9	1.0	2279	2424
5_1	0.072374	0.004779	0.002147	0.000030	0.281621	0.000030	1.467156	0.000066	2.1	1.1	2355	2547
6_1	0.019724	0.000148	0.000621	0.000025	0.281593	0.000025	1.467317	0.000073	3.2	0.9	2301	2460
7_1	0.028868	0.000459	0.000885	0.000034	0.281655	0.000034	1.467239	0.000085	5.1	1.2	2229	2343
8_1	0.033951	0.000884	0.000986	0.000034	0.281605	0.000034	1.467237	0.000093	3.4	1.2	2296	2452
9_1	0.022676	0.000293	0.000744	0.000022	0.281598	0.000022	1.467083	0.000077	3.3	0.8	2300	2459
10_1	0.038035	0.000803	0.001111	0.000035	0.281632	0.000035	1.467325	0.000105	4.0	1.2	2274	2416

Note $^{176}Hf/^{177}Hf_i$ primary ratio is calculated using the decay constant ^{176}Lu $\lambda = 1.867 \times 10^{-11}$; $\varepsilon_{Hf}(t)$ is calculated for the age of 2050 million years; T_{DM} is the model age of the source, calculated with the account of magma melting from the depleted mantle using $^{176}Hf/^{177}Hf = 0.28325$ and $^{176}Lu/^{177}Hf = 0.0384$; T^{C}_{DM} is the model age of the source, calculated from the two-stage model using $^{176}Lu/^{177}Hf = 0.015$, based on the magma melting from the middle continental crust formed from the depleted mantle

Results and Discussion

Hf isotopic composition of zircons from granodiorites of the Talovsky intrusion (sample 8034/287) $\varepsilon_{Hf}(2050)$ range from +2.1 to +5.4 (Table 1). The $T_{Hf}(DM)$ model age, calculated by a one-stage model for the magma melting from the depleted mantle, is 2218-2355 Ma, whereas the calculated $T_{Hf}(DM)^c$ varies from 2324 to 2460 Ma (Table 1).

The positive values of $\varepsilon_{Hf}(2050)$ in zircons from granodiorites of the Talovsky massif ranging from +2.1 to +5.4 point out to a very short crust prehistory of the source of their parental melts with $T_{Hf}(DM) = 2218–2355$ Ma and $T_{Hf}(DM^s) =$ 2324–2460 Ma.

The results calculated from the single-stage model, coincide with the Sm-Nd isotopic data for the same granitoids, where εNd (2050) = +2.4 and $T_{Nd}(DM) = 2284$ Ma (Savko et al. 2014). These characteristics suggest a brief crustal prehistory of the granitoids and their derivation from juvenile Paleoproterozoic sources. This indicates that the generation region of the granitoid magmas could not contain any significant amount of Archean granite-gneiss continental crustal complexes. This fact excludes the presence of the Archaean crust in the Vorontsovsky terrane. Thus, Hf isotopic composition of zircons from granodiorites of the Talovsky intrusion is another independent evidence of the juvenile Paleoproterozoic crust of the Vorontsovsky terrane separating the megablocks of Sarmatia and Volgo-Uralia.

Conclusions

Hf isotopic ratios in zircons from granodiorites of the Talovsky massif, located in the Eastern part of the Vorontsovsky terrane, were studied using the Neptune Plus multi-collector mass spectrometer with NWR 213 laser ablation system. The obtained positive values of $\varepsilon_{Hf}(2050)$ ranging from +2.1 to +5.4 point out to a very brief crustal prehistory of the studied granitoids. This fact excludes any significant contribution of the continental crust to the genesis area of granitoid magmas of the Archean granitic-gneiss complexes.

Thus, Hf isotopic composition of zircons from granodiorites of the Talovsky intrusion is another independent evidence for the juvenile Paleoproterozoic crust of the Vorontsovsky terrane separating the megablocks of Sarmatia and Volgo-Uralia.

Acknowledgements The work was carried out at the UB RAS "Geoanalytik" Center for Collective Use with the financial support of RFBR grant No. 17-05-00618.

References

Black LP, Gulson BL. The age of the Mud Tank carbonatite, Strangways Range, Northern Territory. J. Aust. Geol. Geophys. 1978;3:227–232.

Eby GN. Chemical subdivision of the A-type granitoids: petrogenetic and tectonic implications. Geology. 1992;20:641–644.

Giovanardi T, Lugli F. The Hf-INATOR: A free data reduction spreadsheet for Lu/Hf isotope analysis. Earth Science Informatics. 2017:1–7.

Jackson SE, Norman JP, William LG, Belousova EA. The application of laser ablation-inductively coupled plasma-mass spectrometry to in situ U-Pb zircon geochronology. Chemical Geology. 2004;211:47–69.

Savko KA, Samsonov AV, Larionov AN, Larionova YuO, Bazikov NS. Paleoproterozoic A- and S-Granites in the Eastern Voronezh Crystalline Massif: Geochronology, Petrogenesis, and Tectonic Setting of Origin. Petrology. 2014; 22(3): 205–233.

Savko KA, Samsonov AV, Sal'nikova EB, Kotov AB, Bazikov NS HT/LP Metamorphic Zoning in the Eastern Voronezh Crystalline Massif: Age and Parameters of Metamorphism and Its Geodynamic Environment. Petrology. 2015; 23(6):559–575.

Savko KA, Skryabin VYu. Geokhronologiya i veshchestvennyi sostav gabbrodiorit-tonalitovykh i granodiorit-granitnykh porod Talovskoi intruzii (Voronezhskii kristallicheskii massiv). Vestn. Voronezhskogo un-ta, ser. geol. 2012;2: 95–104. [Savko KA, Skryabin VYu. Geochronology and material composition of gabbrodiorite-tonalite and granodiorite-granite rocks of Talovsky intrusion (Voronezh crystalline massif). Proceedings of Voronezh State University. Series: Geology. 2012; 2:95-104. (in Russ.)].

Terentiev RA, Savko KA. Paleoproterozoic high-Mg low-Ti gabbro-granite series in Eastern Sarmatia: geochemistry and formation conditions. Russian Geology and Geophysics. 2016; 57: 913–938.

Anthropogenic Particles in Contemporary Surface Dirt Sediments in an Urban Environment

Andrian A. Seleznev, Ilia V. Yarmoshenko, Georgy P. Malinovsky, Daria Kiseleva, Lyubov V. Leonova, Elena M. Baglaeva and Ekaterina O. Ilgasheva

Abstract The objective of the study was to characterize morphology and chemical composition of anthropogenic particles in contemporary sediments in an urban environment. The object of the study was the contemporary surface dirt and dust sediments. The study was conducted through the example of Ekaterinburg (Russia). Six typical neighborhoods were chosen for investigations within the residential area of the city. The samples of environmental compartments were collected on these territories. Samples of dirty snow sludge in the winter season, and samples of soil, ground, surface dirt and dust sediment in the summer season were collected in courtyards and adjacent areas of each neighborhood. The anthropogenic particles were selected by visual inspection from the 0.25–1 mm particle-size fraction of the collected environmental compartments. Morphological analysis was conducted by optical microscopy. The chemical composition and structure of particles were determined by scanning electron microscopy. According to their morphological features, structure and elemental content, the analyzed particles were divided into seven different groups: brick, glass, spherical particles, foil, tire fragments, slag and building plaster. The catalog of the particles was compiled.

Keywords Anthropogenic particles · Contemporary anthropogenic sediments · Puddle sediments · Contemporary surface dirt and dust sediments · Dirty snow sludge · Urban environment · Slag · Scanning electron microscopy

A. A. Seleznev (✉) · I. V. Yarmoshenko · G. P. Malinovsky · E. M. Baglaeva
E. O. Ilgasheva
Institute of Industrial Ecology, Ural Branch of the Russian Academy of Sciences, S. Kovalevskoy St., 20, 620219 Ekaterinburg, Russian Federation
e-mail: seleznev@ecko.uran.ru

D. Kiseleva · L. V. Leonova
The Zavaritsky Institute of Geology and Geochemistry of the Ural Branch of the Russian Academy of Sciences, 15 Akademika Vonsovskogo str., 620016 Ekaterinburg, Russian Federation

S. Votyakov et al. (eds.), *Minerals: Structure, Properties, Methods of Investigation*, Springer Proceedings in Earth and Environmental Sciences,
https://doi.org/10.1007/978-3-030-00925-0_33

Introduction

Particles of the anthropogenic origin are a significant part of environmental compartments in an urban territory. The sources of anthropogenic particles in the urban environment are related to the emissions of metallurgical enterprises (Zhu et al. 2013), coal-fired power plants (Iordanidis et al. 2008; Lu et al. 2009), cement plants (Gołuchowska 2001), automobile transport emissions and abrasion of traffic related materials (brake pads and tires) (Crosby et al. 2014; Kim et al. 2009), abrasion of road surfaces and building constructions (Bucko et al. 2011), soil and ground erosion (Kim et al. 2007).

Anthropogenic particles migrate, precipitate from the atmosphere; they are transferred with the wind and atmospheric rainwater runoff and deposited on various surfaces, in the soil (Wang 2013; Meena et al. 2011), snow (Bucko et al. 2011) and on leaves (Gautam et al. 2005), and are finally transported to the sediments of water bodies (Dong et al. 2014). Anthropogenic particles may be found in the backfill material and artificial grounds across urban territories (Ma et al. 2014). Particles of anthropogenic origin are noted to determine magnetic properties of environmental compartments (Yang et al. 2010), especially road (Jordanova et al. 2014) and industrial dust (Meena et al. 2011; Robertson et al. 2003).

One of the depots of particulate matter and pollutants in the urban environment is the contemporary sediments in local depressed zones of microrelief, which integrate the pollution over the territory and time of forming the landscape (Seleznev, Yarmoshenko 2014). The objectives of the study were: (1) to define the morphology and chemical composition of anthropogenic particles in contemporary sediments in an urban area, (2) to form a catalog of particles and to classify them. The study was conducted through the example of the city of Ekaterinburg.

Materials and Methods

The surface facie of contemporary anthropogenic sediments of a residential urban area (in other words, surface dirt and dust sediments or puddle sediments) was chosen as the main object of the study. The samples of the environmental compartments involved in the sediment formation were collected at the experimental sampling sites in residential areas of the city. The experimental site represented a typical block of houses (multistory buildings) constructed in the second part of the 20th century. Experimental sites were chosen in different parts of the city, in different lithogenic substrate areas. Each site consisted of a courtyard and a facade territory adjacent to a network of street roads.

The sampling of the environmental compartments was conducted according to the following scheme. Five samples were collected at each selected site: two samples in the facade part and three samples in the courtyard. In the winter period, samples of undisturbed snow cover were collected in the lawn areas. Samples of dirty snow sludge were collected on pavements and roads, in roadside zones, parking lots and courtyard passages. In the summer season, samples of soil and ground, puddle and dust sediments, surface dirt and road dust were collected in the areas of children's playgrounds, organized and unorganized parking lots and roads, respectively. Each sample of the environmental compartment included 3–5 sub-samples from different localizations within the given micro-landscape zone.

The particle size composition analysis was conducted for each sample. The samples of snow and snow dirt pulp were melted at ambient temperature. The solid mass of the snow and snow dirt pulp samples was divided into the following particle size fractions: >1 mm, 0.25–1 mm, 0.1–0.25 mm, 0.05–0.1 mm, 0.01–0.05 mm, and 0.002–0.01 mm. The particle size fractions of 0.01–0.05 mm and 0.002–0.01 mm were separated by decantation with vacuum filtering from the obtained solutions through filters with the pore sizes of 0.008 mm and 0.002 mm, respectively. The wet sieving with the addition of distilled water allowed separating the particle size fractions: 0.05–0.1 mm, 0.1–0.25 mm, 0.25–1 mm and >1 mm. The same methods of granulometry were applied to separate the samples of soil, ground, dirt and dust sediments and road dust. However, the sequence of fractioning was different: >1 mm, 0.25–1 mm, 0.002–0.01 mm, 0.01–0.05 mm, 0.05–0.1 mm, and 0.1–0.25 mm.

The anthropogenic particles were selected from the fraction of 0.25–1 mm particle size by visual diagnostic methods. The following features of the anthropogenic particles were taken into account: shape, morphology, structure, shine, color, hardness, elasticity, and density.

Each particle was photographed with Axioplan 2 optical microscope (Carl Zeiss) equipped with Olympus C-5060 photo camera. Chemical composition and structure of the particles were determined by JSM-6390LV scanning electron microscope (SEM) (Japanese Electron Optics Laboratory) combined with INCA Energy 350 X-Max 50 energy dispersive X-ray spectroscope (EDS) (Oxford INCAEnergy).

Results

The total of 25 anthropogenic particles were selected for the analysis. Table 1 represents the distribution of the anthropogenic particles selected from the samples of the environmental compartments by the season and the location at the sampling area.

Table 1 Distribution of the anthropogenic particles selected from the samples of the environmental compartments by season and location at the sampling area

Season	Part of site	Number of particles
Winter	Facade	9
	Courtyard	8
Summer	Facade	2
	Courtyard	3
	Courtyard	3

Considering the physical characteristics and elemental composition, the particles were attributed to the seven types of the particulate matter: slag, spheres, foil, brick, tires, plaster, and glass. The composition and characteristics of the above types of particles are presented in Tables 2 and 3, respectively. Figure 1 shows SEM photographs of the obtained types of particles.

Discussion

The presence of anthropogenic particles in the solid sediment is caused by emissions of industrial enterprises, power plants, transport, and by weathering of building materials. The most indicative solid particles for urban areas are slags that enter the atmosphere after burning the fuel with high ash content (fuel and

Table 2 The average composition of different types of anthropogenic particles

Type	Number of particles	Element, mass.%	
		Core	Inclusions
Brick	1	O: 48, Si: 29, Al: 10	Ti, Na, Mg, Fe, P, Ca, K, S < 13
Glass	1	O: 47, Si: 34	Mg, Al, Fe, S, Ca, K, Ni, V < 19
Sphere	9	O: 37–48, Si: 12–36, Ca: 12–15, Fe: 20–24	Al, Cr, Mg, Na, Mn, Ti, P, K, S, Ci < 20
Foil	4	O: 35–46, Si: 14–19, Al: 36–43, Fe: 12–38	Ca, Zn, Br, Ag, Pb < 13
Tires	1	O: 41, Si: 19, Ca: 12, Fe: 11	Mg, Al, Na, S < 17
Slag	8	O: 30–53, Si: 13–47, Fe: 11–23	Ca, Ni, Al, Br, S, V, K, Cl, Mg < 40
Plaster	1	O: 30, Ca: 66	Mg, Al, Fe, Na, Ti < 4

Table 3 Physical characteristics of different types of particles

Type	Size, mm	Short description
Slag	0.4–1	Semitransparent or nontransparent; black, grey, dark brown, dark green, or yellowish color; rounded or angular shape
Sphere	0.4–1	Flattened or irregular shape, corroded surface with caverns, cavities and cracks; black, dark brown color; glass shine, or nontransparent
Brick	0.5–0.7	Red-brown, dark red; rounded
Plaster	0.9	Thin, flattened, fragile, from light gray to white color
Glass	0.5–1	Colorless, yellow, blue, green or nontransparent, rounded, shell-like fracture
Foil	0.8–1	Grey, strong metallic appearance, thin film
Tires	<1	Black, nontransparent, mat, long, with a smooth surface

Fig. 1 The SEM photographs of particles: *1–2* slag particles, *3–5* spheres, *6* brick, *7* glass, *8* plaster, *9* tire, *10* foil

metallurgical slags). The chemical and mineral-phase compositions, structure and properties of slags depend on of the mineral component of the fuel composition, its calorific value, combustion regime and other characteristics. The chemical composition of slags is mainly represented by SiO and Al_2O_3, while inclusions consist of the oxides: FeO, CaO and MgO. Metallurgical slags may contain oxides of Mn, Cr, Ti, V and other metals.

The form, physical properties, and composition of slag particles are related to the conditions of their formation. Spherical solid particles are formed by burning fuel at high temperatures. Vitreous angular particles are formed at a lower temperature of fuel combustion. Apparently, more friable particles are formed at the lowest temperatures.

After deposition on the surface, the slag solid particles are included in the composition of contemporary loose surface sediment. The transfer within solid sediment determines and changes the structure of solid particles, in particular, the degree of shape roundness, and appearance of micro-scratches and chips on the surface of spherical particles. The degree of roundness and the number of chips are proportional to the residence time of the solid particle in the composition of the loose sediment and depend on the conditions of the sediment transport along the microlandscape. On the surface of ferruginous particles, corrosion can occur, which may result in the particle destruction with time. Figure 1(5) shows the initial stage of this process, in which rosette-shaped Fe oxide structures are formed at the surface of the particle.

The fragments of bricks and plaster in the surface sediments are the result of weathering of building materials. Particles of tires and foil in the studied sediments may be attributed to the impact of vehicles (brake pads and car parts wear).

Inclusions on the surface of solid particles are mainly metals and elements of the lithogenic substrate, products of gasoline combustion, abrasion of brake pads, tires, and metallic car details. Particles of slag can have up to 40% inclusions in their composition. Inclusions can cling to the particles due to various physicochemical interactions in the sedimentary cocktail.

Conclusion

In general, the types and physical properties of anthropogenic particles in the surface sediments characterize the anthropogenic impact and the conditions of transport of the sedimentary material and pollutants in an urban area. The systematization of anthropogenic particles and description of their properties can be used in the future to study the state of an urban environment characterized by intense multi-factor anthropogenic impact associated with high population density, significant automobilization, energy production at power plants of various generations and different types of fuel used, and the presence of metallurgy enterprises.

Acknowledgments The study was carried out at the UB RAS "Geoanalytik" Center for Collective Use and represents the scientific background for the project supported by the Russian Science Foundation (grant No. 18-77-10024).

References

Bucko MS, Magiera T, Johanson B, Petrovský E, Pesonen LJ. Identification of magnetic particulates in road dust accumulated on roadside snow using magnetic, geochemical and micro-morphological analyses. Environmental Pollution. 2011;159:1266–1276.

Crosby CJ, Fullen MA, Booth CA, Searle DE. A dynamic approach to urban road deposited sediment pollution monitoring (Marylebone Road, London, UK). Journal of Applied Geophysics. 2014;105:10–20.

Dong C, Zhang W, Ma H, Feng H, Lu H, Dong Y, Yu L. A magnetic record of heavy metal pollution in the Yangtze River subaqueous delta. Science of the Total Environment. 2014;476–477:368–377.

Gautam P, Blaha U, Appel E. Magnetic susceptibility of dust-loaded leaves as a proxy of traffic-related heavy metal pollution in Kathmandu city, Nepal. Atmospheric Environment. 2005;39:2201–2211.

Gołuchowska BJ. Some factors affecting an increase in magnetic susceptibility of cement dusts. Journal of Applied Geophysics. 2001;48;103–112.

Iordanidis A, Buckman J, Triantafyllou AG, Asvesta A. Fly ash-airborne particles from Ptolemais-Kozani area, northern Greece, as determined by ESEM-EDX. International Journal of Coal Geology. 2008;73:63–73.

Jordanova D, Jordanova N, Petrov P. Magnetic susceptibility of road deposited sediments at a national scale e Relation to population size and urban pollution. Environmental Pollution. 2014;189:239–251.

Kim W, Doh S-J, Yu Y. Anthropogenic contribution of magnetic particulates in urban roadside dust. Atmospheric Environment. 2009;43:3137–3144.

Kim W, Doh S-J, Park Y-H, Yun S-T. Two-year magnetic monitoring in conjunction with geochemical and electron microscopic data of roadside dust in Seoul, Korea. Atmospheric Environment. 2007;41:7627–7641.

Lu SG, Chen YY, Shan HD, Bai SQ. Mineralogy and heavy metal leachability of magnetic fractions separated from some Chinese coal fly ashes. Journal of Hazardous Materials. 2009;169:246–255.

Ma MM, Hu SY, Lin H, Cao LW, Wang LS. Magnetic responses to traffic related contamination recorded by backfills: A case study from Tongling City, China, Journal of Applied Geophysics. 2014;107:119–128.

Meena NK, Maiti S, Shrivastava A. Discrimination between anthropogenic (pollution) and lithogenic magnetic fraction in urban soils (Delhi, India) using environmental magnetism. Journal of Applied Geophysics. 2011;73:121–129.

Robertson DJ, Taylor KG, Hoon SR. Geochemical and mineral magnetic characterisation of urban sediment particulates, Manchester, UK. Applied Geochemistry. 2003;18:269–282.

Seleznev AA, Yarmoshenko IV. Study of urban puddle sediments for understanding heavy metal pollution in an urban environment. In: Environmental Technology & Innovation. 2014;1–2:1–7.

Wang XS. Heavy metal pollution in urban topsoils: mineralogical analyses and magnetic characterization. Environ Earth Sci. 2013;70(7):3155–3161.

Yang T, Liu Q, Li H, Zeng Q, Chan L. Anthropogenic magnetic particles and heavy metals in the road dust: Magnetic identification and its implications. Atmospheric Environment. 2010;44:1175–1185.

Zhu Z, Li Z, Bi X, Han Z, Yu G. Response of magnetic properties to heavy metal pollution in dust from three industrial cities in China. Journal of Hazardous Materials. 2013;246–247:189–198.

$CuCrS_2$ Phase in Uakit Iron Meteorite (IIAB), Buryatia, Russia: Preliminary Data

Victor V. Sharygin

Abstract Preliminary data on the occurrence and chemical composition were obtained for phase $CuCrS_2$ found in Uakit iron meteorite (IIAB). This phase may be a potentially new mineral for meteorites. It was observed in small phosphide-sulfide inclusions in kamacite in association with daubreelite, troilite and schreibersite. Chemical composition is $Cu_{0.94}Fe_{0.09}Cr_{0.97}S_{2.00}$. It is suggested that in similarity with a synthetic phase, the Uakit mineral may be trigonal $CuCrS_2$ (*R*3*m* or *R*-3*m*) and may belong structurally to the caswellsilverite group.

Keywords $CuCrS_2$ phase · Caswellsilverite · IIAB iron meteorite · Uakit

Introduction

Uakit iron meteorite (IIAB) was found in the summer 2016 by a gold prospector group during excavation works on the river terrace (stream Mukhtunnyi, left feeder of the Uakit River) in 4 km west of the Uakit settlement (Baunt Evenk district, Republic of Buryatia, Russia). At present, one sample (3.96 kg, Fig. 1) of meteorite was identified. However, the information about finding of larger meteorite mass (≈50 × 50 cm) is known among prospectors around Uakit. The date of fall is unknown. This iron meteorite was approved 28 June 2017 by the Meteorite Nomenclature Committee. Kamacite is the main mineral of the meteorite (>98 vol. %). Minor and accessory primary minerals are represented by schreibersite

V. V. Sharygin (✉)
V.S. Sobolev Institute of Geology and Mineralogy, SB RAS,
3 prospekt Akad. Koptyuga, Novosibirsk 630090, Russia
e-mail: sharygin@igm.nsc.ru

V. V. Sharygin
Novosibirsk State University, 2 Pirogov str., Novosibirsk 630090, Russia

V. V. Sharygin
ExtraTerra Consortium, Institute of Physics and Technology,
Ural Federal University, 19 Mira str., Ekaterinburg 620002, Russia

S. Votyakov et al. (eds.), *Minerals: Structure, Properties, Methods of Investigation*,
Springer Proceedings in Earth and Environmental Sciences,
https://doi.org/10.1007/978-3-030-00925-0_34

Fig. 1 The general view of the Uakit meteorite (IIAB) and a cut-off with troilite-daubreelite nodules. Symbols: Dbr+Tro – troilite-daubreelite nodule

(rhabdite), nickelphosphide, taenite, plessite (taenite + kamacite + tetrataenite), cohenite, tetrataenite, daubreelite, kalininite, troilite, carlsbergite, sphalerite, VN (“uakitite”, IMA 2018-003), cooper and the potentially new mineral $CuCrS_2$. The presence of large sulfide nodules (up to 1 cm) of troilite-daubreelite composition is common of the Uakit meteorite. The appearance of Ni-rich magnetite, pentlandite, heazlewoodite, awaruite-nickel as well as Ni-rich goethite, akageneite, Ni-rich siderite, Ca-Fe-carbonates, gypsum, and unidentified hydrated Fe-rich phosphate and Ca-Fe-sulfate is related to different stages of the terrestrial alteration (Ripp et al. 2017; Lastochkin et al. 2017). Unlike most of IIAB meteorites (Axon et al. 1981), the Uakit meteorite is very interesting in mineralogy and contains a lot of exotic accessory minerals: carlsbergite CrN, sphalerite ZnS, nickelphosphide $(Ni, Fe)_3P$, “uakitite” VN, kalininite $(Zn, Fe)Cr_2S_4$ and cohenite Fe_3C. The detailed mineralogical study of the meteorite gave a possibility to find a new exotic phase $CuCrS_2$ (after “uakitite”). This phase was observed in small phosphide-sulfide inclusions (up to 60 μm) in kamacite.

Materials and Methods

Polished fragments were used for optical examination of meteorite in reflected light. The identification of all minerals was based on energy-dispersive spectra (EDS), back-scattered electron (BSE) images and elemental mapping (EDS system), using a TESCAN MIRA 3MLU scanning electron microscope equipped with an INCA Energy 450 XMax 80 microanalysis system (Oxford Instruments Ltd.) at the V.S. Sobolev Institute of Geology and Mineralogy (IGM), Novosibirsk, Russia. The instruments were operated at an accelerating voltage of 20 kV and a probe current of 1 nA in high-vacuum mode. EDS analyses of the $CuCrS_2$ phase and other minerals were done at an accelerating voltage of 20 kV, a probe current of 1 nA, and accumulation time of 20 s. The following simple compounds and metals were used as reference standards for most of the elements: $Ca_2P_2O_7$ (P), Cr_2O_3 (Cr), pyrite (S), metallic Ti, Fe, Cu, Zn, Mn, Ni, V and others. The correction for matrix effects was done using the XPP algorithm, implemented in the software of the microanalysis system. Metallic Co served for quantitative optimization (normalization to probe current and energy calibration of the spectrometer). In addition some minerals that coexisted with $CuCrS_2$ were analyzed using a microprobe JEOL JXA-8100 (IGM) in wave-dispersive mode (WDS, 20 kV, 50 nA).

Results and Discussion

Three phosphide-sulfide inclusions with $CuCrS_2$ have been found in kamacite. They contain daubreelite, troilite, schreibersite, rarely carlsbergite, kamacite and "uakitite" (Fig. 2). The features of any secondary alteration (appearance of magnetite, etc.) are absent in these isolations. Unlike most of daubreelite-troilite inclusions in the meteorite, they have specific texture as far as daubreelite and troilite do not show "layering". The relationships of minerals within inclusions indicate that $CuCrS_2$ phase was crystallized after troilite and schreibersite (±"uakitite"), but before daubreelite (Figs. 2 and 3).

Unlike under the natural conditions, the $CuCrS_2$ compound is well-known in materials science and has a broad spectrum of useful properties, which are very promising for microelectronics (antiferromagnetic, semiconductor, superionic

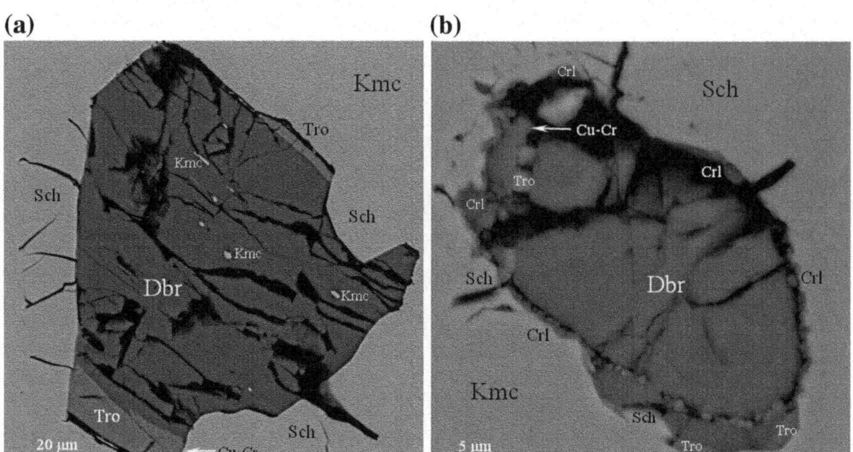

Fig. 2 BSE images of phosphide-sulfide inclusions with $CuCrS_2$ phase in kamacite of the Uakit meteorite. Symbols: Kmc – kamacite; Sch – schreibersite; Dbr – daubreelite; Tro – troilite; Crl – carlsbergite; Cu-Cr – $CuCrS_2$

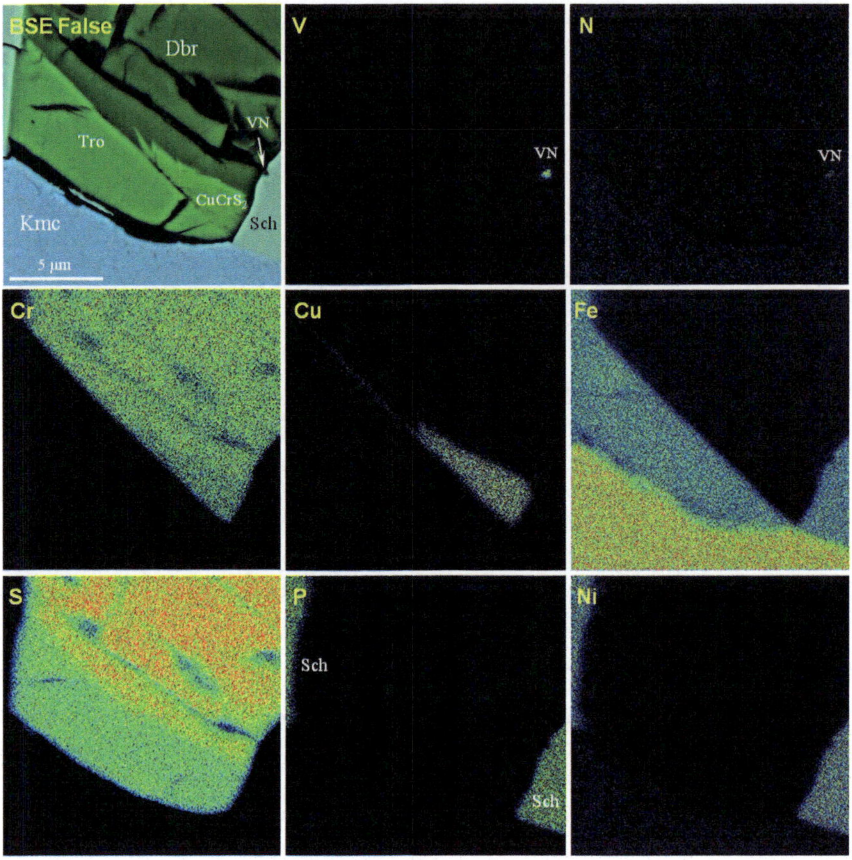

Fig. 3 Elemental maps for part of phosphide-sulfide inclusion with $CuCrS_2$ phase in kamacite (see Fig. 2A), meteorite Uakit. For symbols see Fig. 2

conductor) (Abramova et al. 2004; Almukhametov et al. 2003). The $CuCrS_2$ compound was first synthesized in the 50 s of the last century (Hahn and De Lorent 1957). It belongs to the family of trigonal compounds, covering sulfides, selenides and oxides. The crystal structure of disulfides Me^+CrS_2 (Me^+ = Na, K, Li, Cu, Ag, Au) contains the triple layers of S-Cr-S among which monovalent ions are localized (Ushakov et al. 2013). The crystal structure and properties of $CuCrS_2$ changing at different temperature have been studied in details (Engelsman et al. 1973; Nagard et al. 1979; Fitzgerald, and Al Mukhtar 1980; Wintenberger and Allain 1987; Yakshibaev et al. 1991; Mazalov et al. 2009; Tewari et al. 2011; Vassilieva et al. 2009). In general, the crystal lattice of $CuCrS_2$ (with $a \approx 3.48$ Å and $c \approx 18.70$ Å) is essentially vacant, and only a half of the tetrahedral sites are occupied by Cu+ ions; the environment of Cr^{3+} ions is octahedral. At present, three structural modifications of $CuCrS_2$ may be conventionally distinguished: β (room-temperature), α (high-temperature) and low-temperature. The high-temperature phase of the $CuCrS_2$ (α-$CuCrS_2$) features as a centrosymmetric *R*-3*m* state, with Cr^{3+} ions in octahedral environment and disordered arrangement of Cu in tetrahedral site, which provides high mobility of Cu and superionic conductor property (Almukhametov et al. 1997). The disorder-order phase transition indicates at near 670 K (Almukhametov et al. 2003; Almukhametov et al. 1997). The crystal structure of the room-temperature $CuCrS_2$ (β-$CuCrS_2$) is non-centrosymmetric *R*3*m*, and the right- and left-handed domains are possible in polydomain crystals (Fitzgerald and Al Mukhtar 1980; Abramova et al. 2010). At low temperatures (<100 K) β-$CuCrS_2$ may represent a mixture of two electronic phases: $Cu^{2+}Cr^{2+}S_2$ and $Cu^{+}Cr^{3+}S_2$ (Abramova et al. 2004). Moreover, a crystallographic transition from rhombohedral *R*3*m* to monoclinic *Cm* occurs at 37.5 K (Rasch et al. 2009). Thus, the review of all structural data shows that at room temperature this compound represents an ordered phase $Cu^{+}Cr^{3+}S_2$ (β-$CuCrS_2$).

The natural Cu-Cr-sulfide in the Uakit meteorite slightly differs in chemical composition from the ideal $CuCrS_2$ in lower Cu and Cr at constant admixture of Fe (Table 1). It is quite possible that Fe occupies both octahedral and tetrahedral sites. The compositional similarity of natural and synthetic phases, as well as the structural data for the synthetic compound, suggest that the phase from the Uakit meteorite may be trigonal $CuCrS_2$ (*R*3*m* or *R*-3*m*, Z = 3). It drastically differs from a Cu-thiospinel (cuprokalininite $CuCr_2S_4$) in chemical composition.

It should be noted that Na-analog of $CuCrS_2$, caswellsilverite $NaCrS_2$ (*R*-3*m*, Z = 3, *a* = 3.55 Å, c = 19.5 Å) (Okada and Keil 1982) was previously found in the Norton County enstatite achondrite. Later this mineral was observed in four enstatite chondrites. Thus, the $CuCrS_2$ phase in the Uakit meteorite seems to be a potentially new mineral for meteorites. Structurally this new natural mineral seems to belong to the caswellsilverite group, including caswellsilverite $NaCrS_2$, schöllhornite $Na_{0.3}(H_2O)_1[CrS_2]$ and cronusite, $Ca_{0.2}(H_2O)_2CrS_2$. The latter group is structurally similar to oxides of the delafossite group (delafossite $CuFeO_2$, mcconnellite $CuCrO_2$). In general, the structural features of the natural $CuCrS_2$ need more detailed investigations.

Crystallization conditions for the $CuCrS_2$ phase in the Uakit meteorite are not easy to explain. It should be noted that all Cu in the Uakit meteorite is mainly accumulated in sulfide-rich associations. In addition to $CuCrS_2$ native copper occasionally occurs in some troilite-daubreelite globules with "layered structure" (Ripp et al. 2017). Small amounts of Cu may be concentrated in thiospinels (daubreelite, kalininite) and troilite. Bulk compositions of whole meteorite and kamacite (ICP-MS and LA-ICP-MS) indicate copper concentrations 144–294 ppm (Ripp et al. 2017).

In general, it is evidenced that most Cu is concentrated in the sulfide (±phosphide) liquid after its separation from the metal melt. Different forms for accumulation of Cu (Cu^{o} and $CuCrS_2$) seem to be related to local conditions of crystallization and cooling in particular sulfide inclusions. Relationships of phases in the studied inclusions with $CuCrS_2$ (Figs. 2 and 3) suggest a very slow cooling rate. As a result, schreibersite (±"uakitite") and troilite are crystallized first and then $CuCrS_2$ and daubreelite. In the system Cu-Cr-S (Shabunina and Aminov 1994),

Table 1 Chemical composition (EDS-WDS, wt.%) of minerals from phosphide-sulfide inclusion, meteorite Uakit (see Figs. 2A, 3)

Phase	*n*	Fe	Ni	Co	Cu	Zn	Cr	Mn	S	P	Sum	Formula
Sch	*5*	54.83	29.48	0.12			0.24			15.33	100.01	$Fe_{1.98}Ni_{1.01}Cr_{0.01}P_{1.00}$
Dbr	*6*	19.46	0.00	0.00	0.00	0.20	35.66	0.19	44.47		99.97	$Fe_{1.00}Zn_{0.01}Mn_{0.01}Cr_{1.98}S_{4.00}$
Tro	*4*	62.71	0.11	0.00	0.00	0.00	0.80		36.60		100.22	$Fe_{0.98}Cr_{0.02}S_{1.00}$
$CuCrS_2$	*6*	2.75	0.00	0.00	33.47	0.00	28.02	0.00	35.72		99.96	$Cu_{0.94}Fe_{0.09}Cr_{0.97}S_{2.00}$

phase $CuCrS_2$ (± other Cu-phases) may be in association with the sulfide liquid in the temperature range 1363–790 °C. However, its abundant crystallization begins at 1060–790 °C, when the amount of sulfide melt is getting insufficient.

References

Abramova GM, Vorotynov AM, Petrakovskiy GA, Kiselev NI, Velikanov DA, Bovina AF, Almukhametov RF, Yakshibaev RA, Gabitov EV. Electron transition in intercalated disulfide $CuCrS_2$. Physics of the Solid State. 2004. 46(12): 2225–2228.

Abramova GM, Petrakovskiy GA, Vtyurin AN, Rasch JCE, Krylov AS, Gerasimova JV, Velikanov DA, Boehm VM, Sokolov V. Anomalous Raman phenomenon of $CuCrS_2$. Journal of Raman Spectroscopy. 2010. 41:1775–1778.

Almukhametov RF, Yakshibayev RA, Gabitov EV. X-Ray study of the superionic phase transition in $CuCrS_2$. Ionics. 1997. 3(3–4):296-298.

Almukhametov RF, Yakshibayev RA, Gabitov EV, Abdullin AR, Kutusheva RM. Structural properties and ionic conductivities of $CuCr_{1-x}V_xS_2$ solid solutions. Physica Status Solidi B. 2003. 236:29–33.

Almukhametov RF, Yakshibayev RA, Gabitov EV, Ableev IB. Structural studies and ionic conductivity of $CuCr_{1-x}V_xS_2$ ($0 \leq x \leq 0.3$). Ionics. 1997. 3(3–4):292-295.

Axon HJ, Kinder J, Haworth CW, Horsfield JW. Carlsbergite, CrN, in troilite, FeS, of the Sikhote Alin meteoritic iron. Mineralogical Magazine. 1981. 44:107–109.

Engelsman FMR, Wiegers GA, Jellinek F, van Laar B. Crystal structures and magnetic structures of some metal(I) chromium(III) sulfides and selenides. Journal of Solid State Chemistry. 1973. 6:574–582.

Fitzgerald AG, Al Mukhtar JH. An electron microscope study of copper chromium sulphide ($CuCrS_2$) crystals. Physica Status Solidi A: Applied Research. 1980. 59:233–239.

Hahn H, De Lorent C. Über ternäre chalkogenide des chroms mit einwertigem kupfer und silber. Zeitschrift für Anorganische und Allgemeine Chemie. 1957. 290: 68–81.

Lastochkin EI, Ripp GS, Izbrodin IA, Khromova EA, Sharygin VV. Mineral composition of meteorite Uakit (Republic of Buryatia). Proceedings of the IV Russian Young Scientists Conference, 2017, GIN SB RAS, Ulan-Ude, pp. 146–148 (in Russian).

Mazalov LN, Sokolov VV, Kryuchkova NA, Vovk EI, Filatova IY, Abramova GM. X-ray photoelectron spectroscopic studies of the charged state of 3*d* metal ions in $CuCr_{1-x}V_xS_2$ (x= 0-0.4). Journal of Structural Chemistry. 2009. 50:439-445.

Nagard NL, Collin G, Gorochov MO. Etude structurale et propriétés physiques de $CuCrS_2$. Materials Research Bulletin. 1979. 14:1411–1417.

Okada A, Keil K. Caswellsilverite, $NaCrS_2$: a new mineral in the Norton County enstatite achondrite. American Mineralogist. 1982. 67:132–136.

Rasch JCE, Boehm M, Ritter C, Mutka H, Schefer J, Keller L, Abramova GM, Cervellino A, Löffler JF. Magnetoelastic coupling in the triangular lattice antiferromagnet $CuCrS_2$. Physical Review B. 2009. 80:104431(1–6).

Ripp GS, Sharygin VV, Izbrodin IA, Ragozin AL, Khromova EA. Mineralogy and geochemistry of iron meteorite Uakit (IIAB), Buryatia. Proceedings of the 200th Anniversary Meeting of the Russian Mineralogical Society, 2017, Saint-Petersburg, V. 2, pp. 311–313 (in Russian).

Shabunina GG, Aminov TG. Interactions in the system Cu-Cr-S. (Russian) Journal of Inorganic Chemistry. 1994. 39:1505–1509 (in Russian).

Tewari GC, Tripathi TS, Kumar P, Rastogi AK, Pasha SK, Gupta G. Increase in the thermoelectric efficiency of the disordered phase of layered antiferromagnetic $CuCrS_2$. Journal of Electronic Materials. 2011. 40(12):2368–2373.

Ushakov V, Kukusta DA, Yaresko AN, Khomskii DI. Magnetism of layered chromium sulfides $MCrS_2$ (M = Li, Na, K, Ag, and Au): A first-principles study. Physical Review B. 2013. 87:014418.

Vassilieva IG, Kardash TY, Malakhov VV. Phase transformations of $CuCrS_2$: structural and chemical study. Journal of Structural Chemistry. 2009. 50:288–295.

Yakshibaev RA, Akmanova GR, Almukhametov RF, Konev VN. Ionic conductivity and diffusion in $CuCrS_2$-$AgCrS_2$ mixed conductors and their alloys. Physica Status Solidi A: Applied Research. 1991. 124:417–426.

Wintenberger M, Allain Y. Structure magnétique hélicoidale de $CuCrS_2$. Solid State Communications. 1987. 64:1343–1346.

Vibrational Spectroscopy of Kenyaite and Magadiite in the Southern Urals

Mikhail V. Shtenberg, Vladimir A. Popov, Svetlana M. Lebedeva, Rimma T. Zainullina and Mikhail A. Rassomakhin

Abstract Spectroscopic (infrared, Raman) data and chemical composition are reported for the kenyaite ($Na_2Si_{22}O_{41}(OH)_8 \cdot 6H_2O$) and magadiite ($NaSi_7O_{13}(OH)_3 \cdot H_2O$). Kenyaite dense white nodules were found in the dumps of exploration drill holes on the locality of "cacholong" at the Quarken district of the Orenburg region. The kenyaite nodules occurred in the disturbed black shale with the small magadiite amounts. Apparently, this is the first finding of kenyaite and magadiite in Russia.

Keywords Kenyaite · Magadiite · Infrared · Raman · Spectroscopy · Southern Urals

Introduction

Hydrous sodium silicates: kenyaite ($Na_2Si_{22}O_{41}(OH)_8 \cdot 6H_2O$) and magadiite ($NaSi_7O_{13}(OH)_3 \cdot H_2O$) were first found and described in Magadi Lake beds (Kenya) in 1967 by Eugster (1967). It is noted in the work that kenyaite forms small nodules in the magadiite layer, and the part of these nodules has a quartz (chert) core. Later, magadiite was found in altered volcanic rocks in Trinity County, California, USA (Eugster et al. 1967; McAtee et al. 1968). In addition, the scheme of the transformation of magadiite-kenyaite-quartz is given in (Eugster 1967). Recrystallization under pressure above 100 °C yields quartz and heating to 700 °C in air produces tridymite + quartz. Later, this scheme of the transformation kenyaite

M. V. Shtenberg (✉) · V. A. Popov · S. M. Lebedeva · R. T. Zainullina
Institute of Mineralogy, Ural Branch of RAS, 456317 Ilmensky Reserve, Miass, Chelyabinsk Region, Russia
e-mail: shtenberg@mineralogy.ru

V. A. Popov
e-mail: popov@mineralogy.ru

M. A. Rassomakhin
Ilmen State Reserve, 456317 Ilmensky Reserve, Miass, Chelyabinsk Region, Russia

S. Votyakov et al. (eds.), *Minerals: Structure, Properties, Methods of Investigation*, Springer Proceedings in Earth and Environmental Sciences,
https://doi.org/10.1007/978-3-030-00925-0_35

and magadiite into quartz was confirmed in the field (Eugster 1969). Magadiite and kenyaite were also obtained in laboratory (Beneke et al. 1983; Lagaly et al. 1975), and magadiite was synthesized long before its discovery in nature (McCulloch 1952).

Our studies started with the study of samples from the locality of "cacholong" (51°55′55.49″N, 59°35′22.63″E) at the Quarken district of the Orenburg Region. Some of the samples are presented by a dense white mineral aggregate with an "opoka effect" (adherence of the tongue to the sample) due to a microporosity and good wettability. In the area of the manifestation of cacholong, the sedimentary rocks of the Carboniferous age (Paleozoic) are developed: limestones, sandstones, siltstones and carbonaceous shales intruded by various subvolcanic dikes. Mineralization in sedimentary and volcanic strata is associated with metamorphism (regional and local metasomatism). Metasomatic transformations are partly expressed by the formation of flints and chalcedony nodules, among which there are original white concretions (Fig. 1).

Materials and Methods

The infrared spectra were obtained using Nicolet 6700 Thermo Scientific Fourier spectrometer in spectral range 400–4000 cm^{-1} with 4 cm^{-1} resolution and 32 scans. The pellets were prepared by compressing $\sim$2 mg of the sample with KBr (530 mg). Raman spectra were recorded on a Horiba Jobin Yvon HR spectrometer with a He–Ne laser (632.8 nm, 20 mW) and the Olympus BX41 microscope.

Fig. 1 White kenyaite nodules in dumps of exploration mines

The spectra were obtained by accumulating 20 scans of 10 s each in 180° geometry in a range of 100–2000 cm^{-1} from 5 μm sections. Registration and processing of spectra were carried out using Labspec v.5 software. The chemical composition of kenyaite was calculated from gravimetric SiO_2-analysis, photometric Na-determination, and potentiometric titration curves.

Results and Discussion

The studied nodules have different sizes (1–20 cm) and shapes (from isometric to elongated, complex and flattened), sharp or gradual (diffuse) contacts with the host shales. Nodules are partly monomineral, and some of them are zonal. In the transition zones of zonal concretions, there are aggregates with signs of co-crystallization of quartz, kenyaite, and magadiite. Some kenyaite nodules along the periphery have magadiite rims. Mineral nodules are developed metasomatically by the minerals of the host shale - carbonates, chlorites, mica, clay minerals and carbonaceous formations.

Kenyaite forms white or light gray dense concretions (Fig. 2a). The hardness of the mineral is about 4 of the Mohs scale. The refractive index is about 1.48, birefringence is low (0.010). In monomineral aggregates in the sections, split kenyaite individuals can be seen, the structure is fine-grained, homogeneous (Fig. 2b). The chemical composition of kenyaite is close to the theoretical values (wt %): SiO_2 85.1; Na_2O 3.01; MgO 0.32; CaO 0.43; H_2O^- 3.45; H_2O^+ 7.12 (Σ 99.43).

Figure 3 shows the infrared spectrum of kenyaite of the K2 sample compared to the published data. The main absorption bands in the 1000–1200 cm^{-1} region and the double peak in the 800 cm^{-1} region is common to many silicates, including quartz. Characteristic for the spectrum of kenyaite are small absorption bands with

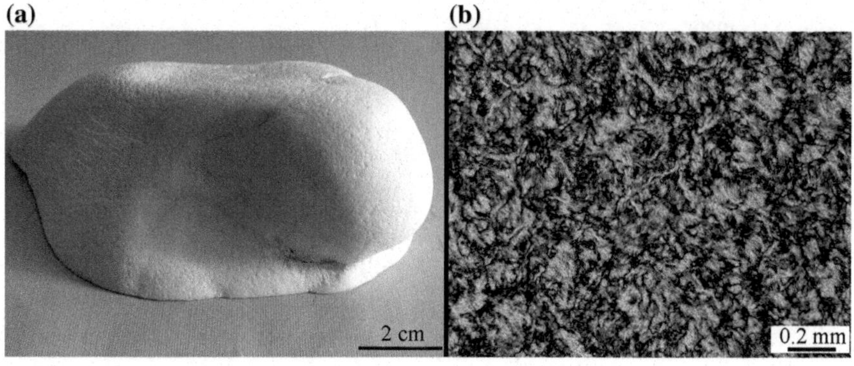

Fig. 2 Monomineral kenyaite nodule (sample K2): **a** – general view; **b** – aggregate of split kenyaite individuals (thin section, with analyzer)

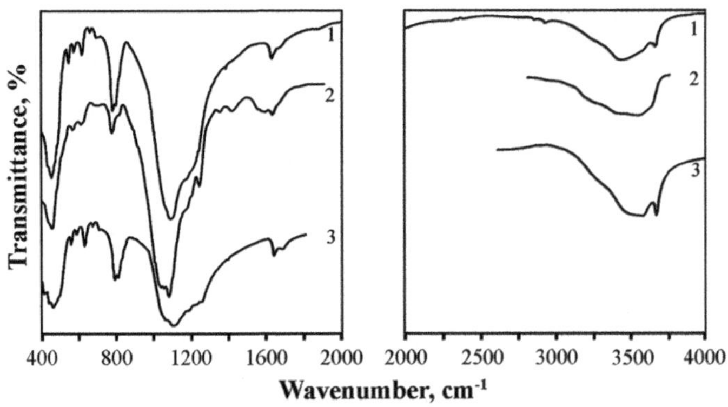

Fig. 3 Infrared transmission spectra of kenyaite: *1* – sample K2; *2* – sample from Magadi Lake, Kenya (Chukanov 2014); *3* – synthesized kenyaite (Huang et al. 1999)

maximum of 544 cm^{-1}, 574, 618, 660 and 693 cm^{-1}, related to symmetric stretching vibrations of Si-O-Si. The bands 1630 cm^{-1} and 1672 cm^{-1} are associated with the deformation vibrations of molecular water, bands 3440 cm^{-1} and 3660 cm^{-1} refer to the stretching vibrations of hydroxyl groups (Huang et al. 1999). The different degree of resolution of the bands in the spectra can be explained by the variation in the chemical composition and structural perfection of the minerals in the samples.

The Raman spectrum of kenyaite is shown in Fig. 4 and compared with the spectrum from (Huang et al. 1999). The characteristic bands of the Raman spectrum and their interpretation are given in Table 1. A comparison of the Raman spectra indicated that the position of the bands recorded at different points in the sample

Fig. 4 Raman spectra of kenyaite: *1* – sample K2; *2* – synthesized kenyaite (Huang et al. 1999)

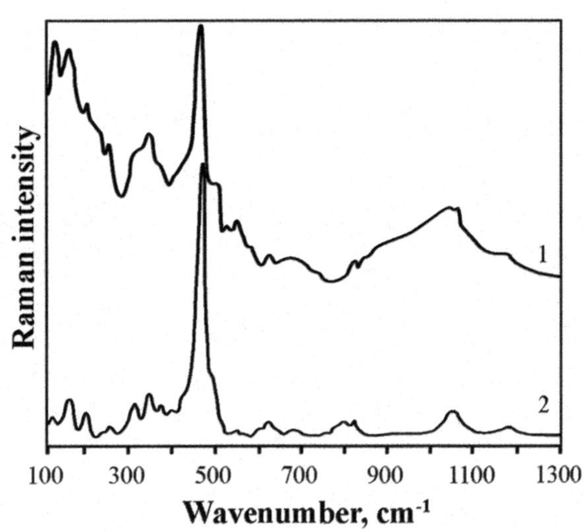

Table 1 Vibrational data (cm^{-1}) for kenyaite and magadiite spectra and their assignments

Kenyaite		Magadiite		Assignments (McCulloch 1952)
Infrared	Raman	Infrared	Raman	
		3659		ν(OH)
3645		3566		
3437		3440		
		1662		δ(HOH)
1633		1628		
1210	1180s	1235		ν_{as}(Si-O-Si)
1180		1172	1187vw	
1098	1061w	1079		ν(Si-O$^-$)
	1049w		1049w	
	819w	821	820w	ν_s(Si-O-Si)
797				
784		782	789w	
693	690m	706		
660		669	686w	
618	620w	621	632vw	
574		577		
544	549w	545		
	497s		492s	
	464vs		464vs	
455		444	442s	
		413		
	379s		373w	δ(Si-O-Si), δ(O-Si-O) ν(Na-O)
	348m		336w	
	201vw		252vw	
			187m	
	164w		160w	
	131w			

Note: ν_{as} – asymmetric stretching vibrations; ν_s – symmetric stretching vibrations; δ – bending vibrations, vs – very strong, m – medium, w – weak, vw – very weak, s – shoulder

studied corresponds to kenyaite, confirming the identification by the results of X-ray and infrared analysis. Our X-ray analysis data confirmed that the K2 sample consists of kenyaite (Shtenberg et al. 2017). The small deviations in the parameters of the obtained reflexes from published data are due to the heterogeneity of the composition. Additional reflections may be due to a small amount of quartz.

The spectrum of kenyaite is characterized by a strong band in the region of 464 cm^{-1}, corresponding to symmetric stretching vibrations of Si-O-Si, a group of bands in the region of 800–1180 cm^{-1}, responsible for asymmetric stretching vibrations of Si-O-Si and stretching vibrations of Si-O bonds, as well as a group of bands from 100 to 400 cm^{-1}, which can be attributed to the deformation vibrations of Si-O and M-O oscillations. A strong band in the Raman spectrum of 464 cm^{-1} is also manifested in several other silicates, e.g., in quartz.

Magadiite forms inclusions in kenyaite aggregates or peripheral rims on them (Fig. 5a). There are also inclusions of magadiite in host shales. The mineral has a low hardness of about 2 of the Mohs scale, a low average refractive index of about 1.48, and low birefringence. The white color of the mineral is caused by reflection from the planes of perfect cleavage along one direction, indicating a layered structure. In the anatomical picture of concretion, the "shadow" of the original substituted shale is sometimes visible (Fig. 5b).

Figure 6 shows the IR spectra of the K1 sample and the spectra of the magadiite from various sources. Apparently, the spectra are very similar. Interpretation of the absorption bands is presented in Table 1. The spectrum of magadiite differs from

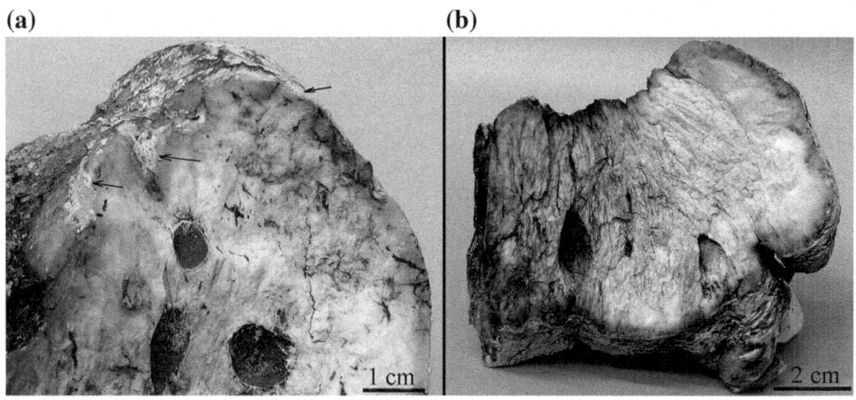

Fig. 5 Kenyaite nodule with magadiite rim (shown by arrows) **(a)** and relics of replaced shale **(b)**

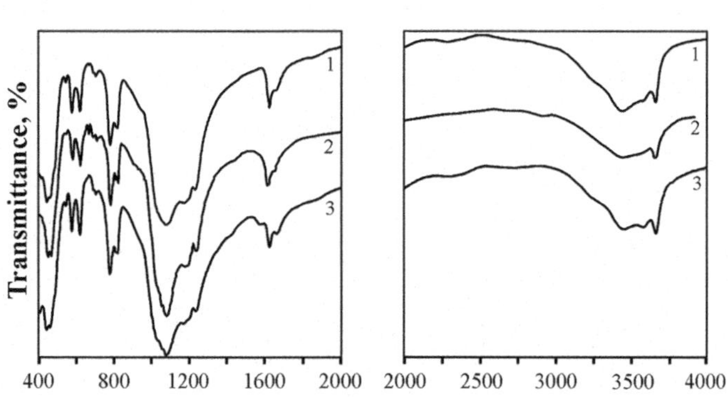

Fig. 6 Infrared transmission spectra of magadiite: *1* – sample K1; *2* – sample from Kanem region, Chad (Chukanov 2014); *3* – sample from Chad Lake, Kenya; data are from IR HR Inorganics database (Nicolet Instrument Corporation)

kenyaite spectrum by a greater degree of band resolution in the range 1000–1200 cm^{-1}, where three lines with maximum of 1235 cm^{-1}, 1172 and 1080 cm^{-1} are clearly traced. The bands in the 500–600 cm^{-1} region are more intense. The shape of the bands in the region of stretching and deformation vibrations of water indicates a complex structure of the entry of molecular water and hydroxyl groups. As in the case of infrared spectra, the Raman spectra of magadiite and kenyaite differ in the distribution of the lines in intensity and in the small shift of the maximum.

Conclusions

According to the published data (Eugster 1969), in the sediments with kenyaite and magadiite, makatite, kanemite, thrones, thermonatrite, gaylussite can be associated. The study of the locality of "cacholong" is going to be continued including the detailed toponymic study and searching for vein systems with mineral fill chambers (what should be done when performing the "Alpine-type living principle").

Acknowledgments The authors are grateful to V.N. Udachin, E.V. Belogub, E.D. Zenovich, T.V. Semenova who contributed to the implementation of research. The study is supported by a grant from the Russian Foundation for Basic Research (№ 16-35-60045).

References

Beneke K., Lagaly G. Kenyaite-synthesis and properties. Am. Miner. 1983;68:818–826

Chukanov NV. Infrared spectra of mineral species Extended library. Dordrecht: Springer Netherlands: Imprint: Springer, 2014. 1726 p. https://doi.org/10.1007/978-94-007-7128-4

Eugster HP. Hydrous sodium silicates from Lake Magadi, Kenya: precursors of bedded chert. Science. 1967;157(3793):1177–1180. https://doi.org/10.1126/science.157.3793.1177

Eugster HP. Inorganic bedded cherts from the Magadi area, Kenya. Contr. Mineral. Petrol. 1969;22:1–31. https://doi.org/10.1007/bf00388011

Eugster HP., Jones BF., Sheppard RA. New hydrous sodium silicates from Kenya, Oregon and California: possible precursors of chert. (abstr.). Geol. Soc. Amer. Ann. Meet. 1967:60

Huang Y., Jiang Z., Schwieger W. Vibrational spectroscopic studies of layered silicates. Chem. Mater. 1999;11:1210–1217. https://doi.org/10.1021/cm980403m

Lagaly G., Beneke K. Weiss A. Magadiite and H-magadiite: I. Sodium magadiite and some of its derivatives. Am. Miner. 1975;60:642–649

McAtee JL., House R., Eugster HP. Magadiite from Trinity county, California. Am. Miner. 1968;53:2061–2069

McCulloch L. A new highly silicious soda–silica compound. J. Am. Chem. Soc. 1952;74(10): 2453–2456. https://doi.org/10.1021/ja01130a003

Shtenberg MV., Popov VA., Lebedeva SM., Zenovich ED., Bozhko NA. Keniyait i magadiit na Yuzhnom Urale. Mineralogiya. 2017;3(1):16–22. [Shtenberg MV., Popov VA., Lebedeva SM., Zenovich ED., Bozhko NA. Kenyaite and magadiite in the Southern Urals. Mineralogy. 2017;3(1): 16–22. (In Russ.)]

Isotopic ($\delta^{13}C$, $\delta^{18}O$, Rb-Sr and Sm-Nd) Studies of Carbonates from the Vorontsovskoe Gold Deposit (Northern Urals)

Elena I. Soroka, Oksana B. Azovskova, Mikhail Yu. Rovnushkin, Mikhail E. Pritchin, Irina V. Smoleva, Natalia G. Soloshenko, Maria V. Streletskaya and Maria V. Zaitceva

Abstract Carbonate vein mineralization of the ore-bearing rocks of the Vorontsovskoe gold deposit (Northern Urals) represented by calcite and dolomite is considered in the article. Carbonate breccia is one of the major gold collectors on the deposit. Ore-bearing rocks of the Vorontsovskoe deposit (carbonate breccias, volcano-sedimentary rocks) are altered by hydrothermal processes. The lightening of C and O isotopic composition is observed in vein carbonates of ore-bearing rocks of the deposit. Isochron derived age of calcites is 429.8 ± 2.9 Ma, initial $^{87}Sr/^{86}Sr$ 0.7073537 ± 0000071, MSWD 0.60. The obtained $\delta^{13}C$, $\delta^{18}O$, Rb-Sr and Sm-Nd isotopic data of carbonates of the ore-bearing rocks of the Vorontsovskoe gold deposit indicates mostly crustal nature of mineralizing fluid. Metamorphogenic calcite may inherit the carbonate Rb-Sr isotope system of more ancient sediments trapped in the cement of the ore-bearing breccias of the Devonian limestones. However, the isotopic values of vein carbonates of the altered volcanic rocks may be affected by a magmatic fluid during mineralization.

Keywords Vein carbonates · Calcite · Dolomite · Carbonate breccia · $\delta^{13}C$ · $\delta^{18}O$ · Rb-Sr and Sm-Nd isotopic data · Vorontsovskoe gold deposit

---

E. I. Soroka (✉) · O. B. Azovskova · M. Yu. Rovnushkin · M. E. Pritchin
N. G. Soloshenko · M. V. Streletskaya · M. V. Zaitceva
A.N. Zavaritsky Institute of Geology and Geochemistry of the Ural Branch
of the Russian Academy of Sciences, 15 Vonsovskogo Street, Ekaterinburg, Russia
e-mail: soroka@igg.uran.ru

I. V. Smoleva
Institute of Geology of the Komi Science Centre of the Ural Branch
of the Russian Academy of Sciences, 54 Pervomayskaya Street, Syktyvkar, Russia

S. Votyakov et al. (eds.), *Minerals: Structure, Properties, Methods of Investigation*,
Springer Proceedings in Earth and Environmental Sciences,
https://doi.org/10.1007/978-3-030-00925-0_36

Introduction

The Vorontsovskoe deposit belongs to the gold-arsenic-sulfide formation and corresponds to the Carline type according to a number of mineralogical and geochemical characteristics (Murzin et al. 2010, 2016). It is localized in the marginal part of the Auerbach andesitoid volcano-plutonic complex, represented by volcanogenic-sedimentary rocks with lenses of Devonian limestone (D_1), and is confined to the Krasnoturyinsk ore field (Grabezhev et al. 2014). Intrusive rocks of the ore field are represented by diorites and granodiorites. The dikes of intermediate and basic chemical composition (from dolerites and pyroxene-plagioclase porphyrites to quartz diorites) are widespread. There are gold-bearing skarns and quartz veins formed near the contact with the intrusion. Carbonate breccia is one of the major gold collectors on the deposit. Ore-bearing rocks of the Vorontsovskoe deposit (carbonate breccias, volcano-sedimentary rocks) are altered by hydrothermal processes; carbonate and quartz-carbonate veins are widely developed, often with sulfide and realgar mineralization. The interpretation of C, O, Rb, Sr, Sm and Nd isotopic data of vein carbonates can clarify their formation and fluid origin during the ore mineralization.

Materials and Methods

Stable isotopes of C, O ($\delta^{13}C$, $\delta^{18}O$) of carbonate samples of rock veins and limestones of the Vorontsovskoe deposit were analyzed in the "Geonauka" Center for Collective Use of the Institute of Geology of the Komi SC, UB RAS. The carbonate monofractions were decomposed using the phosphoric acid. The measurement of their carbon and oxygen isotopic composition by the Continuous-flow isotope ratio mass spectrometry (CF-IRMS) in continuous helium flow mode was carried out using Thermo Fisher Scientific analytical complex, which included Gas Bench II sample preparation and input system connected to DELTA V Advantage mass spectrometer. The values of $\delta^{13}C$ are reported in per mil relative to the PDB standard, $\delta^{18}O$ - per mil relative to the SMOW standard. NBS 19 and NBS 18 international reference materials were used for calibration. The $\delta^{13}C$ and $\delta^{18}O$ determination error was $\pm 0.1‰$ (1σ).

The mineral composition of carbonates was determined by powder X-ray diffractometry using Shimadzu XRD-7000 diffractometer in the "Geoanalytic" Center for Collective Use of the Institute of Geology and Geochemistry by O. Galakhova.

The study of Sm-Nd, Rb-Sr isotopic system was conducted for 6 calcite monofractions hand-picked with the use of a binocular loupe in the "Geoanalytic" Center for Collective Use of the Institute of Geology and Geochemistry, UB RAS. The procedure of chemical sample preparation for the measurement of Sm, Nd, Rb and Sr concentrations and isotopic composition consisted of open acid sample decomposition and chromatographic separation of Rb and Sr as well as the REE sum followed by separation of Sm and Nd. The decomposition of samples was carried out by a mixture of HF and HNO_3 in the ratio of 5:1 and 1:1 depending on the sample composition in Savillex^{TM} vials. Before the decomposition step, ^{149}Sm-^{150}Nd tracer was added to the mixture based on the optimal mixing conditions and error minimization. Chromatographic separation was carried out by gradient elution of 2.3 N and 3.9 N HCl on a column packed with Bio-Rad AG ® 50 W-X8 Cation Exchange Resin (200–400 mesh) and the subsequent separation of Sm and Nd from the REE sum using the column packed with LN7A resin by gradient elution with 0.3 N and 0.7Nn HCl. The concentration and isotopic composition of Sm, Nd, Rb and Sr were determined by ICP-mass spectrometry by ELAN-9000 and Neptune Plus ICP-MS spectrometers. The mass bias was corrected on Neptune Plus using Eu internal standard. Eu solution was added to the test samples immediately before measurement. $^{151}Eu/^{153}Eu$ correction ratio was obtained by repeated measurement of the reference Nd solution (Neptune test solution) with the addition of the Eu solution ($^{151}Eu/^{153}Eu = 0.917$). Typical errors in the analysis of the isotopic composition of Sm, Nd, and isolated mixtures of samples with a tracer for $^{147}Sm/^{144}Nd$ and $^{143}Nd/^{144}Nd$ ratios did not exceed 0.00056 and 0.000031% ($\pm 2\sigma$), respectively, and were monitored by analyzing BCR-2 certified reference material. Isochron parameters calculation were carried out using ISOPLOT 3-v3.71_r5 software (DePaolo 1988).

Results and Discussion

$\delta^{13}C$ and $\delta^{18}O$ of the carbonate veins of the Vorontsovskoe deposit along with their mineral composition are given in Table 1.

The lightening of C and O isotopic composition is observed in vein carbonates of ore-bearing rocks of the deposit (Fig. 1): from calcite of marbled limestone having $\delta^{13}C = 1.17‰$ and $\delta^{18}O = 23.36‰$ (typical for marine carbonates), to calcites from tuff siltstone with δ^{13} C (−2.66 − 2.53‰) and dikes of the basic composition with $\delta^{18}O$ (12.95 − 18.61‰). According to (Murzin et al. 2010; Zamyatina and 2016) the carbonates characterized by the consistent lightening of C and O isotope composition in the series "ore-bearing calcareous breccias – jasperoid – skarn - quartz veins" are formed during the process of gold mineralization. Some of the vein carbonates (samples 17/15, 5/15, 3/16), as expected from the results of electron paramagnetic resonance (EPR) spectroscopy (Soroka et al. 2018), are of metamorphic origin, and so is calcite from the carbonate breccias (samples 72-1, 62-9, 62-14).

The results of the Rb-Sr and Sm-Nd isotopic analysis are given in Table 2 and Table 3, respectively.

The isochron was obtained using the samples of calcite of the marbleized carbonate breccias and from vein on a contact with altered limestone (samples 72-1,

Table 1 $\delta^{13}C$ and $\delta^{18}O$ isotopic data of the carbonate veins of the Vorontsovskoe deposit

Sample	Mineral, wt%	Rock	$\delta^{13}C$, ‰ PDB	$\delta^{18}O$, ‰ SMOW
2–15	Clc, 100	Tuff siltstone	−2.66	13.24
3–15	Dlt, 73; Qtz, 14; Clc, 11; Klt, 2	Tuff siltstone	−2.53	15.97
6–15	Mg-Clc, 81; Qtz, 17; Chl, 2; Pyr	Contact with dike of Px-pophyrite and tuff siltstone	−1.20	14.69
7–15	Clc, 97; Qtz,3	Tuff siltstone	0.45	17.51
9–15	Qtz, 58; Dlt, 29; Clc, 9; Klt 4	Contact of limestone with volcanites	−1.31	14.89
10–15	Clc, 93; Qtz, 6; Dlt 2	Contact of limestone with volcanites	0.05	15.12
12–15	Clc, 80; Dlt, 16; Qtz, 4	Contact of limestone with volcanites	−1.20	13.44
14–15	Dlt, 49; Clc, 39; Qtz, 12	Contact of limestone with dike	−4.51	12.95
17–15	99 Clc, 1 Qtz	Limestone	1.17	23.36
1a/16	Clc, 99; Qtz,1	Limestone in contact with dike	0.9	17.83
3/16	Clc, 97; Dlt, 3	Limestone	1.41	24.39
4/16	Clc, 95; Dlt, 3	Limestone	−2.96	15.28
5/16	Clc, 99; Dlt, 1	Limestone	1.41	22.19
6/16	Clc, 90; Qtz, 1	Limestone	1.11	24.21
7a/16	Clc, 92; Chl, 6; Qtz, 2	Limestone	1.06	20.21
7б/16	Clc, 92; Chl, 6; Klt, 1	Limestone	1.02	19.34
72–1	Clc	Carbonate breccia	1.46	17.19
62–9	Clc	Carbonate breccia	1.30	21.12
62–14	Clc	Carbonate breccia	–	–

Note Clc—calcite; Dlt—dolomite; Qtz—quartz; Klt—kaolinite; Chl—chlorite

10-15, 62-14) (Fig. 2). Calculated age was 429.8 ± 2.9 Ma, initial $^{87}Sr/^{86}Sr$ 0.7073537 ± 0000071, MSWD 0.60.

$^{87}Sr/^{86}Sr$ values given in Table 2 indicate crustal fluid sources. $^{87}Sr/^{86}Sr$ (0.706112 − 0.706532) correspond to the vein calcite of tuff siltstones. According to Murzin et al. (Murzin et al. 2010), $^{87}Sr/^{86}Sr$ = 0.7048-0.7079 in carbonates of ore-bearing rocks and ores of the Vorontsovskoe deposit implies the presence of two isotopic reservoirs: metamorphogenic fluid balanced with limestone and magmatogenic fluid.

A trend of C lightening from calcite of carbonate breccias to vein carbonate of altered volcanites is shown in Fig. 3a. All points of the $^{87}Sr/^{86}Sr$ – $\delta^{18}O$ diagram in Fig. 3b form an almost vertical trend from calcite of carbonate breccias to vein carbonate of volcanites. It should be noted that the studied calcite has a value of $\delta^{13}C$, close to sea limestone, and according to EPR analysis (Soroka et al. 2018),

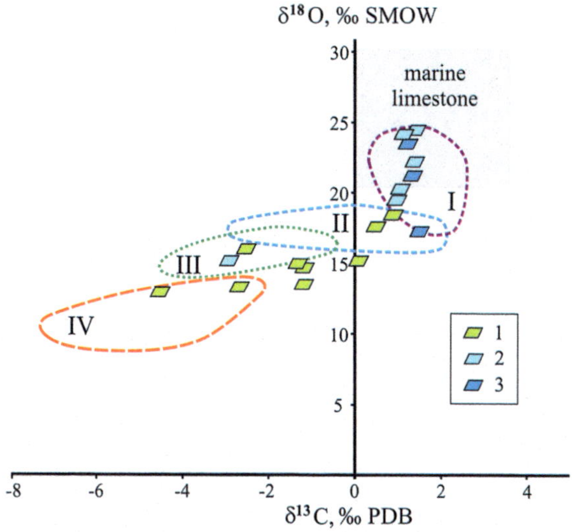

Fig. 1 A diagram of carbon and oxygen isotopic composition in carbonates of the Vorontsovskoe deposit: 1—vein carbonates of altered volcanites and dikes; 2—carbonate of marbled limestones; 3—carbonate of ore-bearing calcareous breccias. Areas I–IV (Zamyatina and Murzin 2016): I—marbled limestones; II—carbonate of ore-bearing calcareous breccias; III—carbonate of jasperoids; IV—carbonate of quartz veins of the Peschansk gold-ore deposit. The fields of a global magmatic reservoir and carbonate of marine limestone are also shown in the diagram

Table 2 Results of Rb-Sr isotopic analysis of the vein carbonates of the Vorontsovskoe deposit

Sample	Rb, ppm	Sr, ppm	$^{87}Rb/^{86}Sr$	±2σ	$^{87}Sr/^{86}Sr$	±2σ
72–1	14.73	271.59	0.1569	0.0019	0.708319	0.000017
10–15	0.44	123.74	0.0102	0.0002	0.707417	0.000017
62–14	3.23	154.92	0.0603	0.0007	0.707720	0.000017
62–9	2.74	149.72	0.0530	0.0006	0.707877	0.000019
2–15	0.17	196.94	0.0025	0.0001	0.706112	0.000006
7–15	0.38	170.33	0.0064	0.0001	0.706532	0.000013

Table 3 Results of Sm-Nd isotopic analysis of the vein carbonates of the Vorontsovskoe deposit

Sample	Sm, ppm	Nd, ppm	$^{147}Sm/^{144}Nd$	±2σ	$^{143}Nd/^{144}Nd$	±2σ	εNd
72–1	5.066	26.23	0.11678	0.00005	0.511933	0.00005	−3.0
62–9	1.470	7.64	0.11630	0.00062	0.512119	0.00060	0.7
62–14	0.902	4.56	0.11951	0.00002	0.511983	0.00003	−2.0
2–15	1.098	6.70	0.09915	0.00002	0.511903	0.00003	−3.5
7–15	2.767	12.13	0.13797	0.00001	0.511899	0.00002	−3.6
10–15	0.907	4.00	0.13716	0.00142	0.512672	0.00128	11.5

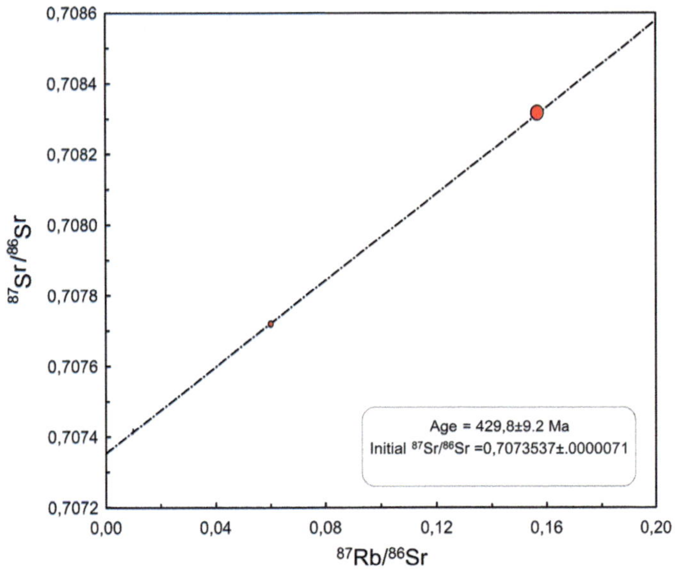

Fig. 2 Rb-Sr diagram of calculated calcite age

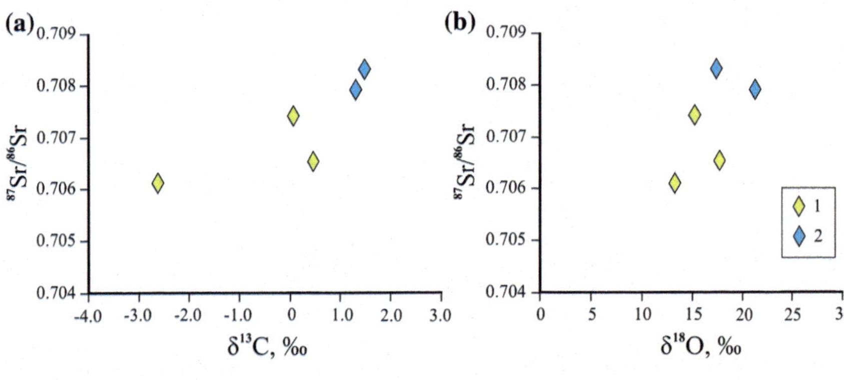

Fig. 3 Carbon, oxygen and strontium isotope correlations from carbonates of the Vorontsovskoe deposit. 1—vein carbonates of the altered volcanites and dikes; 2—carbonate of ore-bearing calcareous breccias

the Mn^{2+} band width in carbonates of marbleized carbonate breccias corresponds to that of non-defective crystals of metamorphosed limestone.

The calculated age of the studied calcite is 429.8 ± 2.9 Ma. It is more ancient than brecciated Devonian limestone of the Vorontsovskoe deposit. According to Murzin et al. (2010), isotopic dating (SHRIMP) of zircon from ore-bearing carbonate breccia cement gave the concordant age of 518.5 ± 3.7 Ma. It is assumed that zircon firstly got into the clay fraction of limestone, and then into the cement of breccias as the erosion result of more ancient sediments. The calculated calcite age of 429.8 ± 2.9 may indicate that newly formed carbonate has inherited Rb-Sr isotopic system of the Silurian volcanogenic-sedimentary deposits of the Krasnoturyinsk ore field (Grabezhev et al. 2014).

εNd values calculated for the age of 430 Ma range from −3.6 to 11.5 for the calcite of carbonate breccias and vein calcites (Table 3). ^{147}Sm/^{144}Nd vary from 0.09915 to 0.13797, and ^{143}Nd/^{144}Nd from 0.511899 to 0.512672. According to Krupenin et al. (2016), the Riphean limestones of the Southern Urals have negative values of εNd characteristic of the rocks that originated from the reprocessing of crust formation. The oceanic water might have been a probable source of limestone Nd with 0.21–0.23 values for the sedimentary carbonates. The high positive value of εNd = 11.5 of the vein carbonates corresponds to carbonate from limestone contact with altered tuff siltstone. The isotopic composition of granitoids and vein carbonate of the Krasnoturyinsk ore field (Grabezhev et al. 2014) indicates the magmatic nature of the fluid with ^{87}Sr/^{86}Sr = 0.70503-0.70520, U-Pb age of zircon from quartz diorite is 407.7 ± 1.6 Ma (MSWD 1.5).

Conclusions

Thus, the obtained δ^{13}C, δ^{18}O, Rb-Sr and Sm-Nd isotopic data of carbonates of the ore-bearing rocks of the Vorontsovskoe gold deposit indicates mostly crustal nature of mineralizing fluid. Besides, metamorphogenic calcite may inherit the carbonate Rb-Sr isotope system of more ancient sediments trapped in the cement of the ore-bearing breccias of the Devonian limestones. However, the isotopic values of vein carbonates of the altered volcanic rocks may be affected by a magmatic fluid during mineralization.

Acknowledgements The work was carried out at the UB RAS "Geoanalitik" Center for Collective Use with the financial support by the scientific program of UB RAS (grant No. 18-5-5-54).

References

DePaolo DJ. Neodinium isotope geochrmistry: an introduction. Heidelberg: Springer-Verlag; 1988

Grabezhev AI, Ronkin YuL, Puchkov VN, Gerdes A, Rovnushkin MYu. Krasnotur'inskoe medno-skarnovoe rudnoe pole (Severnyi Ural): U-Pb vozrast rudokontroliruyuschikh dioritov i ikh mesto v skheme metallogenii regiona. DAN. 2014;456(4):1–5. [Grabezhev AI, Ronkin YuL, Puchkov VN, Gerdes A, Rovnushkin MYu. Krasnotur'insk skarn copper ore field, Northern Urals: the U–Pb age of ore controlling diorites and their place in the regional metallogeny. Doklady Earth Sciences. 2014;456(2):641–645. (In Russ.)]

Kruhenin MT, Kuznetsov AB, Konstantinova GV. Sr-Nd sistematika I raspredelenie RZE v tipovykh magnezitovykh mestorozhdeniaykh nizhnego rifeya Yuzhno-Uralskoi provintsii. Litosfera. 2016;5:58–80. [Kruhenin MT, Kuznetsov AB, Konstantinova GV. Sr-Nd Systematics and REE distribution in the typical magnesite deposits of Lower Riphean of Southern Ural province. Lithosphere. 2016;5:58–80. (In Russ.)]

Murzin VV, Naumov EA, Azovskova OB, Varlamov DA, Rovnushkin MYu, Pirajno F. The Vorontsovskoe Au-Hg-As ore deposit (Northern Urals, Russia): Geological setting, ore mineralogy, geochemistry, geochronology and genetic model. Ore Geology Reviews. 2016; 85:271–298. https://doi.org/10.1016/j.oregeorev.2016.10.037

Murzin VV, Sazonov VN, Ronkin YuL. Model formirovaniya Vorontsovskogo zolotorudnogo mestorozhdeniya na Urale (Karlinskii tip): novye dannye i problem. Litosfera. 2010;6:66–73. [Murzin VV, Sazonov VN, Ronkin YuL. A formation model of the Vorontsovskoe gold deposit in the Urals (Carline type). Lithosphere. 2010; 6:66–73. (In Russ.)]

Soroka EI, Pritchin ME, Azovskova OB, Rovnushkin MYu, Lutoev VP, Smoleva IV. Phiziko-khimicheskie issledovaniya zhilnykh karbonatov Vorontsovskogo zolotorudnogo mestorozhdeniya. Vestnik Permskogo Universiteta. Geologiya. 2018;17(1):41–52. [Soroka EI, Pritchin ME, Azovskova OB, Rovnushkin MYu, Lutoev VP, Smoleva IV. Physical and chemical studies of the vein carbonates of the Vorontsovskoe gold deposit. Bulletin of Perm University. Geology. 2018;17(1):41–52. (In Russ.)]

Zamyatina DA, Murzin VV. Istocyniki veschestva i fluyida pri formirovanii zoloto-sulfidnogo orudeneniya Auerbakhovskogo volcano-plutonichrskogo poyasa na Stvernom Urale. Litosfera. 2016;1:169–177. [Zamyatina DA, Murzin VV. Substance and fluid sources in gold sulfide mineralization of the Auerbakh volcanic-plutonic belt in the Northern Urals. Lithosphere. 2016;1:169–177. (In Russ.)]

Composition and Microscopic Features of Background Cosmic Dust from Peat

Vladimir A. Tselmovich

Abstract A technique has been developed that makes it possible to separate from peat and identify a significant number of particles of background cosmic dust (CD), including presumably interstellar particles. Comparative studies of peat sections from the Yaroslavl region and from the site of the fall of the Tunguska space body (TSB) have been carried out. The background magnetic component of the CD contains mainly magnetite microspheres and particles of native metals. Most often we managed to detect native Fe and FeCr alloy. Studying these particles by other methods will yield in new fundamental results on the origin of the CD, and layers with a high CD content can be used as stratigraphic frames. The microparticles isolated from the site of the TSB fall with traces of high-temperature action are very different from the background component of the CD and can be attributed to the TSB substance.

Keywords Cosmic dust · Microprobe analysis · Peat · Native iron · Magnetite · Microsphere

Introduction

Origin of cosmic dust. The origin of cosmic dust is extremely diverse: it includes the remains of disintegrated comets, and particles of matter ejected by the Sun and brought to us by the force of light pressure. The presence of dust clouds at a height significantly exceeding 100 km is established. These are the so-called “noctilucent clouds”, consisting of cosmic dust. Under the influence of terrestrial attraction, a significant part of these cosmic dust particles slowly settles to the ground. The presence of such cosmic dust was found on high snow tops.

V. A. Tselmovich (✉)
Borok Geophysical Observatory, The Branch of Shmidt’s Institute of Physics of the Earth, Borok, bldg. 142, Nekouz District, Yaroslavl Region 152742, Russia
e-mail: tselm@mail.ru

S. Votyakov et al. (eds.), *Minerals: Structure, Properties, Methods of Investigation*, Springer Proceedings in Earth and Environmental Sciences,
https://doi.org/10.1007/978-3-030-00925-0_37

In addition to the slowly settling space dust, "falling stars"—hundreds of millions of meteors every day—fall into the Earth's atmosphere. They have a speed of tens (hundreds) kilometers per second. Meteors are burned by friction against particles of air before they reach the surface of the earth. The products of their combustion also settle on the ground. However, particles smaller than 50 μm reach the Earth's surface without heating. Among the meteors there are also exceptionally large specimens that reach the surface of the Earth. The fall of a large Tunguska meteorite is known to be at 5 am on June 30, 1908. Unfortunately, nobody could find the meteorite itself. Expeditions to the site of the fall of the Tunguska cosmic body continue to this day. Expedition to the Tunguska State Nature Reserve (Krasnoyarsk Territory, Evenk Municipal District, Vanavara settlement) took place from July 16 to August 3, 2017 (http://tunguska.ru/forum/index.php?topic=1091). The hypothesis of Academician V.I. Vernadsky, who considered possible the falling of not a meteorite, but an enormous cloud of cosmic dust, which was traveling at colossal speed, is interesting (Vernadsky 1932). By this hypothesis Academician Vernadsky confirmed the appearance in these days of a large number of glowing clouds, moving at high altitude at a speed of 300–350 km per hour. This hypothesis could also be explained by the fact that the trees surrounding the meteorite crater remained standing, while the ones located further were blown up by a blast wave.

Materials and Methods

The search for tiny space treasures requires the collection of dust from clean environments with low levels of terrestrial particles. One such environment is peat deposits. A particle pre-concentration is required for sampling CD. The most effective methods for solving this problem are magnetic ones.

Determination of the dynamics of the arrival of the CD on the Earth's surface was carried out on the basis of the investigation of peat deposits (Tselmovich et al. 2016a, b). The search for horizons enriched with mineral matter (terrigenous or cosmogenic) was carried out using the parameter Irs—the residual saturation magnetization in terms of the power of the core being studied. Then, using the "TESCAN VEGA 2" SEM—EDS, the composition and genesis of mineral particles in horizons with increased values of Irs was determined. Note that the implementation of quantitative determinations of the mineral and, especially, of the cosmogenic substance in the horizons of peat deposits is a complex and ultimately unresolved problem. Nevertheless, the Irs parameter values are certainly related to the amount of mineral matter contained in the peat samples. The combination of petrophysical and micromineralogical methods made it possible to determine the horizons enriched with a cosmogenic substance, and to draw conclusions about changes in its quantity. In this case, the top layers of peat contain a large number of anthropogenic pollutants. They give a strong magnetic signal. In this layer, the cosmic dust is more difficult to identify. However, it is of interest in the environmental aspect, since it allows the composition and morphology of mineral particles-the pollutants of the atmosphere to be diagnosed.

Of particular interest was the microprobe analysis of regions with low values of magnetic characteristics (for example, the Irs remanent magnetization). It turned out that in these zones there is practically no terrigenous magnetic material. Magnetite (Mt) microspheres with a detrital surface, as well as native Fe and other native metals, were found. These findings were attributed to CD and micrometeorites (MM). A very small amount of peat (about 8 cm^3) was used for layer analysis, which made it impossible to isolate a sufficient number of particles, so it was decided to work out a technique for separating microparticles from more peat (from 1 L). As a result of the experiment, about 5 mg of magnetic and 40 mg of non-magnetic substances of various origins were isolated.

For comparative analysis, a substance was used from the peat bogs of the Yaroslavl region and from the peat bog at the site of the fall of the Tunguska cosmic body.

Results and Discussion

The magnetic substance of both objects was similar and is represented mainly by Mt microspheres, native Fe and intermetallides (FeCr, FeW, FeAl) presumably of cosmic origin. However, such particles can form in other processes, so it is important to have different arguments in favor of this hypothesis. These include the following:

Mt microspheres have a well developed detritus surface and can be referred to as "cosmic" spheres, which have been repeatedly described in hundreds of scientific publications. They were formed during the ablation process during the flight of meteorites through the atmosphere, the mechanism of formation of such balls in the melting core of troilite is known. They could also form particles of native Fe through the melting crust. The mechanisms of these processes are described (Tselmovich 2013, 2015). Explanation of the cosmogenic nature of native Fe is given in a study (Pechersky et al. 2017), in which microprobe studies on the diagnosis of CD were performed by V. A. Tselmovich.

Particles of Fe can be conditionally divided into two groups in size: up to 50 microns and more than 50 microns. It is known that particles up to 50 microns are not heated while passing through the atmosphere (Folco and Cordier 2015). Indeed, all fine particles have sharp, angular shapes, while particles larger than 50 μm have obvious melting characteristics. These arguments also work in favor of the hypothesis of their cosmic origin. The particles of native Fe could also have a volcanic origin. However, there are serious arguments in favor of the cosmic origin of these particles.

1. In the analyzed layers (horizons 50, 60, 70, 100 and 120 cm from the Mokeikha-Zybinsk peat deposit, Yaroslavl region), there is approximately the same amount of Fe particles, relatively uniformly arriving "from above". Volcanic iron

arrives impulsively, and it would be recorded as a marker of a catastrophic phenomenon.

2. There are no active volcanoes in the area of this peat bog. The detected particles of this size are large and they quickly fall out of the ash clouds, so the version of the volcanic material is rejected.

3. Some of the Fe particles have very characteristic tracks of several microns in size (Fig. 1). The tracks are micron-sized holes that permeate particles at different angles, which reflects the rotation of relatively massive particles when they are "fired" by high-speed microparticles. Similar tracks were observed by the author during the analysis of the meteorite 7428 (San Julio De Morayra) from the collection of the Vernadsky State Geological Museum. Such structures could arise when colliding with swarms of high-speed microparticles. Figure 2a, b, c shows a micrograph of magnetic microparticles isolated from the peat of the Mokeikha-Zybinsk deposit (Yaroslavl Region, horizons 60, 70 and 120 cm.) Similar results were obtained by the author on horizons 50, 60, and 100 cm as well on peat deposits of the Kemerovo Region and Khabarovsk Territory (Tselmovich et al. 2016a; b), from peat from the place of the fall of the Tunguska Space Body (TSB), Fig. 3a–e. A significant number of Mt beads, native Fe, and FeCr intermetallides have been found at all horizons. This indicates a single mechanism of accumulation of the background fossil of CP in different geographical locations and environments. However, in the area of the coral funnel, discovered by E. V. Dmitriev at the site of the fall of the TSB (Dmitriyev 2012), along with the background CD, particles were found that can be attributed to the TSB itself due to their special morphology and composition. Thus, a thin nickel film was deposited on the silicate stick of vegetable origin present in the funnel (an example is shown in Fig. 4a, b). Such a structure could arise due to the high temperature and

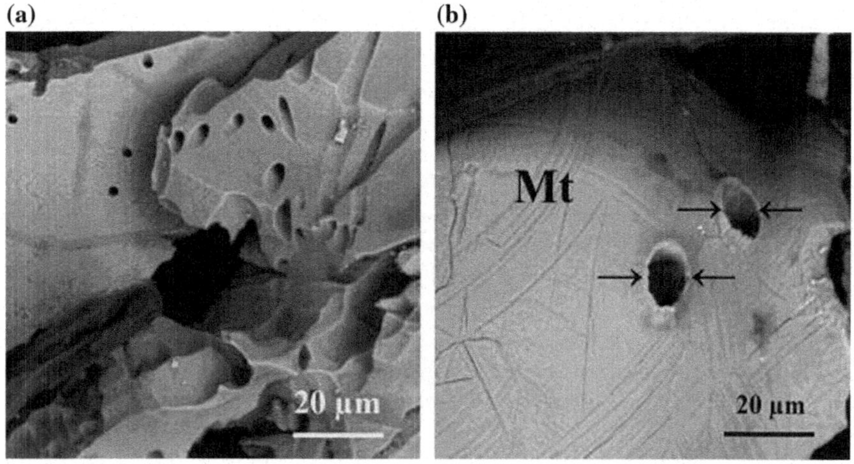

Fig. 1 Holes in Mt particle found in the peat bog from the Kemerovo region (**a**); for comparison: holes in the particle from the peat shreibersite meteorite San Julio De Moreira (**b**)

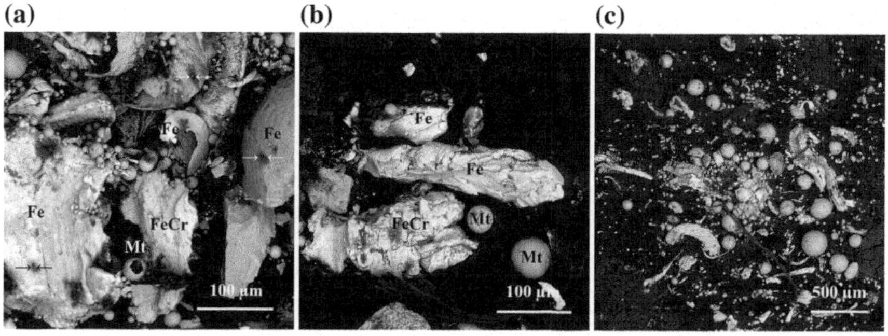

Fig. 2 Mt beads and Fe from various horizons of the peat deposit: 60 cm horizon: Mt, Fe (**a**); Horizon 70 (**b**); Horizon 120 cm: Mt, Fe, FeCr (**c**)

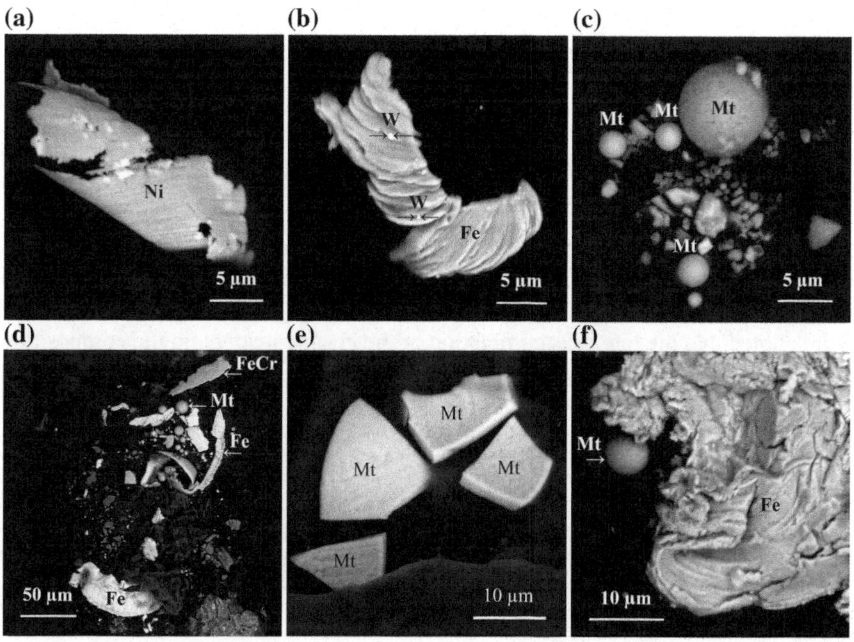

Fig. 3 CD from peat from the place of falling of TSB; Ni flakes (**a**); Fe flakes (**b**); small ablative Mt microspheres and micrometeorite (**c**); particles of Fe, FeCr (**d**); fragments of a hollow Mt micrometeorite (**e**); Fe and Mt microspheres (**f**)

hydrophilic properties of nickel. The discovery of melted tungsten microspheres in the gland indicates the high temperature (Fig. 4c). These findings convincingly speak in favor of the hypothesis of V. I. Vernadsky, who believed that the cloud of cosmic dust, which was traveling at colossal speed, could fall. Perhaps it cloud flew along with the comet.

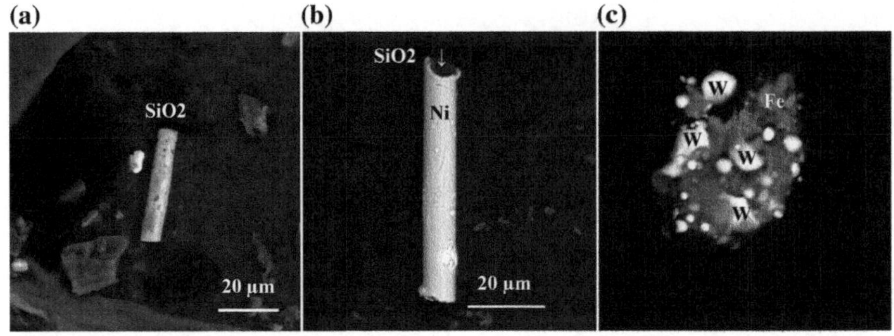

Fig. 4 Microparticles from the site of the fall of the TSB ("coral" funnel, found by E. V. Dmitriev), which are indicators of the high-temperature process. A quartz stick of vegetable origin (**a**); thin Ni film fused to a quartz stick with a drop in the TSB (**b**); melted microspheres W in an iron matrix (**c**)

Conclusions

It is shown that the magnetic background component of cosmic dust extracted from peat has a similar composition in remote peat sections of the sections selected both at the site of the TSB fall and in the peat deposit of the Yaroslavl Region. This fact is of great importance for identifying the microparticles that fell to the ground during the Tunguska catastrophe. Such particles are separated from the coralloid funnel. They carry traces of high-temperature impact and they can be attributed to the material of TSB.

It is noted that the rate of accumulation of the cosmic matter and the composition of its main magnetic components in the Holocene varies little both in the depth of the peat sections and in the location of the peat bog. This fact can be used in climatic reconstructions: in a cold climate, the ratio of the terrigenous substance to the CD varies in favor of the CD.

Acknowledgements The work was carried out within the budget topic "Spatial-temporal structure of the ancient geomagnetic field and petrophysics of magnetic minerals as indicators of geological and geophysical processes", State Registration No. AAAA-A17-117040610183-6. Microprobe studies were carried out with the support of the Russian Foundation for Basic Research, project No. 16-05-00703a.

References

Dmitriyev YeV. Tungusskaya kometa: neobychnaya istoriya odnoy intriguyushchey nakhodki. LENAND: Moscow; 2012. S. 282–289. [Dmitriev EV. Tunguska comet: an unusual story of one intriguing find. LENAND: Moscow; 2012. P. 282-289. (in Russ.)]

Folco L and Cordier C. Micrometeorites. EMU Notes in Mineralogy. 2015;15:253–297

Pechersky DM, Kuzina DM, Markov GP, Tselmovich VA. Native iron in the Earth and space. Izvestiya Physics of the Solid Earth. 2017;5:44–62

Tselmovich VA. Kosmicheskiye shariki na poverkhnosti Chelyabinskogo meteorita. Asteroidy i komety. Chelyabinsk, 20–22 iyunya 2013g. Chelyabinsk “Kray Ra”. 2013;140-147. [Tselmovich VA. Space balls on the surface of the Chelyabinsk meteorite. Materials and reports of the International Scientific and Practical Conference “Asteroids and comets. The Chelyabinsk event and the study of the fall of the meteorite into Lake Chebarkul”. Chebarkul, June 21-22, 2013;140–147. (in Russ.)]

Tselmovich V.A. Vozmozhnosti mikroskopicheskoy diagnostiki kosmicheskoy pyli v torfe. “Meteority, asteroidy, komety»”. Materialy konferentsii i shkoly molodykh uchenykh “CHEBARKUL’ 2015”. “TETA”: Chelyabinsk; 2015. P. 193–196. [Tselmovich VA. Possibilities of microscopic diagnostics of cosmic dust in peat. Materials of the international conference and school of young scientists “CHEBARKUL 2015”. “TETA”: Chelyabinsk; 2015. P. 193–196. (in Russ.])

Tselmovich VA, Kurazhkovskii AYu, Kazansky AYu, Shchetnikov AA. Research of the holocene events. EXPERIMENT In GeoSciences: Mineralogy.Petrology.Geochemistry. Crystallography. Geophysics. Isotopy. 2016a;22(1):21–23

Tselmovich VA, Kurazhkovskii AYu, Kazansky AYu, Shchetnikov AA, Blyakharchuk TA, Amelin II. Catastrophic events in the Holocene and their registration in peat deposits. Proceedings of the 11th Intl School and Conference “Problems of Geocosmos”. Oct 03–07 2016. St.Petersburg, Russia. 2016b;91–98

Vernadsky VI. Ob izuchenii kosmicheskoy pyli. Mirovedeniye. 1932;5:32–41. [Vernadsky VI. On the study of cosmic dust. World Science. 1932;5:32–41 (in Russ.)]

A New Self-consistent Interatomic Potential Set for Structure Modeling of Monazites and Xenotimes with Variable Composition

Amina S. Ulanova, Ekaterina I. Marchenko and Nikolai N. Eremin

Abstract A new semi-empirical self-consistent pairwise potential model for rare-earth monazites and xenotimes modeling has been developed. Its transferability and reliability have been demonstrated by testing the potentials toward the prediction of structural, thermodynamical and mechanical properties of all minerals under consideration. The partial ionic charge model with a Morse function is used, and it allows different monazite and xenotime-based solid solutions, including those with plutonium impurities, to be modeled.

Keywords Atomistic model · Monazite · Xenotime · Rare-earth elements · Plutonium · Crystal structure

Introduction

Monazite and xenotime (rare earth orthophosphates, $MePO_4$) are common natural phosphate minerals. The typical structure of La-Gd phosphates is a monoclinic monazite type (space group $P2_1/n$), while for YPO_4 and Tb-Lu compounds a tetragonal xenotime structure ($I4_1/amd$) is realized (Fig. 1). $TbPO_4$ and $DyPO_4$ compounds are dimorphic and can be obtained also with a monazite structure under certain synthesis conditions (Boakye et al. 2008). Natural monazites are known to be a complex isomorphic mixture of the lighter lanthanides (La-Gd) with significant amount of the heavier rare earths and radioactive actinides. Due to their properties, these compounds are considered as a promising durable matrix for high-level radioactive waste (Ewing et al. 1995).

A. S. Ulanova (✉)
Department of Natural Sciences, Lomonosov MSU, Dushanbe, Tajikistan
e-mail: ulanova-amina@mail.ru

E. I. Marchenko · N. N. Eremin
Department of Geology, Lomonosov Moscow State University, Moscow, Russia
e-mail: neremin@geol.msu.ru

S. Votyakov et al. (eds.), *Minerals: Structure, Properties, Methods of Investigation*, Springer Proceedings in Earth and Environmental Sciences,
https://doi.org/10.1007/978-3-030-00925-0_38

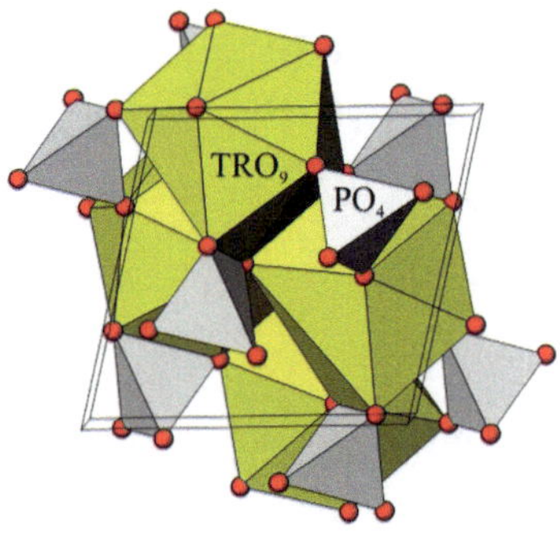

Fig. 1 Left—[001] view of monazite crystal structure (group $P2_1/n$); right ≈ [110] view of xenotime crystal structure (group $I4_1/amd$). Unit cells are emphasized

Materials and Methods

Study of various properties of these minerals is preferably to be performed by semi-empirical structural modeling than by quantum chemical calculations: very close properties of all rare-earth monazites allow the principle of transferability of interatomic potentials parameters to be successfully used. But, as it was noted in our recent review, the application of the method of interatomic potentials for modeling crystalline structures of monazites is rather irrelevant and unsuccessful (Grechanovsky et al. 2016). Recently (Eremin et al. 2017) we have performed a comparative analysis of existing models of interatomic potentials for semi-empirical simulations of monazites. It was shown that none of the previously published models is able to describe simultaneously the structural, elastic and thermodynamic properties of these compounds with high accuracy. That is why we have developed a new set of partially ionic interatomic interaction potentials (Eremin et al. 2017) for the structural simulation of monazites of seven light lanthanides (La-Gd). This set provides an excellent description of the crystal structures of monazites and reproduces the elastic and thermodynamic properties of compounds with high accuracy due to the justified crystal-chemical charge model. We have used coulombic interaction of atomic effective charges (q(Me) = 1.6 e_0, q(P) = 1.2 e_0, q(O) = −0.7 e_0) as well as short-range (up to 15 Å) Morse potential (Eq. 1) in the developed potential model:

$$V(r) = D_M \cdot [\exp(-2\alpha(r - r_0)) - 2\exp(-\alpha(r - r_0))]. \tag{1}$$

D_M, α and r_0 parameters were optimized in Eremin et al. (2017) using experimental La-Gd monazite values from (Ni et al. 2005). We also expanded a list of potentials and added Me-O interaction potentials for heavier lanthanides (Tb-Lu), Y and Pu. Me-O potential parameters for Tb-Lu and Y were optimized using $MePO_4$ xenotimes structural information from (Ni et al. 2005). Pu-O interaction parameters

Table 1 Parameters of Morse pair potentials for monazite and xenotime modeling with the following effective atomic charges: q(Me) = 1.6 e_0, q(P) = 1.2 e_0, q(O) = −0.7 e_0

Interaction	D_M, eV	σ, Å^{-2}	r_0, Å	R_{max}, Å
O-O	0.018716	1.374805	3.645169	15.0
P-O	0.446152	3.263861	1.609157	15.0
La-O	0.099973	3.456018	2.662300	15.0
Ce-O	0.117341	3.335213	2.631052	15.0
Pr-O	0.133622	3.278962	2.607182	15.0
Nd-O	0.173034	3.563096	2.562213	15.0
Sm-O	0.190876	3.737843	2.518803	15.0
Eu-O	0.193661	3.399594	2.518195	15.0
Gd-O	0.198682	3.172605	2.515892	15.0
Pu-O	0.119590	3.701095	2.596788	15.0
Y-O	0.318939	1.908249	2.540487	15.0
Tb-O	0.231571	2.008679	2.572061	15.0
Dy-O	0.306464	1.898378	2.552877	15.0
Ho-O	0.356088	1.837013	2.539438	15.0
Er-O	0.402338	1.792183	2.526532	15.0
Tm-O	0.441546	1.758439	2.517015	15.0
Yb-O	0.472769	1.729378	2.509278	15.0
Lu-O	0.505421	1.700125	2.501459	15.0

were optimized using $PuPO_4$ monazite structural data (Jardin et al. 2008). Final values of (Eq. 1) potentials for Me-O interaction are presented in Table 1. Calculations were performed using GULP 4.5 software (Gale and Rohl 2003).

Results and Discussion

Developed potentials allowed us to reach excellent agreement with experimental crystal structures and satisfactory reproduction of thermodynamic data as is shown in Table 2. Thus, the unit cell metric parameter deviation for all monazites from their experimental values did not exceed 0.5% along with precise description of β monoclinic angle in all cases. As is seen from Fig. 2, the $LaPO_4$ calculated entropy temperature dependence does not deviate from the experimental values more than 3% in the entire studied temperature interval (from 200 K to 1600 K). For all monazites, a good agreement is also observed between the measured bulk and shear moduli and their theoretical estimates.

As can be seen from Table 3, the developed potential set also perfectly reproduces the crystal structures of xenotimes of heavier lanthanides (Tb-Lu) and Y.

Table 2 Comparison of the experimental and calculated data for Ln-, Gd-, Ce- and Pu-monazites

Parameter	Experiment	Calculation	Experiment	Calculation
$LaPO_4$			$CePO_4$	
a, Å	6.831	6.809	6.788	6.764
b, Å	7.071	7.082	7.016	7.029
c, Å	6.503	6.514	6.465	6.476
β,°	103.27	103.27	103.43	103.43
K, GPa	99	109.74	–	112.58
G, GPa	53	53.98	–	55.73
S_{298}, J/(mol·K)	108.7	111.02	119.97	110.46
$GdPO_4$			$PuPO_4$	
a, Å	6.644	6.611	6.7641	6.7433
b, Å	6.841	6.855	6.9984	7.0040
c, Å	6.328	6.350	6.4536	6.4620
β,°	103.97	103.97	103.64	103.64

Experimental structural data are from Ni et al. (2005), bulk and shear moduli for $LaPO_4$—from (Morgan and Marshall 1995), and S_{298} values—from (Thiriet 2005)

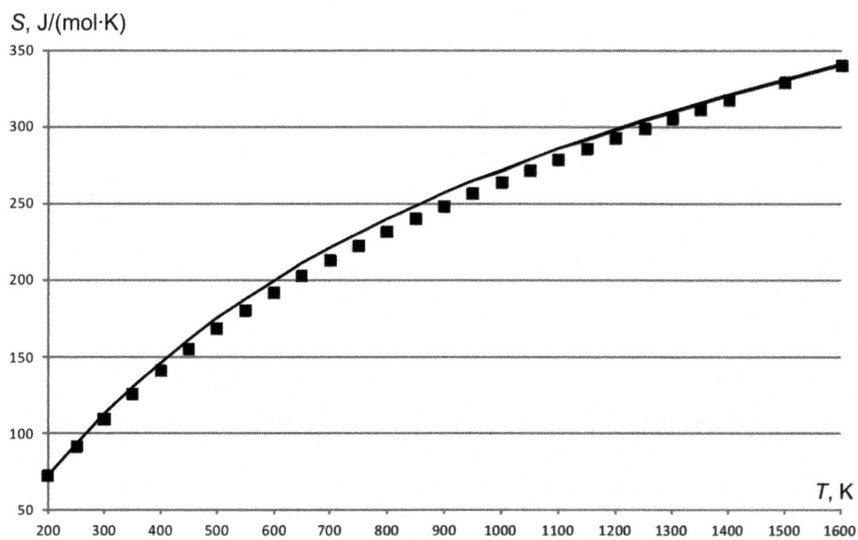

Fig. 2 $LaPO_4$ calculated entropy temperature dependence in 200 K–1600 K temperature interval. Squares—experimental data from (Gavrichev et al. 2008), solid line—calculation results

Table 3 The results of xenotime crystal structure modeling in comparison with the experimental data from (Ni et al. 2005)

Xenotime	Unit cell parameter	Experiment, Å	Calculation, Å
$TbPO_4$	*a*	6.9319	6.9320
	c	6.0606	6.0606
$DyPO_4$	*a*	6.9051	6.9048
	c	6.0384	6.0384
YPO_4	*a*	6.8947	6.8949
	c	6.0276	6.0276
$HoPO_4$	*a*	6.8772	6.8776
	c	6.0176	6.0175
$ErPO_4$	*a*	6.8508	6.8514
	c	5.9968	5.9967
$TmPO_4$	*a*	6.8294	6.8302
	c	5.9798	5.9796
$YbPO_4$	*a*	6.8093	6.8089
	c	5.9639	5.9637
$LuPO_4$	*a*	6.7828	6.7855
	c	5.9467	5.9465

Conclusions

We have developed the consistent set of the partially ionic potentials for interatomic interaction with effective ionic charges $q(\mathrm{Me}) = 1.6\ e_0$, $q(\mathrm{P}) = 1.2\ e_0$, $q(\mathrm{O}) = -0.7\ e_0$ for the modeling of monazites of all lanthanides, yttrium and plutonium (III). This set provides an excellent description of monazite and xenotime crystal structures and allows available experimental S_0 (298 K), bulk and shear moduli values to be reproduced with a high accuracy. This set can be used to evaluate the thermodynamic stability of monazite and xenotime based phases of variable composition as well as for molecular dynamic calculations of their radiation resistance.

References

Boakye EE. et al. Synthesis and phase composition of lanthanide orthophosphate nanoparticles $LnPO_4$ (Ln = La, Gd, Tb, Dy, Y) and solid solutions for fiber coatings. J. Am. Ceramic Soc. 2008;91(12):3841–3849.

Eremin NN., Marchenko EI, Ulanova AS. Razrabotka modeli mezhatomnykh potentsialov i atomisticheskoye modelirovaniye kristallicheskikh struktur monatsitov legkikh redkozemelnykh elementov. Vestnik filiala MGU v Dushanbe. 2017; 1(3): 94–107. http://msu.tj/files/vestnik/vestnik_en_3_1_2017.pdf.

Ewing RC., Weber WJ., Clinard FW. Radiation effects in nuclear waste forms for high-level radioactive waste. Prog. Nuclear Energy. 1995;29:63.

Gale JD. Rohl AL. The General Utility Lattice Program (gulp). Mol. Simul. 2003;29(5):291–341.

Gavrichev KS, Ryumin MA, Tyurin AV, Gurevich VM, Komissarova LN. Refined heat capacity of $LaPO_4$ in the temperature range 0–1600 K. Thermochimica Acta. 2008;474:47–51.

Grechanovsky AE., Urusov VS., Eremin NN. Molecular dynamics study of self-radiation damage in mineral matrices. J. Struct. Chem. 2016;57(6):1243–1262. https://doi.org/10.1134/S0022476616060263.

Jardin R. et al. The high-temperature behaviour of $PuPO_4$ monazite and some other related compounds, J. Nucl. Mat. 2008;378:167–171. https://doi.org/10.1016/j.jnucmat.2008.05.011.

Morgan PED, Marshall DB. Ceramic composites of monazite and alumina. J. Am. Ceram. Soc. 1995;78(6):1553–1563.

Ni Y, Hughes JM, Mariano AM. Crystal chemistry of the monazite and xenotime structures. Amer. Miner. 2005;80:21–26.

Thiriet C.et al. The low temperature heat capacity of $LaPO_4$ and $GdPO_4$, the thermodynamic functions of the monazite-type $LnPO_4$ series. J. Chem. Thermodynamics. 2005;37: 131–139.

Symmetry Point Groups and Topological Entropies of Polyatomic Convex Clusters

Yury L. Voytekhovsky

Abstract The H_S topological entropy of all 2907 convex 4- to 9-atomic polyhedral clusters has been calculated from the point of different symmetrical positions of the atoms. It shows a general trend to drop with growing symmetry of clusters with many local exceptions. The H_V topological entropy of the same clusters has been calculated from the point of different valences (chemical bonds) of the atoms. It classifies the variety of clusters in more details. The relationships between the two entropies are discussed.

Keywords Polyhedral convex cluster · Symmetry point group · Topological entropy

Introduction

A general theory of convex polyhedra is given in (Grünbaum 1967). In the series of papers we consider a special problem on the combinatorial variety of convex n-hedra rapidly growing with n. In (Voytekhovsky and Apatity 2008; Voytekhovsky 2014) all combinatorial types of convex 4- to 12-hedra and simple (only 3 facets/edges meet at each vertex) 13- to 16-hedra have been enumerated and characterized by automorphism group orders (a.g.o.'s) and symmetry point groups (s.p.g.'s). Asymptotically, almost all n-hedra (and n-acra, *i.e.* n-vertex polyhedra, because of duality) seem to be combinatorially asymmetric (i.e. primitive triclinic). A method of naming any convex n-acron by a numerical code arising from the adjacency matrix of its edge graph has been suggested in (Voytekhovsky 2016). The combinatorial types of convex n-acra with the $\min_n$ and $\max_n$ names and some asymptotical (as $n \to \infty$) relations between the latter have been found in (Voytekhovsky 2017a, b). Here we consider the topological entropies as additional characteristics of convex n-acra.

Y. L. Voytekhovsky (✉)
Saint-Petersburg Mining University, 2, 21st Line, 199106 Saint-Petersburg, Russia
e-mail: woyt@geoksc.apatity.ru

S. Votyakov et al. (eds.), *Minerals: Structure, Properties, Methods of Investigation*, Springer Proceedings in Earth and Environmental Sciences,
https://doi.org/10.1007/978-3-030-00925-0_39

Obviously, convex n-acra can also be interpreted as atomic clusters with atoms located in vertices and the edges considered as chemical bonds. It is interesting to know, if the topological entropy correlates with the a.g.o.'s of atomic clusters. If so, it can be taken as a continuous approximant of the discrete s.p.g.'s. On the other hand, there are convex n-acra with different numbers of edges, as a whole, and different valences of the vertices, in particular. It follows from the general theory of systems that their complexity mostly depends on relationships between the elements (e.g., chemical bonds of the atoms) rather than on the number of the elements themselves (e.g., the number of vertices equivalent under the automorphism group). Does the topological entropy effectively fix the complexity of the convex n-atomic clusters? The paper discusses these questions.

Statistical Entropy and Its Properties

The concept of entropy has been suggested in thermodynamics by Clausius in 1865. Its statistical interpretation has been performed by Boltzmann in 1872. Afterwards, Shannon (1948) and Halphen (1957) have independently found the formula

$$H = -\sum_{i=1}^{n} p_i \log p_i$$

within the framework of the mathematical theory of communication and population statistics, respectively. In any case, this is the convolution of some distribution of probabilities p_i with an obvious restriction $p_1 + \ldots + p_n = 1$. The H function is bounded by $H_{\min} = 0$, if one of $p_i = 1$ (the others are 0's), and $H_{\max} = \log n$, if any $p_i = 1/n$. Its schematic graphs for two (arcs with $H_{\max} = \lg 2$) and three (surface with $H_{\max} = \lg 3$) probabilities are given over the barycentric diagram $p_1 + p_2 + p_3 = 1$ in Fig. 1. It is easy to see that small changes of the probabilities p_i at the corners of a diagram affect big changes of H, while the same changes of p_i in the central part of a diagram do not affect H that much.

Entropy H_S of Convex n-acra

The above formula allows us to characterize the topology of convex n-acra in different ways. For example, let us define the entropy H_S from the point of different symmetrical positions of the vertices. In this case, for any n-acron, numbers n_i of vertices in different symmetrical positions and, afterwards, the probabilities $p_i = n_i/n = n_i/\Sigma n_i$ should be calculated. It follows from the above that, for given n, $H_{S,\max}$ is attained by n-acra with any two vertices non-equivalent under the automorphism group, i.e. by combinatorially asymmetric n-acra (i.e. those, which cannot be

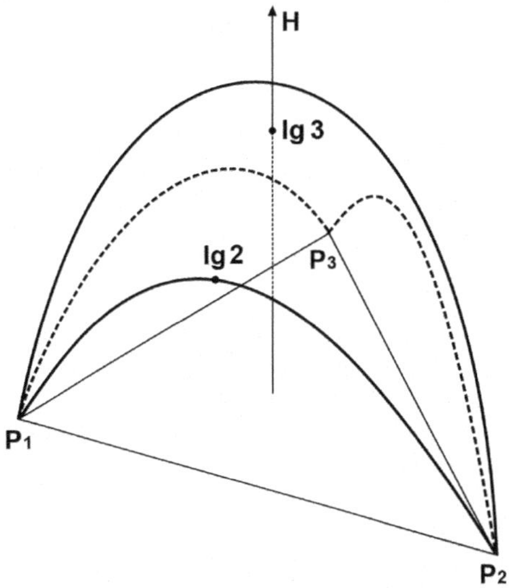

Fig. 1 The graph of $H(p_1, p_2, p_3)$. Hereinafter in nites, as decimal logarithms are used

transformed to symmetric convex n-acra by the continuous transformations), n 7. At the same time, $H_{S,\min}$ is attained by vertex-transitive n-acra. These include regular (Platonic) and semi-regular (Archimedian) polyhedra, as well as the infinite series of prisms and antiprisms. The further question is how H_S depends on a.g.o.'s and s.p.g.'s of n-acra with growing n. The edge graphs of all 2907 convex 4- to 9-acra and their a.g.o. and s.p.g. statistics (Table 1) have been extracted from (Voytekhovsky and Apatity 2008).

The lexicographically ordered sequences of the vertices numbers in different symmetry positions for all convex 4- to 9-acra and related s.p.g.'s are as follows. **4-acron**. 4: $\bar{4}3m$(tetrahedron). **5-acra**. 14: $4\ mm$ (tetragonal pyramid), 23: $\bar{6}m2$ (trigonal bipyramid). **6-acra**. 1122: m, 15: $5\ m$ (pentagonal pyramid), 222: 2, $mm2$, 6: $\bar{6}m2$ (trigonal prizm), $m\bar{3}m$ (octahedron). **7-acra**. 1111111: 1, 11122: m, 1222: 2, m, $mm2$, 124: $mm2$, 133: $3\ m$, 16: $6\ mm$, 25: $\overline{10}m2$. **8-acra**. 11111111: 1, 111122: m, 11222: 2, m, 1124: $mm2$, 1133: $3\ m$, 2222: 2, $mm2$, 17: $7\ m$, 224: $mm2$, $2/m$, 26: $\bar{3}m$, $\bar{6}m2$, $6/mmm$, 44: mmm, $\bar{4}2m$, $\bar{4}3m$, 8: $\bar{8}2m$, $m\bar{3}m$. **9-acra**. 111111111: 1, 1111122: m, 111222: m, 12222: 2, m, $mm2$, 1224: $mm2$, 144: $mm2$, $4\ mm$, 18: $8\ mm$, 27: $\overline{14}m2$, 333: 3, $3\ m$, 36: $3\ m$, $\bar{6}m2$. The data have been used to calculate H_S (Fig. 2). The entropy H_S shows the general trend: the higher a.g.o., the lower H_S. But, there are a lot of local exceptions. The two 5-acra contradict to the trend. Some n-acra with the same a.g.o.'s (and even s.p.g.'s) have different H_S, while some n-acra with the same H_S have different s.p.g.'s and even a.g.o.'s. Moreover, some n-acra with higher a.g.o.'s also have higher H_S.

Table 1 Symmetry statistics of convex 4- to 9-acra, V – vertices, F – facets

п.г.а.	т.г.с.	V	4	5		6				7					8							9							
		F	4	5	6	5	6	7	8	6	7	8	9	10	6	7	8	9	10	11	12	7	8	9	10	11	12	13	14
1	1										2	3	2			3	22	48	44	21	2	2	48	237	533	662	449	164	16
2	2						1				1	2		1		2	5	5	8	1	1		5	8	22	10	25	3	7
	m							1		1	2	4	4			4	11	17	19	12	4	4	17	48	71	87	74	46	18
3	3																									1			
4	mm2							1	1	1		2	2	1	1	2		1	2	4	3	2	1		7	5	10	4	5
	2/m																		1										
6	3m										2			2				2					2			2			1
8	4mm			1																				2				2	
	mmm																1												
	$\bar{4}2m$																2		1		1								
10	5m						1																						
12	$\bar{6}m2$				1	1												1					1			1			2
	$\bar{3}m$																				1								
	6mm										1																		
14	7m																1												
16	$\bar{8}2m$																		1										
	8mm																							1					
20	$\overline{10}m2$													1															
24	6/mmm																				1								
	$\bar{4}3m$		1																		1								
28	$\overline{14}m2$																												1
48	$m\bar{3}m$								1						1														
Σ			1	1	1	1	2	2	2	2	8	11	8	5	2	11	42	74	76	38	14	8	74	296	633	768	558	219	50
			1	2		7				34					257							2606							

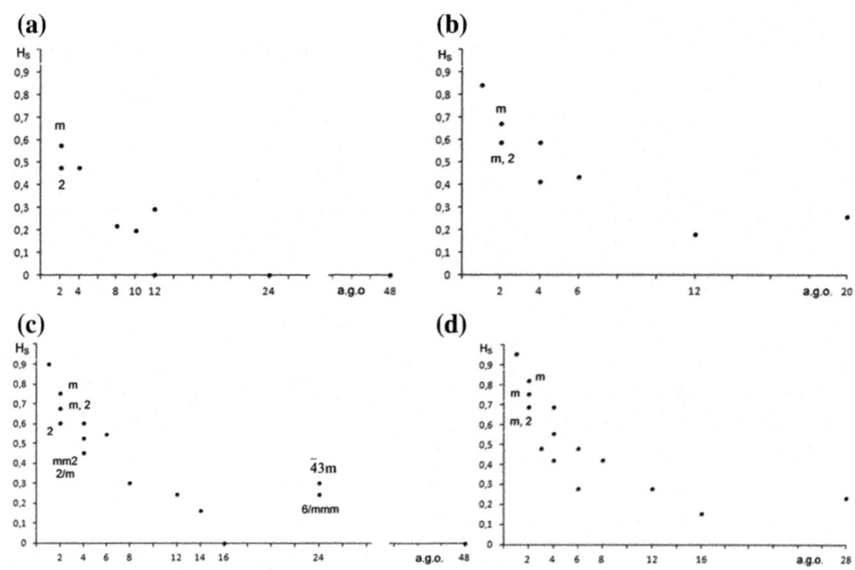

Fig. 2 Entropy H_S of convex 4- to 6-acra (A, 10 in total), 7-acra (B, 34), 8-acra (C, 257), and 9-acra (D, 2606) versus a.g.o. The s.p.g.'s are given to the dots if they do not follow from the Table 1

Entropy H_V of Convex *n*-acra

It seems that the entropy H_S characterizes a 'disorder' more than a 'complexity' of *n*-acra. The first parameter is quite well characterized by s.p.g's. From this point of view, combinatorially asymmetric *n*-acra are maximum disordered, while *n*-acra with $H_S = 0$ are maximum ordered. At the same time, there are *n*-acra with the same s.p.g.'s, but vertices of different valences. We presume them to be of different complexity, which is not fixed by H_S. To distinguish between them, we suggest the entropy H_V considering different valences of vertices of *n*-acra: $p_i = v_i/n$. For example, there are 7 combinatorially asymmetric 7-acra (Table 1) of the same entropy $H_S = H_{\max} = lg\,7$. But almost all of them are unique as for valences of their vertices (Voytekhovsky 2016, Fig. 3): 232, 3211, 331 (two 7-acra), 412, 43, and 511. Hereinafter each sequence records the numbers v_i of *i*-valent vertices from v_3 to $v_{\max}$. Obviously, the entropy H_V differs for the six classes. In the same way, the combinatorially asymmetric 8-acra (140 in total, Table 1) can be divided into 31 classes (Voytekhovsky and Apatity 2008). But, as 0's and permutations of the indexes v_i do not change H_V, combinatorially asymmetric 8-acra can be divided into 12 classes of different H_V. The lexicographically ordered sequences of the numbers of the vertices with different valences for convex 4- to 9-acra and related s.p.g.'s have been extracted from (Voytekhovsky and Apatity 2008) and are as follows.

4-acron. 4: $\bar{4}3m$(tetrahedron). **5-acra**. 23: $\bar{6}m2$ (trigonal bipyramid), 41: *4 mm* (tetragonal pyramid). **6-acra**. 06: $m\bar{3}m$ (octahedron), 222: *mm2*, 24: *mm2*, 321: *m*, 42: *2*, 501: *5 m* (pentagonal pyramid), 6: $\bar{6}m2$ (trigonal prizm). **7-acra**. 052: $\overline{10}m2$, 133: *3 m*, 151: *m*, 2221: 2, 2302: *mm2*, 232: *1*, *mm2*, 2401: *mm2*, 25: 2, *mm2*, 3031: *3 m*, 313: *m*, 3211: *1*, *m*, 331: *1*, *m*, 412: *1*, 2, 4201: *m*, *mm2*, 43: *1*, 2, *m*, *3 m*, 511: *1*, *m*, 6001: *6 mm*, 61: *m*, *mm2*.

8-acra. 044: $\bar{4}2m$, 0602: *6/mmm*, 062: *mm2*, 08: $\bar{8}2m$, 1331: *m*, 1412: *m*, 143: *1*, *m*, 1511: *1*, *m*, 161: *1*, *m*, 206: $\bar{3}m$, 2141: *mm2*, 2222: 2, *mm2*, 22301: *m*, 224: *1*, 2, *m*, *mm2*, 23111: *1*, 2321: *1*, *m*, 24002: *mm2*, 2402: *1*, *mm2*, 24101: *1*, 242: *1*, 2, *m*, *2/m*, 2501: *1*, *m*, *mm2*, 26: 2, *m*, $\bar{6}m2$, 3113: *m*, 31211: *1*, 3131: *1*, *m*, 3212: *1*, *m*, 32201: *1*, *m*, 323: *1*, *m*, 33011; *1*, 3311: *1*, 34001: *m*, 341: *1*, *m*, 4004: $\bar{4}3m$, 4022: *1*, *mm2*, 40301: *m*, 404: *1*, *mm2*, $\bar{4}2m$, 4121: *1*, *m*, 4202: *1*, 2, 42101: *1*, *m*, 422: *1*, 2, *m*, *mm2*, 4301: *1*, *m*, *3 m*, 44: *1*, 2, *m*, *mmm*, $\bar{4}2m$, 503: *1*, *m*, *3 m*, 5111: *1*, 52001: *m*, 521: *1*, *m*, 602: 2, *m*, 6101: *1*, 62: *1*, 2, *m*, *mm2*, 70001: *7 m*, 701: *m*, 8: *mm2*, $m\bar{3}m$.

9-acra. 036: $\bar{6}m2$, 0441: *mm2*, 0522: *mm2*, 054: *m*, *4 mm*, 0603: $\bar{6}m2$, 0621: *1*, *mm2*, 07002: $\overline{14}m2$, 072: 2, *mm2*, 0801: *mm2*, 09: $\bar{6}m2$, 1251: *m*, 1332: *1*, 135: *1*, *m*, 1413: *m*, 14211: *1*, *m*, 1431; *1*, *m*, 15102: *m*, 1512: *1*, *m*, 15201: *1*, *m*, 153: *1*, *m*, 16011: *1*, *m*, 1611: *1*, *m*, 17001: *m*, 171: *1*, *m*, 2142: 2, *m*, 21501: *m*, 216: *m*, 2223: 2, *m*, 22311: *1*, *m*, 224001: 2, 2241: *1*, *m*, *mm2*, 2304: *mm2*, 23121: *1*, *m*, 23202: 2, *m*, 232101: *1*, *m*, 2322: *1*, 2, *m*, *mm2*, 23301: *1*, *m*, 234: *1*, 2, *m*, *mm2*, 240201: 2, 2403: *1*, *m*, 241011: *1*, 24111: *1*, 242001: *1*, 2, *m*, 2421: *1*, 2, *m*, 250002: *mm2*, 25002: *1*, *mm2*, 250101: *1*, 2502: *1*, 2, *m*, 25101: *1*, *m*, 252: *1*, 2, *m*, 260001: *mm2*,

2601: *1*, *2*, *m*, *mm2*, 27: *1*, *2*, *m*, *mm2*, 3033: *m*, *3 m*, 3051: *1*, *m*, 31221: *1*, *m*, 31302: *1*, 313101: *1*, 3132: *1*, 31401: *1*, *m*, 315: *1*, *m*, 32031: *1*, *m*, 32112: *1*, *m*, 321201: *1*, *m*, 3213: *1*, *m*, 322011: *1*, *m*, 32211: *1*, *m*, 323001: *1*, *m*, 3231: *1*, *m*, 33021: *1*, *m*, 33102: *1*, *m*, 331101: *1*, 3312: *1*, *m*, 33201: *1*, *m*, 333: *1*, *m*, *3*, *3 m*, 340011: *1*, *m*, 34011: *1*, *m*, 341001: *1*, *m*, 3411: *1*, *m*, 35001: *1*, *m*, 351: *1*, *m*, 40212: *2*, 402201: *mm2*, 4023: *2*, *m*, 40311: *1*, *m*, 404001: *4 mm*, 4041: *1*, *2*, *m*, *mm2*, 41022: *m*, 4104: *1*, *m*, 41121: *1*, *m*, 41202: *1*, *m*, 412101: *1*, 4122: *1*, *2*, *m*, *mm2*, 41301: *1*, *m*, 414: *1*, *2*, *m*, *mm2*, 4203: *1*, *2*, *m*, *mm2*, 42111: *1*, 422001: *1*, *2*, *m*, 4221: *1*, *2*, *m*, *mm2*, 43002: *1*, *2*, 430101; *1*, *m*, 4302: *1*, *2*, *m*, 43101: *1*, *m*, 432: *1*, *2*, *m*, *mm2*, 440001: *1*, *m*, *mm2*, 4401: *1*, *2*, *m*, *mm2*, 45: *1*, *2*, *m*, *4 mm*, 50031: *m*, 5013: *1*, *m*, 50211: *1*, *m*, 503001: *m*, 5031: *1*, *m*, 5112: *1*, *m*, 51201: *1*, *m*, 513: *1*, *m*, 52011: *1*, *m*, 521001: *1*, *m*, 5211: *1*, *m*, 53001: *1*, *m*, 531: *1*, *m*, 6021: *1*, *2*, *m*, 6102: *1*, *2*, *m*, 61101: *1*, 612: *1*, *2*, *m*, 620001: *m*, *mm2*, 6201: *1*, *2*, *m*, 63: *1*, *2*, *m*, *3 m*, $\bar{6}m2$, 7011: *1*, *m*, 71001: *1*, *m*, 711: *1*, *m*, 800001: *8 mm*, 8001: *m*, *mm2*, 81: *1*, *m*, *mm2*. The data have been used to calculate the entropy H_V (Fig. 3). The main feature of H_V is that it classifies the variety of convex 4- to 9-acra in more details than H_S with $H_S \geq H_V$ for any *n* and s.p.g.

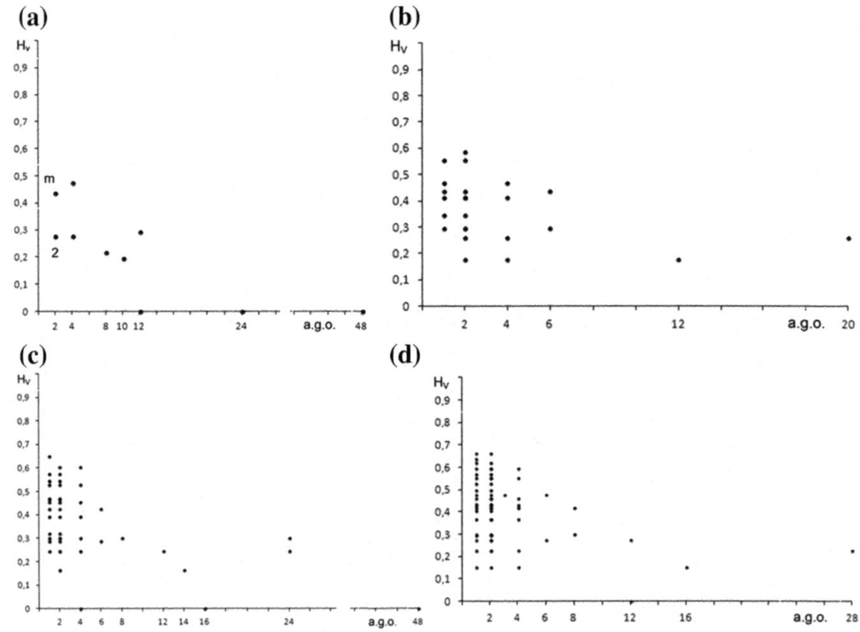

Fig. 3 Entropy H_V for the same classes of convex *n*-acra as in Fig. 2

Discussion

The relationships between the entropies H_S and H_V in a general case can be formulated in two statements.

Statement 1. $H_S \geq H_V$ for any convex n-acron, *i.e.* for any n and s.p.g.

Proof. First of all, the statement is true for all convex 4- to 9-acra (Table 2). $H_S > H_V$ mostly for n-acra of low symmetry, while $H_S = H_V$ mostly for n-acra of high symmetry with the transition classes of a.g.o.'s from 2 to 12. Careful consideration of n-acra has allowed to establish the following. Let us take any n-acron with vertices of different symmetry positions. Obviously, vertices equivalent under the automorphism group have the same valences. The question is if the non-equivalent vertices have different valences or not. $H_S = H_V$ if so, $H_S > H_V$ if not. In the latter case, decrease in the variety of valences is resulted in reduction of H_V (if compared with H_S) in accordance with the general properties of the entropy H. The rigorous proof is given in Appendix 1.

Table 2 Relationships between H_S and H_V for convex 4- to 9-acra

п.г.а.	т.г.с. V	4	5		6				7					8							9							
	F	4	5	6	5	6	7	8	6	7	8	9	10	6	7	8	9	10	11	12	7	8	9	10	11	12	13	14
1	1									>	>	>			>	>	>	>	>	>	>	>	>	>	>	>	>	>
2	2					>				>	>		=		>	>	>	>	>	=		>	>	>	>	>	>	>
	m						>		>	>	>	>			>	>	>	>	>	>	>	>	>	>	>	>	>	>
3	3																								=			
4	mm2						>	=	>		≥	≥	>	>	>		=	>	>	≥	>	>		>	≥	≥	≥	≥
	2/m																	=										
6	3m									>			=				>					≥			=			=
8	4mm		=																				>				≥	
	mmm															=												
	$\bar{4}2m$															=		=		=								
10	5m					=																						
12	$\bar{6}m2$			=	=												=					=			>			=
	$\bar{3}m$																			=								
	6mm									=																		
14	7m															=												
16	$\bar{8}2m$																	=										
	8mm																						=					
20	$\overline{10}m2$												=															
24	6/mmm																			=								
	$\bar{4}3m$	=																		=								
28	$\overline{14}m2$																											=
48	$m\bar{3}m$							=						=														

Note > means $H_S > H_V$ for all n-acra in the class; = means $H_S = H_V$ for all n-acra in the class; ≥ means both types of n-acra are in the class

Statement 2. Entropy H_V never reaches maximum $lg\ n$.

Proof. In other words, we should prove that there are no n-acra with all vertices of different valences. Actually, even more strict statement takes place: any convex n-acron has at least 4, or 3 and 2, or 3 pairs of vertices of the same valences. The rigorous proof is given in Appendix 2.

Conclusions

It follows from general considerations that the H_S topological entropy is hardly interpreted in initial terms (in our case, a.g.o.'s and s.p.g.'s of convex n-acra even for given n). The H_S value fixes any n-acron on a scale between $H_{S,\min} = 0$ and $H_{S,\max} = lg\ n$. But, small changes of the probabilities p_i at the corners of a field of definition (Fig. 1) affect big changes of H_S, while the same changes of p_i in the central part of the field of definition do not affect H_S that much. $H_{S,\max} = lg\ n$ is attained by combinatorially asymmetric convex n-acra ($n \geq 7$). $H_{\min} = 0$ is attained by regular and semi-regular n-acra (all the cases are enumerated) as well as the infinite series of prisms and antiprisms (even $n \geq 4$). Between the two bounds, the H_S entropy of convex 4- to 9-acra shows a general trend: the higher a.g.o., the lower H_S. But, there are a lot of exceptions.

The entropy H_S characterizes a 'disorder' rather than a 'complexity' of convex n-acra. The first one is quite well characterized by s.p.g's. The second one should distinguish n-acra of the same s.p.g. and different numbers of edges, for example, the overwhelming majority of combinatorially asymmetric n-acra. To do this, the topological entropy H_V is suggested, which considers the valences of vertices of n-acra. It classifies the variety of convex 4- to 9-acra in more details. It is proved that H_V can reach 0 as minimum (for regular and semi-regular polyhedra, as well as the infinite series of prisms and antiprisms), but never $lg\ n$ as maximum, because there are no convex n-acra with all vertices of different valences. It is also proved that $H_S \geq H_V$ for any convex n-acron. $H_S = H_V$ if the vertices non-equivalent under the automorphism group also have different valences, and $H_S > H_V$ if not.

Appendix 1

Consider the sequences of numbers v_i of different valences for convex 5- to 9-acra (related H_V are in parentheses) ordered by the algorithm to follow: … p … q … (H_1) → … $p - 1$ … $q + 1$ … (H_2), where $1 \leq p \leq q$. **5-acra**. 23 (0,292) → 14 (0,217). **6-acra**. (Hereinafter 0's and permutations of v_i are omitted in the sequences as they do not affect H_V.) 222 (0,477) → 123 (0,439) → 24 (0,276) → 15 (0,196) → 6 (0). **7-acra**. The main trend: 1222 (0,587) → 1123 (0,555) → 223 (0,469) → 133 (0,436) → 124 (0,415) → 34 (0,297) → 25 (0,260) → 16 (0,178); offshoot: 124 (0,415) → 115 (0,346). **8-acra**. The main trend: 11123

(0,649) → 1223 (0,574) → 1133 (0,545) → 233 (0,470) → 224 (0,452) → 134 (0,423) → 44 (0,301) → 35 (0,287) → 26 (0,244) → 17 (0,164) → 8 (0); offshoots: 2222 (0,602) → 1223 (0,574); 1133 (0,545) → 1124 (0,527) → 1115 (0,466); and 134 (0,423) → 125 (0,391) → 116 (0,319). **9-acra**. The main trend: 11223 (0,661) → 2223 (0,595) → 1233 (0,569) → 333 (0,477) → 234 (0,461) → 144 (0,419) → 135 (0,407) → 45 (0,298) → 36 (0,276) → 27 (0,230) → 18 (0,152) → 9 (0); offshoots: 11223 (0,661) → 11133 (0,636) → 11124 (0,620); 1233 (0,569) → 1224 (0,553) → 1134 (0,528) → 1125 (0,499) → 1116 (0,435); 234 (0,461) → 225 (0,432); and 135 (0,407) → 126 (0,369) → 117 (0,297).

The above sequences could be ordered in different ways. We have followed the rule of a "slow down" to include as many sequences in the main trends, as possible. With no exception, the above algorithm causes $H_1 > H_2$. To prove the inequality in a general case (for any $1 \leq p \leq q$ and n), we should show that

$$-(p/n)\ln(p/n)-(q/n)\ln(q/n) > -[(p-1)/n]\ln[(p-1)/n]-[(q+1)/n]\ln[(q+1)/n].$$

If $p \to 1$, then $[(p-1)/n]\ ln\ [(p-1)/n] \to 0$. Hence, for $p = 1$ we get an obvious inequality $(q + 1)\ (1 + 1/q)^q > 1$. For $2 \leq p \leq q$ we should prove the inequality $p^p/(p-1)^{p-1} < (q+1)^{q+1}/q^q = f(q)$. Consider $f(q)$ as a continuous function and use a logarithmic derivative $df/dq = ln(1 + 1/q) \times (q+1)^{q+1}/q^q > 0$. That is, $f(q)$ grows with growing $q = p, p + 1, p + 2$, etc.

Let us show that the above inequality takes place even for the minimum argument $q = p$, i.e. $p^p/(p-1)^{p-1} < (p+1)^{p+1}/p^p$ or $1 < (p+1)^{p+1}\ (p-1)^{p-1}/p^{2p} = f(p)$. Again, consider $f(p)$ as a continuous function and use a logarithmic derivative $df/dp = ln(1 - 1/p^2) \times (p+1)^{p+1}\ (p-1)^{p-1}/p^{2p} < 0$. That is, $f(p)$ drops with growing $p = 2, 3, 4$, etc. Indeed, $f(2) = 1.6875$, $f(3) = 1.404\ldots$, $f(4) = 1.287\ldots$, $f(5) = 1.223\ldots$, $f(6) = 1.182\ldots$ Nevertheless, if $p \to \infty$, then $\lim f(p) = lim\ (p+1)^{p+1}\ (p-1)^{p-1}/p^{2p} = lim\ (1 + 1/p)^p\ (1 - 1/p)^p\ [1 + 2/(p-1)] = e \times e^{-1}$ $1 = 1$. That is, $f(p)$ tends to 1 from above, i.e. $f(p) > 1$ for any p. Thus, $H_1 > H_2$ for any $1 \leq p \leq q$ and n.

Appendix 2

Assume that a convex polyhedron exists with all the facets being different (*i.e.* of different number of edges). Let us consider its Schlegel diagram on a facet with a maximum number of edges (k-lateral facet, Fig. 4a). More precisely, let us consider how its corona (*i.e.* a set of facets touching it edge-to-edge) is built. After $(k-1)$-, $(k-2)$- … 4-, and 3-lateral facets being attached to k-lateral one in any order, 3 more edges are free. And we can conclude that our initial assumption is wrong. Obviously, in the above case, 3 same (*i.e.* of the same number of edges), or 2 and 1,

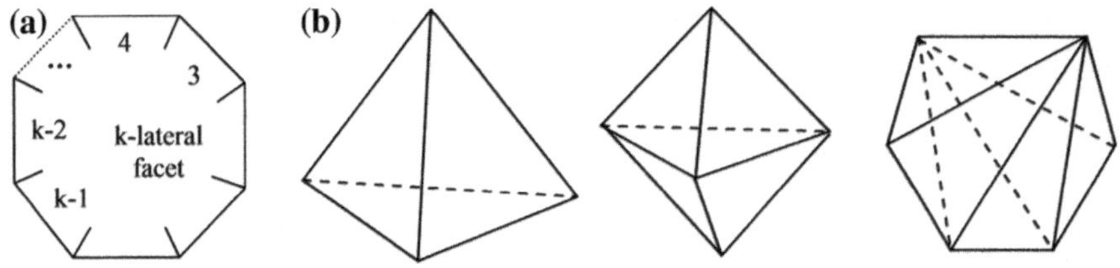

Fig. 4 The Schlegel diagram on *k*-lateral facet (**a**) and the limit convex 4-, 5-, and 6-acra (**b**)

or 3 different facets can be attached to them. As any (*i.e.* 3- to *k*-lateral) facet is used, 4 same, or 3 and 2, or 3 pairs of same facets will result on a polyhedron.

Now, assume that not all $k - 3$ types of the facets are submitted in the corona. After the facets of each type being attached by one to *k*-lateral facet, more than 3 edges are free. To complete the corona, one should choose more than 3 facets from their less than before $(k - 3)$ variety. Obviously, both reasons may not reduce the frequency of occurrence of the facets in the corona: 4 same, or 3 and 2, or 3 pairs of same facets. Finally, because of the duality, any convex *n*-acron has at least 4, or 3 and 2, or 3 pairs of vertices of same valences. The limit cases are: a tetrahedron, a trigonal dipyramid, and a 6-acron of *mm2* s.p.g. (Figure 4b).

References

Grünbaum B. Convex Polytopes. New York: Springer; 1967.

Halphen E. L'analyse intrinsèque des distributions de probabilité. Publ Inst Stat Univ Paris. 1957; 6; 2; 77–159.

Shannon CE. A mathematical theory of communication. Bell Syst Tech J. 1948; 27; 379–423; 623–656.

Voytekhovsky YL. How to name and order convex polyhedra. Acta Cryst. 2016; A72; 5; 582–585. https://doi.org/10.1107/S2053273316010846.

Voytekhovsky YL. Convex polyhedra with minimum and maximum names. Acta Cryst. 2017; A73; 3; 271–273. https://doi.org/10.1107/S2053273317004053.

Voytekhovsky YL. Accelerated scattering of convex polyhedra. Acta Cryst. 2017; A73; 5; 423–425. https://doi.org/10.1107/S2053273317009196.

Voytekhovsky YL. E.S. Fedorov's algorithm of the generation of the combinatorial diversity of convex polyhedra: the latest results and applications. J Struct Chem. 2014; 55; 7; 1293–1307.

Voytekhovsky YL, Stepenshchikov DG. Kombinatornaya kristallomorfologiya. Kn. 4: Vypuklye polyedry. T. 1: 4- … 12-edry. Apatity: KSC RAS; 2008. [Voytekhovsky YL, Stepenshchikov DG. Combinatorial Crystal Morphology. Book 4: Convex Polyhedra. Vol. 1: 4- to 12-hedra. Apatity: KSC RAS; 2008. (In Russ.)].

System of Mineralogy Revisited

Yury L. Voytekhovsky

Abstract What is the system of mineralogy? That is, what is the best mathematical idea to describe the growing diversity of mineral species? Some biomineral analogues (the concepts of individuals, species, ontogeny, phylogeny, the system of species) originating from Linnaeus's "Systema Naturae" are briefly considered. They are due to the common structural and evolutionary approaches in recent biology and mineralogy. Mineralogy is successful in the structural analysis, while biology is used in evolutionary considerations of individuals and species. The concepts of tolerance spaces and structures (with various order relations) are suggested to describe the system of mineralogy.

Keywords Mineral individuals and species · Ontogeny · Phylogeny · System of mineralogy

Introduction

Biology and mineralogy synchronously developed fundamental categories beginning from the determination of individuals and species "Systema Naturae" (Linnaei 1735). Biology had priority in some fields, and mineralogy in others. For instance, the theory of symmetry with its terminology (asymmetry, dissymmetry, antisymmetry, color and curvilinear symmetry, homology, fractal symmetry, etc.) is their shared platform, but is mathematically more developed for crystals due to the nature of their organization (Rozhnov 2013). Both sciences developed the notions of the ideal (ideal crystals of stoichiometric composition, ideal bauplan) and actual individuals and species, distorted in ontogeny under the influence of the environment (heterochrony, heterotopy, allometry (Häckel 1874)), and phylogeny. These concepts were developed in biology and successfully used in mineralogy (Grigoriev 1961; Grigoriev and Zhabin 1975; Zhabin 1979). In the interpretation of the system

Y. L. Voytekhovsky (✉)
Saint-Petersburg Mining University, 2, 21st Line, Saint-Petersburg, Russia199106
e-mail: woyt@geoksc.apatity.ru

S. Votyakov et al. (eds.), *Minerals: Structure, Properties, Methods of Investigation*, Springer Proceedings in Earth and Environmental Sciences,
https://doi.org/10.1007/978-3-030-00925-0_40

of species, biologists and mineralogists have diverged particularly widely, but even here there are meaningful analogues. After the deaths of D.P. Grigoriev (1909–2003), A.G. Zhabin (1934–2007), and N.P. Yushkin (1936–2012), these categories were more intensely discussed in biology and paleontology. To identify these terms more precisely, it seems useful to consider the biomineral analogues, using examples of the categories of individual, species, ontogeny, phylogeny, and the system of species.

Individuals

"In the 18th and partly in 19th centuries, two major concepts were developed related to the objects of this science: the concept of mineral individuals and of mineral species. The former, i.e. the concept of mineral individuals or, as it was termed at that time, "individuals of the mineral world," "mineral entities," and "mineral objects," is formulated approximately as follows: an individual is a natural separation of homogenous chemical matter physically delineated from other objects by natural separating surfaces. A natural crystal is an individual, as it is delineated by crystalline facets, and every mineral grain or any other homogenous separation delineated from the adjacent media by bounding surfaces is an individual as well." (Grigoriev 1961, p. 12). This formulation (also see (Grigoriev and Zhabin 1975; Zhabin 1979)) contains virtually all the problematic issues of a formal definition of a mineral individual. Like 200 years ago, we continue formulating "approximately."

Three issues of the above definition are discussed below. How should we understand "an individual is a natural separation"? What is an appropriate measure for the human role in the origin of mineral individuals? B. V. Chesnokov and his coauthors established several new minerals formed in the spoil tips of the Chelyabinsk Coal Basin. The Commission on New Minerals and Mineral Names (CNMMN) no longer registers such names, because spoil heaps are manmade objects. What about mineral individuals within a human organism? Are these considered as formed in nature, or somewhere else, with human assistance or without? It has been already established that they include pathological and functional minerals. A human organism in the former case can be compared with an autoclave or a test tube, and, in the latter case, with the natural environment. Is the man a part of nature or have humans already torn off the umbilical cord connecting them with the environment? Similarly speaking, should biological clones grown in laboratories be denied the right to have the status of individuals in their own right? Perhaps, the status of individuals of *Homo sapiens* born from a surrogate mother from embryos developed in a laboratory should be reconsidered?

Mineralogists generally agree that the concept of mineral individuals is understood intuitively, without a strict definition. I do not agree with this, because in science, a concept cannot be considered as clearly defined prior to the formal definition. Clarity in science means a completely finalized definition. In mineralogy, definitions are often problematic. We can see that, even at the starting point, a specialized aspect of the problem gradually becomes ethical. How to understand that an "individual is a separation of a homogenous chemical substance…"? What is the degree of homogeneity acceptable for a mineral individual? It is well known how strongly the chemical composition of different zones and pyramids of growth, of which crystals are composed, can vary. "Because crystallization of minerals never happens in an unchangeable environment, … diachronously formed zones have different chemical composition and different crystallized structure … Along with chronologically different parts of a mineral (growth zones), different parts of the crystal are formed during crystallization … due to the deposition of the substance on different facets or different areas of other mineral surfaces (in grains). These parts of an individual F. Becke named … growth pyramids … Different content of admixture in growth pyramids of different facets depends on the effect of the degree of structural similarity of mixing components on the ability of the facet to absorb the admixture and the external conditions, in which the crystal was formed …" (Grigoriev 1961, pp. 56, 60).

Strictly speaking, because the mineral is not homogenous, mineral individuals cannot be described using chemical terminology, although this is widely used. It is known when one growth pyramid is formed by tantalite (Fe, Mn)Ta_2O_6, another by columbite (Mn, Fe)Nb_2O_6, and the list of these analogues can be continued. But how can two mineral species exist together in one individual? And if they survive together in one body with perfect crystallography, perhaps, this indicates the entity of the species? D. P. Grigoriev was puzzled by this problem and constantly mentioned it in his lectures. Such "centaurs" are not known in biology. At least one of the concepts (individual or species) should be redefined in mineralogy to remove this contradiction.

How to understand that an "individual is a segregation …, physically delimited from other individual by natural demarcating surfaces"? Note the logical contradiction of this phrase known even to Aristotle. A mineral individual is defined by a "natural demarcating surface" with another individual, which at the time when the definition is made is also undefined. The segregation, delimitation, the finite nature of a mineral individual should more correctly be defined through a distortion of its immanent feature, for instance, through the enclosed surface of broken chemical bonds of the crystalline lattice. The term "surface" in the physical context is also conventional and the term "crystalline lattice" seems more reliable. However, there is a perceived tendency to embrace by the term "mineral" even those natural forms which have never been crystalline or ceased to be crystalline (Nickel 1995).

In addition, the term "crystalline lattice" loses its meaning as we approach nanosized structures. From which fragment can it be recognized? Does the mineral individual begin from that moment? Apparently, this level can be recognized by *X*-ray methods, which appears to be unbiased. On the other hand, it seems incorrect to define a fundamental concept through an analytical procedure. "The birth of minerals is the first moment of their formation. Of three stages of mineral genesis, the origin is the least studied" (Grigoriev 1961, p. 19). Even after 50 years, this remains true. Formally, it should be correct to distinguish "mineral individuals" with the constitutional attributes; "mineral embryos" that overcame the potential barrier and acquired the ability to grow; and "mineral separations" with uncertain prospects of growth. Thus, the category of a mineral individual at present cannot be considered strictly defined for these reasons and some other reasons. The habit of defining it through a "physically evident," which we inherited from our predecessors is based on the observations of macroscopic samples and leads to a number of contradictions.

Species

The category of a mineral species is also contradictory. "The concept of mineral species was formulated based on the tasks of classification of mineral individuals. The latter are not unlimitedly diverse and, while their properties and shapes vary widely, they coincide to various extents in their composition and crystalline structure. This gave rise to the concept of a mineral species as of a combination of mineral individuals, chemically and structurally identical" (Grigoriev 1961, p. 12). Three issues have to be mentioned in this connection. Classification of mineral species is discussed below. With regard to the type of crystalline structure used as the specific character, it should be said that the spatial group of symmetry is presently accepted as such a character. This is a fundamental basis because of the Fedorov-Schönflies theory and convenient because it gives a large, but reasonable number of classes (230). When necessary, the structural classification can be made more or less detailed.

The question of chemical composition as a specific character is full of conventions. "A mineral is usually a crystalline substance with defined limits of composition (limits are more frequently set up not by the nature, but votes of members of CNMNM MMA – *Auth.*), formed by geological processes … The main criterion of a new mineral is that one of the main positions in the structure should be occupied by a chemical component differing from a chemical component on an equivalent position in a known mineral" (Nickel and Mandarino 1987, p. 53). Here one can see a trend toward an increase in the number of new mineral species, for which it is sufficient to have a new (or predominance of a known – *Auth.*) chemical element at least in one structural position, without recognition of whether or not a solid solution with other minerals exists in this position. However, the latter does not change anything for the "compete hard solutions without regular structural arrangement," including for "solid solution with structural arrangement partial solutions" (Nickel 1992). The "rule of 50%" accepted by CNMMN MMA is bizarre and lethal to the category of the mineral species. The formal increase in the mineral diversity of the planet, still tiny compared to the biological diversity, does not change due in part to the fact that the category of a mineral species has been lowered to the level of an artificial logical construction. A mineral species in the current understanding is not a structural block of the universe. A true mineral species should be separated from others by natural (established in the nature and confirmed by laboratory synthesis and theoretical calculations) disruptions of the ability to mix crystalline solid solutions.

A solution was proposed long ago. "The following fundamental innovations have been proposed: … a new way to study minerals forming so-called series, which involves description of the series as of a single species… Minerals forming continuous change in their properties with a change in their composition are called "a series", and are described as a single species. In these cases, a series is a natural mineralogical unit, as its arbitrary subdivision does not allow interpretation of various parts of the series. Plagioclase and spinel represent examples of such series" (Dana et al. 1946, pp. 8, 13). At present, a radical step forward would be to identify a mineral species as a mineral series delimited only by natural interruptions of mixture. Unfortunately, because of the complexity of some mineral series, the above author made a step backward. "The complexity of some multicomponent systems makes us consider the useful subdivision of large series into particular ones with defined boundaries of composition and their separate description", although this interpretation is "not suitable as a way of understanding of some usual and important series, for example, the calcite group. At the same time, some descriptions are written in a way that emphasizes the connections between the species included in the series… The descriptive mineralogy at present is overloaded with small names, while it would be much more useful for science, if there were fewer names and if the nomenclature was simplified. The systematization of the nomenclature in this way is a natural expression of the concept of variability of the mineral composition, which replaces the older understanding of species as of certain phases in the main composition of generally permanent composition" (Dana et al. 1951, pp. 7–8).

In the Russian literature, the idea of mineral series found support a long time ago. Bonshtedt-Kupletskaya (Bonshtedt-Kupletskaya 1975) noted that the newer the mineral species, the more of them are structural analogues of previously known minerals and intermediary members of the mineral series. Mozgova (1985) described homologous series of structurally regulated phases, which under certain condition are transformed into homogenous regions of compositions of nonstoichiometric compounds (of solid solutions). The definition of the category of a mineral species as a mineral series may be questionable. For example, a series may only include solid solutions, and a series can only be interpreted as solid solutions of isovalent replacement following the scheme "atom for atom, ion for ion" (Kukharenko 1970, p. 203). We are more inclined to accept Bokii's (1975) approach, which did not formalize the use of the concept of isomorphism and, consequently, the concepts of the isomorphic mixture, solid solution, and mineral series. It cannot be overlooked that the diversity of mineral species is becoming more diverse and more connected. It looks more and more like a single system, in certain conditions and due to various mechanisms revealed itself through mineral species. Their synthetic consideration in the most complex assemblages is significant compared to the existing analytical approach, when for the recognition of mineral species, even the continuous series of compositions of isostructural solid solutions are subdivided.

Ontogeny

The ontogeny of mineral individuals and aggregates is a well developed theory studied in Geology Departments of universities (Grigoriev 1961; Grigoriev and Zhabin 1975; Zhabin 1979). There are many biomineral parallels, but also nuances. According to the definition, a mineral species is described as a set of individuals with a crystalline structure of one of 230 groups of Fedorov-Schönflies and of fixed (within certain limits) chemical composition. Quantitative variations of chemical composition, structure, and external shape of a mineral individual depend on the environment in which it was formed, and change in ontogeny. However, this does not lead to evolutionary changes in mineral species. There are no concepts of ancestors and descendants in mineralogy. In this context, concepts of heterochrony, heterotopy, and allometry are not transferrable to mineralogy. With some extent of convention, they are applied for the description of variations of space and time of the origin, growth tempo of parts of a mineral individual. The following definitions need to be accepted. Heterochrony is a change in the time of the origin of parts of the individual. Heterotopy is a change in the place of origin and development of parts of the individuals. Allometry is an irregular growth of parts of the individuals. Variants of allometry: (1) ontogenetic: established in individual ontogeny or in the comparison of different-age specimens of the same species; (2) intraspecific: established by comparison of the individuals of the same age, but different size; (3) interspecific: established by comparison of mean values of characters for individuals of different species.

The following mineralogical analogues to the above biological concepts are recognized. The examples of heterochrony and heterotopy are diachronous and even absent in some structural positions, including appearance of facets of various simple forms on crystals always growing in a changing natural environment (e.g., in cooling melts and solutions, continuously changing chemical composition due to the missing crystallizing phases). Ontogenetic allometry is irregular growth of pyramids of respective facets of crystals, differently oriented in the growth-supporting medium. The intraspecific allometry is irregular growth of pyramids of respective facets in crystals of one mineral species differently oriented in the growth-supporting medium. The interspecific variability is irregular growth of individuals of different species in one paragenesis (Popov and Popova 1996). On the other hand, a great variety of rock textures resulted from different mechanisms of crystallization of melts, distribution of mineral grain cores in space and their growth rates, allows a concept of population allometry, which seems to be currently absent in biology.

Phylogeny

This field is considerably more poorly developed compared to mineral ontogeny; hence, I will discuss it in greater detail to show possible pathways of future research. The current mathematical graph theory has a sufficient base for an effective description of complex mineral series, such as the calcite group. According to Betekhtin (1950), Kostov (1971), Berry et al. (1983), and Essen (1987), natural solid solutions are not restricted in the pairs calcite-rhodochrosite, rhodochrosite-siderite, and siderite-magnesite (Fig. 1, on the left). The comparison of the size of ions suggests (Dana et al. 1951, pp. 174–175) an uninterrupted miscibility in the pairs smithsonite-siderite and smithsonite-magnesite (Fig. 1, center), and based on the observations of the solutions of natural phases in the pair smithsonite-rhodochrosite (Fig. 1, right). The miscibility in the spinel group is shown after Sokolov (1948) in Fig. 2 (left). Irvine (1965) has stated that during synthesis at above 860 °C, a continuous solid solution of magnetite-hercynites appears (Fig. 2, center), above 1400 °C – magnetite-chromite (Fig. 2, on the right). The mineral series of spinelids is expanded due to ganite $ZnAl_2O_4$. There are data on its solid solution with chromites (Weiser 1967; Osokin 1979), spinel, and hercynites (Spry and Scott 1986). These also include rare brunogeierite $GeFe_2O_4$ and cuprospinel $CuFe_2O_4$ (Bonshtedt-Kupletskaya 1975, p. 15). The series of natural garnets is even more interesting. The data of the above authors and Sobolev (1964) on the ratio of

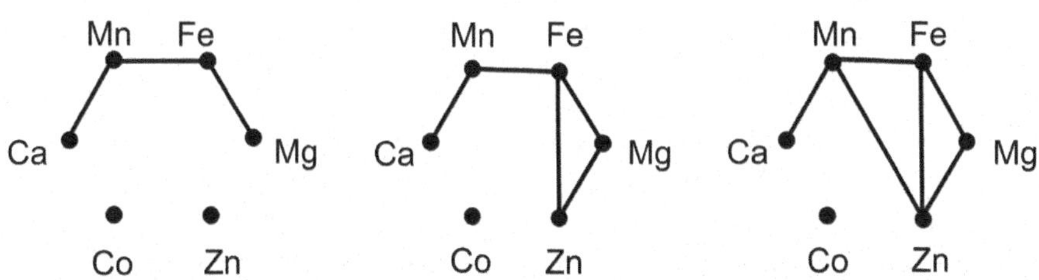

Fig. 1 Ratios of miscibility in the calcite group: (Ca) calcite, (Mn) rhodochrosite, (Fe) siderite, (Mg) magnesite, (Zn) smithsonite, (Co) spherocobaltite. The graph vertices show theoretical limiting compositions while edges show continuous (natural and/or synthetic) series

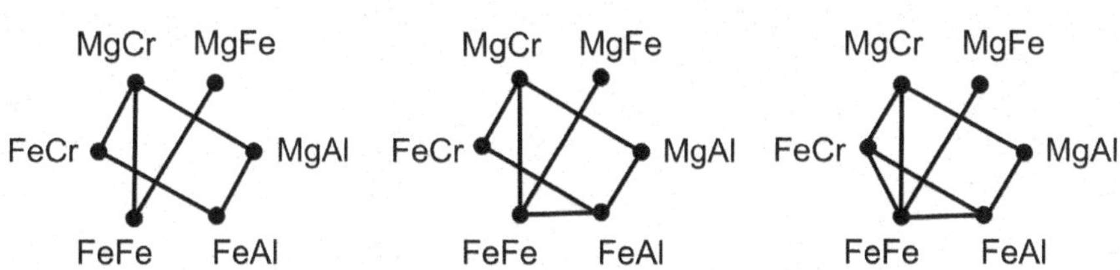

Fig. 2 Ratios of miscibility in the spinel group: (MgCr) picrochromite, (MgFe) magnesioferrite, (MgAl) spinel, (FeAl) hercynite, (FeFe) magnetite, (FeCr) chromite

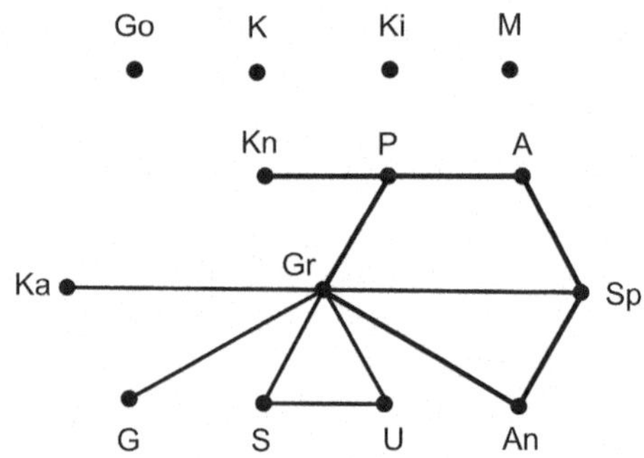

Fig. 3 Ratios of miscibility in the garnet group: (A) almandine, (An) andradite, (G) gibshite, (Go) goldmanite, (Gr) grossular, (K) calderite, (Ka) katoite, (Ki) kimceite, (Kn) knorringite, (M) majorite, (P) pyrope, (S) shorlomite, (Sp) spessartine, (U) uvarovite

miscibility are shown in Fig. 3. This mineral series probably is one of the most complex regarding the binary ratios of miscibility.

Thus, the description of the configuration of the mineral species (series according to Dana et al.) does not cause any fundamental problems. How can this be used? "The comprehensive study of the genesis of minerals should be considered as a combination of two sections of genetic mineralogy, ontogeny, i.e. studies of genesis of mineral individuals and aggregates and phylogeny, i.e. study of the genesis of mineral species and parageneses" (Grigoriev 1961, p. 17). "Phylogeny of minerals is a section of genetic mineralogy devoted to the studies of processes of genesis of mineral species and formation of paragenesis: physicochemical conditions of species and parageneses, reasons of crystallization, thermobarometry, isomorphism, typomorphism, typical chemical qualities, and so on" (Zhabin 1979, p. 12). In biology, phylogeny is understood as historical development (evolution) of species, whereas the habitats and populations are considered in the corresponding sections of ecology. In this sense, the physicochemical conditions of existence (areas of stability), reasons for crystallization, thermobarogeochemistry, isomorphism, typomorphism, typical chemical qualities, and other features of mineral species and parageneses are not a subject of phylogeny.

Only paragenesis can evolve in changing conditions, i.e. separate mineral species in it can appear or disappear. A mineral species in a modern understanding cannot evolve. In these conditions, it is either stable (with fixed chemical composition and crystalline structure), or unstable (problem of metastability is beyond the scope of this discussion). The proposed understanding of a mineral species as a mineral series (after Dana et al.) for the first time allows discussion of phylogeny of mineral species. Its configuration is changing according to a driven change in the physical and chemical parameters. It is shown for calcite and spinel in Figs. 4 and 5.

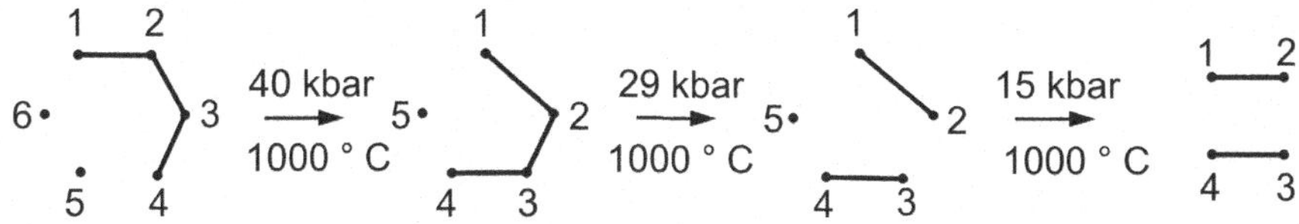

Fig. 4 Phylogeny of calcite in an environment of decreasing pressure

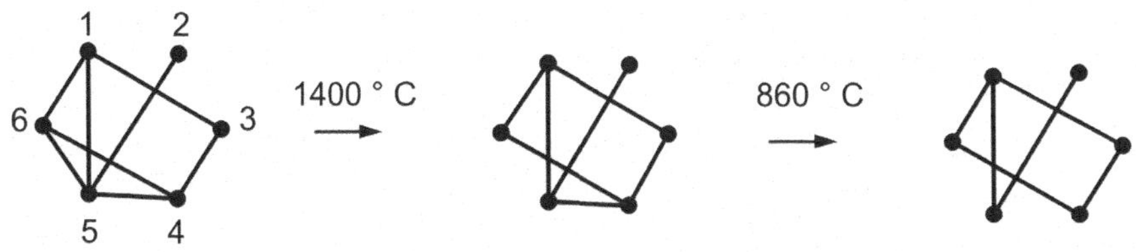

Fig. 5 Phylogeny of spinel in an environment of decreasing temperature

System

"Mineralogy … from the end of the 18th century changes its character—mineralogists focus on mineral descriptions and discovery of new mineral species and varieties. Hence, it was necessary to develop principles of classification of minerals and, at that time and also in the first decade of the 19th century, many mineralogists worked on the development of various "mineralogical systems." The old method of studies and thinking, which was referred to by Hegel as "metaphysical", was mainly connected with objects as something finite and unchangeable. Remains of that method persist in men's minds and have a historical background. Objects should have been studied before researching processes. At first, it is necessary to know what a particular object is and, only then, to study processes happening to this object" (Grigoriev 1961, p. 10). This quotation has two problems. The first is in the text about the relationship of the old "metaphysical" (structural) and new "genetic" methods. The genetic method is not much more progressive than the structural method, as it can never replace the knowledge of "what the image is". This knowledge needs constant updating. This currently happens with the concepts of mineral individuals, species, and the entire mineral system. The genetic method is an escape to a different thinking level. There are many such methods: evolutionary "what came from what," causative "what resulted in what" (modern Argentinian philosopher M. Bunge distinguishes about ten varieties due to subtle differences in the cause and effect quantifiers).

The second is on classification of increased abundance of minerals and formulation of various mineralogical systems. The innate desire of a researcher is to systematize objects of a study, i.e. create their system. However, the logically simplest systems have retained classification dividing one entirety into non-overlapping classes. Examples include Linnaeus' "Systema Naturae" and

Dana's "System of Mineralogy" and others. This box contains diamonds, another box contains topazes; here are horses, there are sheep… with no intermediates. What can possibly be simpler? But the system is not necessarily a classification as well. This constitutes a large resource for the natural history science. Note that the basis of any classification (reflection, symmetrical, and transitive) is an equivalent, but there are also other relations. For instance, the tolerance (reflexive, symmetrical, and nontransitive) relations give rise to a tolerance space, whereas a relation of order (strict, nonstrict, and quasiorder, see Shreider (1971)) gives rise to a structure.

Smirnova (1979) has shown that the system of mineral species (in a modern understanding) can be interpreted as a tolerance space with a relation of partial similarity of chemical composition. At the level of mineral series, a system of minerals based on the character of partial similarity of configuration can be interpreted as tolerance metaspace. In this context, "meta" means that the mineral series are given as graphs without their actual contents. Finally, the evolution of the mineral series can be portrayed as interrupted lines or disappearance of the apices on graphs. In this case, the mineral system can be interpreted as developing and essentially complete partially ordered system (interpreted by a containment graph). Petrov and Krasnova (2010) interpret a system of minerals as a structure with a relationship of strict lexicographic order based on the examples of multirank chemical formulae of minerals. The latter are defined as "successions of symbols of chemical elements arranged in the order of decrease in their atomic weights in the composition of minerals" (p. 5). At first glance, this is a very formal interpretation. However, the alphabet here is the periodic system of elements, which fundamentally distinguishes the R-catalog from all mineralogical dictionaries composed using the same lexicographic principle. Krivovichev and Charykova (2013) suggested one more structure with a relationship of non-strict lexicographic order under the name of "classification of mineral systems".

Conclusions

The history of biology and mineralogy shows that the analysis of biomineral analogues is useful for both these sciences as it helps to expose gaps in knowledge and mutually enriches them by theoretical approaches and practical techniques. The analogues are based on the shared functional morphological and evolutionary approaches to define basic categories and analyze objects at the level of individuals, species, and in general, systems. The mineral individuals, species, and systems of minerals require further interpretations and more rigid definition. The mineral series are more likely proto-elements than mineral species as they are currently understood. The proposed synthetic approach adds to the prevailing analytical approach and is better correlated with natural causes. The boundaries of mineral species should be naturally recognized in nature by discontinuity in the miscibility of solid solutions. The new interpretation initiated a new fundamental question of structures (configuration) of mineral species in these conditions. The new understanding of a mineral species allows examination of its phylogeny, and changes in its structure during the directed change of the physicochemical conditions. The current understanding of the phylogeny of a mineral species (areas of stability) is inaccurate. Most likely it has a biological analogue in the "ecology of a species." The currently accepted terminology talks about the phylogeny of the paragenesis of a mineral species, rather than phylogeny of a mineral species.

The concept of a mineral system becomes more and more ambiguous, but this does not result in a loss of "definiteness." The use of various logical interpretations is caused by the human need for deeper knowledge. Statements, such as "this is the most correct" mineral system, simply cannot exist. The future prospects are in combination of various interpretations. Should the transition from various classifications to tolerance spaces and structures be more comprehensive? The answer is yes and no. No, because a subsequent concept does not necessarily encompass the preceding concept. For instance, a classification is an example of a tolerance space, although the latter is not an example of a structure. Yes, because each new concept of a mineral system adds something to the general pattern of logical properties and mineralogical meanings of its intrasystemic relationships. The future of mineralogy is in coordination of different interpretations of the system of mineral species, supplementing and logically refining each other in respect to mineralogical meanings. It is particularly important to show the mineral system as a tolerance space with their partial similarity in chemical composition and/or crystalline structure.

References

Berry LG, Mason B, Dietrich RV. Mineralogy: concepts, descriptions, determinations. San Francisco: W.F. Freeman; 1983

Betekhtin AG. Mineralogiya. Moscow: Gosgeolizdat; 1950. [Betekhtin AG. Mineralogy. Moscow: Gosgeolizdat; 1950. (In Russ.)]

Bokii GB. Granitsy primeneniya ponyatiya "izomorfizm". In: Izomorfizm v mineralakh. Moscow: Nauka; 1975. P. 7–14. [Bokii GB. Delineation of application of the concept "isomorphism". In: Isomorphism in minerals. Moscow: Nauka; 1975. P. 7–14. (In Russ.)]

Bonshtedt-Kupletskaya EM. Novye mineraly – chleny isomorfnykh ryadov, structurnye analogi izvestnykh mineralov. In: Izomorfizm v mineralakh. Moscow: Nauka; 1975. P. 14–25. [Bonshtedt-Kupletskaya EM. New minerals – members of isomorphic rows, structural analogues of known minerals. In: Isomorphism in minerals. Moscow: Nauka; 1975. P. 14–25. (In Russ.)]

Dana JD, Dana ES, Palache C, Berman H, Frondel C. The System of Mineralogy. Vol. 1. New York: J. Wiley & Sons; 1946

Dana JD, Dana ES, Palache C, Berman H, Frondel C. The System of Mineralogy. Vol. 2. New York: J. Wiley & Sons; 1951

Essen EJ. Karbonatnye tverdye rastvory i vzaimnaya rastvorimost ikh konechnykh chlenov primenitelno k termobarometrii. In: Karbonaty: mineralogiya i khimiya. Moscow: Mir; 1987. P. 105–127. [Essen EJ. Carbonate solid solutions and mutual solubility of their end members with reference to geological thermobarometry. In: Carbonates: mineralogy and chemistry. Moscow: Mir; 1987. P. 105–127. (In Russ.)]

Grigoriev DP. Ontogeniya mineralov. Lvov: University Press; 1961. [Grigoriev DP. Ontogeny of minerals. Lvov: University Press; 1961. (In Russ.)]
Grigoriev DP, Zhabin AG. Ontogeniya mineralov: individy. Moscow: Nauka; 1975. [Grigoriev DP, Zhabin AG. Ontogeny of minerals: individuals. Moscow: Nauka; 1975. (In Russ.)]
Häckel E. Anthropogenie oder Entwickelüngsgeschichte des Menschen. Leipzig: W. Engelmann; 1874
Irvine TN. Chromian spinel as a petrogenetic indicator. Pt 1. Theory. Can J Earth Sci. 1965; 6–7 (2): 648–672
Kostov I. Mineralogiya. Moscow: Mir; 1971. [Kostov I. Mineralogy. Moscow: Mir; 1971. (In Russ.)]
Krivovichev VG, Charykova MV. Klassifikatsiya mineralnykh sistem. St Petersburg: SPb State University; 2013. [Krivovichev VG, Charykova MV. Classification of mineral systems. St Petersburg: SPb State University; 2013. (In Russ.)]
Kukharenko AA. Problema isomorfisma v mineralogii. Proc Russ Miner Soc. 1970; 2: 200–213. [Kukharenko AA. The problem of isomorphism in mineralogy. Proc Russ Miner Soc. 1970; 2: 200–213. (In Russ.)]
Linnaei C. Systema naturae sive regna tria naturae systematice proposita per classes, ordines, genera, species. Leyden: T. Haak; 1735
Mozgova NN. Nestekhiometriya i gomologicheskie serii sulfosolei. Moscow: Nauka; 1985. [Mozgova NN. Nonstoichiometry and homologous series of sulfosalts. Moscow: Nauka; 1985. (In Russ.)]
Nickel EH. The definition of a mineral. Can Miner. 1995; 33: 689–690
Nickel EH. Solid solutions in mineral nomenclature. Can Miner. 1992; 30: 231–234
Nickel EH, Mandarino JA. Procedures involving the IMA commission on new minerals and mineral names, and guidelines on mineral nomenclature. Can Miner. 1987; 25: 353–377
Osokin AS. Ob aktsessornykh zinc-soderzhashchikh khromitakh. In: Novye dannye po mineralogii Cu-Ni i kolchedannykh rud Kolskogo poluostrova. Apatity: Kola Sci Centre; 1979. P. 89–96. [Osokin AS. On accessory zinc-containing chromites. In: New data on the mineralogy of Cu-Ni and pyrite ores of the Kola peninsula. Apatity: Kola Sci Centre; 1979. P. 89–96. (In Russ.)]
Petrov TG, Krasnova NI. R-slovar-katalog khimicheskich sostavov mineralov. St Petersburg: Nauka; 2010. [Petrov TG, Krasnova NI. R-dictionary-catalogue of chemical compounds of minerals. St Petersburg: Nauka; 2010. (In Russ.)]
Popov VA, Popova VI. Paragenezisy form kristallov mineralov. Miass: Inst of mineralogy RAS; 1996. [Popov VA, Popova VI. Parageneses of shapes of mineral crystals. Miass: Inst of mineralogy RAS; 1996. (In Russ.)]
Rozhnov SV. Morfogenez: simmetriya i asimmetriya. Moscow: Paleont Inst RAS; 2013. [Rozhnov SV. Morphogenesis: symmetry and asymmetry. Moscow: Paleont Inst RAS; 2013. (In Russ.)]
Shreider YA. Ravenstvo, skhodstvo, poryadok. Moscow: Nauka; 1971. [Shreider YA. Equality, similarity, order. Moscow: Nauka; 1971. (In Russ.)]
Smirnova NL. O sisteme mineralov. Urovni. Bull Moscow State University. Ser Geol. 1979; 2: 59–63. [Smirnova NL. On the system of minerals. Levels. Bull Moscow State University. Ser Geol. 1979; 2: 59–63. (In Russ.)]
Sobolev NV. Parageneticheskiye tipy granatov. Moscow: Nauka; 1964. [Sobolev NV. Paragenetic types of garnets. Moscow: Nauka; 1964. (In Russ.)]
Sokolov GA. Khromity Urala, ikh sostav, usloviya kristallizatsii i zakonomernosti rasprostraneniya. Proc Inst Geol Sci. 1948; 97(12). [Sokolov GA. Chromites of the Urals, their composition, crystallization conditions and distribution patterns. Proc Inst Geol Sci. 1948; 97 (12). (In Russ.)]
Spry PG, Scott SD. The stability of zincian spinels in sulfide systems and their potential as exploration guides for metamorphosed massive sulfide deposits. Econ Geol. 1986; 81: 1446–1463
Weiser T. Zink- und Vanadium-führende Chromite von Outokumpu, Finland. Neues Jahrb Miner Monatsh. 1967; 7–8: 234–243
Zhabin AG. Ontogeniya mineralov: agregaty. Moscow: Nauka; 1979. [Zhabin AG. Ontogeny of minerals: aggregates. Moscow: Nauka; 1979. (In Russ.)]

Determination of Rare Earth Elements in Rock-Forming Clinopyroxenes Using NexION 300S Mass Spectrometer with NWR 213 Laser Ablation System

Maria V. Zaitceva, Sergei Votyakov and Vladimir R. Shmelev

Abstract The determination of rare earth elements (REE) with a high (tens of microns) locality and low detection limits (up to 0.01–0.1 ppm) in rock-forming minerals, in particular pyroxenes, is a rather complex methodological task. The comparison of the analytical data for minerals with low REE content obtained using high-sensitive single-collector mass spectrometers and excimer 193 nm laser ablation systems with those obtained using more common and relatively cheap quadrupole mass spectrometers and conventional solid-state 213 and 266 nm laser ablation systems remains relevant. The determination of REE content in clinopyroxenes from clinopyroxenite of the Nizhny Tagil massif (Platinum belt, the Urals) using the optimized operational parameters of quadrupole NexION 300S mass spectrometer with the NWR 213 laser ablation system was carried out in the laboratory of the Institute of Geology and Geochemistry; the reached REE concentration detection limit was 0.01 ppm. Our results obtained for clinopyroxenes were found to be in good agreement with those obtained using the Agilent 7500S mass spectrometer with the MicroLas GeoLas Q-plus 193 nm ArF excimer laser ablation system (Kanazawa University, Japan). Our methodological work will be continued in order to reduce the REE detection limits.

Keywords Rare earth elements · Pyroxenes · Rock-forming minerals · Laser ablation · Quadrupole mass spectrometer · Concentration

Introduction

The determination of rare earth elements (REE) with a high (tens of microns) locality and low detection limits (up to 0.01– 0.1 ppm) in rock-forming minerals, in particular pyroxenes, is a rather complex methodological task. Its solution has

M. V. Zaitceva (✉) · S. Votyakov · V. R. Shmelev
A.N. Zavaritsky Institute of Geology and Geochemistry, Ural Branch of the Russian Academy of Sciences, 15 Vonsovskogo Str., Ekaterinburg, Russia
e-mail: zaitseva.mv1991@gmail.com

S. Votyakov et al. (eds.), *Minerals: Structure, Properties, Methods of Investigation*, Springer Proceedings in Earth and Environmental Sciences,
https://doi.org/10.1007/978-3-030-00925-0_41

become possible using modern high-sensitive single-collector mass spectrometers and laser ablation systems with excimer 193 nm lasers (Ishida et al. 2004). However, it is quite difficult to operate such equipment and thus it is available only in a few geochemical analytical centers. The comparison of the analytical data for minerals with low REE content obtained in such centers equipped with modern instruments, with those obtained using more common and relatively cheap mass spectrometers and conventional systems with solid-state lasers emitting at the wavelengths of 213 and 266 nm still remains relevant (Kil et al. 2011).

The aim of the study was to obtain and compare the REE content in pyroxenes using the NexION 300S mass spectrometer and the NWR 213 laser ablation system with those obtained using the Agilent 7500S mass spectrometer with the MicroLas GeoLas Q-plus 193 nm ArF excimer laser ablation system (Kanazawa University, Japan).

Materials and Methods

The clinopyroxenes from two clinopyroxenite samples (No 1636 and 1780, the Nizhny Tagil massif, Platinum belt, Urals) were studied. 1636 clinopyroxene sample had an increased (0.21) ferruginosity and high (4–5 wt.%) content of Al_2O_3, while 1780 clinopyroxene sample was characterized by low (0.08) ferruginosity and low (0.9 wt.%) Al_2O_3 content. NIST SRM 612 glass reference material having REE content of 36-40 ppm was used for instrument calibration (Norman et al. 1996).

The operational parameters of the PerkinElmer NexION 300S inductively coupled plasma mass spectrometer and the ESI NWR 213 laser ablation system located in class ISO7 cleanroom of the Institute of Geology and Geochemistry of the Ural Branch of RAS (Zaitseva et al. 2016) were optimized using the NIST SRM 612 reference glass to obtain the maximum stable signal (Norman et al. 1996). The optimized operational parameters of the mass spectrometer were as follows: Ar sample flow rate —0.94 l/min, RF power —1100 W, dwell time —10 ms, the number of scan cycles —1, the number of replicates — 500. The optimized parameters of the laser ablation system were as follows: laser energy — 10.5–11.5 J/cm^2, pulse repetition frequency —10 Hz, laser spot diameter —50 microns, He flow rate —400 ml/min, laser operating time — 50 s, laser warming time before the measurement —20 s. The length of the connecting tube between the mass spectrometer and the laser ablation system was 1.5 m.

The data processing and concentration calculation (Longerich et al. 1996) were performed using the GLITTER V4.4 software. SiO_2 was used as an internal standard; NIST SRM 612 reference glass was used as an external standard for the standard-sample bracketing (SSB) and was measured after every 10 craters. The reached REE concentration detection limit was 0.01 ppm.

Results

The REE concentrations in the NIST SRM 612 reference glass varied from 36 to 40 ppm (1σ error 4 ÷ 8%). The measured REE concentrations in 1636 and 1780 clinopyroxene samples varied in the range of 0.2 ÷ 15 ppm (1σ error 5 ÷ 15%) and 0.01 ÷ 1.5 ppm (1σ error 14 ÷ 100%), respectively (Fig. 1). Our results of the REE composition have been compared in Fig. 1 with those obtained in the laboratory of the Kanazawa University (Japan). As can be seen from the chondrite-normalized REE distributions, the elemental trends and concentration levels are quite comparable; the closest agreement of the results has been found for the 1636 sample.

Conclusions

The determination of REE content in clinopyroxenes from clinopyroxenite of the Nizhny Tagil massif (Platinum belt, the Urals) using the NexION 300S mass spectrometer with the NWR 213 laser ablation system was carried out in the laboratory of the Institute of Geology and Geochemistry; the reached REE concentration detection limit was 0.01 ppm. Our results obtained for the elemental trends and concentration levels in clinopyroxenes were found to be in good agreement with those obtained using the Agilent 7500S mass spectrometer with the 193 nm ArF excimer laser ablation system (Kanazawa University, Japan). This allows us to conclude that at the moment the studies of REE in rock-forming minerals at

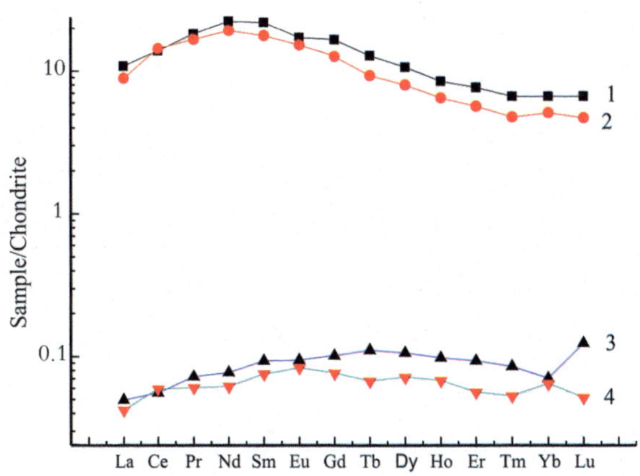

Fig. 1 The comparison of REE content in clinopyroxenes No. 1636 (1, 2) and No. 1780 (3, 4) from the Nizhny Tagil massif obtained in the laboratories of the Institute of Geology and Geochemistry of the Ural Branch of RAS, Ekaterinburg (1, 3) and of the University of Kanazawa, Japan (2, 4). REE concentrations are given in ppm and normalized to those of the chondritic reservoir according to (Taylor and McLennan 1985)

concentration levels of 0.1 ppm are possible using the NexION 300S mass spectrometer with the NWR 213 ablation system. The methodological work will be continued in our laboratory in order to reduce the REE detection limits.

Acknowledgments The work was carried out at the UB RAS "Geoanalitik" Center for Collective Use and financially supported by RSF grant No. 16-17-10283.

References

Ishida Y, Morishita T, Arai S, Shirasaka M. Simultaneous in-situ multi-element analysis of minerals on thin section using LA-ICP-MS. The science reports of the Kanazawa University. 2004; 48:31–42.

Kil Y, Shin HS, Oh HY, et al. Geosci J. 2011;15:177–183. https://doi.org/10.1007/s12303-011-0012-1.

Longerich HP, Jackson SE, Günther D. Laser ablation inductively coupled plasma mass spectrometric transient signal data acquisition and analyte concentration calculation. Journal of Analytical Atomic Spectrometry. 1996;11:899–904.

Norman MD, Pearson NJ, Sharma A, Griffin WL. Quantitative analysis of trace elements in geological materials by laser ablation ICP-MS: instrumental operating conditions and calibration values of NIST glasses. Geostandards Newsletters. 1996;20:247–261.

Taylor SR, McLennan SM. The continental crust, its composition and evolution: an examination of the geochemical record preserved in sedimentary rocks. Oxford: Blackwell; 1985.

Zaitseva MV, Pupyshev AA, Shchapova YuV, Votyakov SL. Metodicheskie aspekty U/Pb datirovaniya tsirkonov na mnogokollektornom mass-spektrometre s induktivno-svyazannoi plazmoi Neptune Plus s pristavkoi dlya lazernoi ablyatsii NWR 213. Analitika i kontrol'. 2016;20(2):121–137. https://doi.org/10.15826/analitika.2016.20.2.008. [Zaitseva MV, Pupyshev AA, Shchapova YuV, Votyakov SL. Methodical aspects of U/Pb dating of zircons using Neptune Plus multi-collector mass spectrometer with inductively coupled plasma and NWR 213 laser ablation system. Analytics and Control. 2016;20(2):121-137. https://doi.org/10.15826/analitika.2016.20.2.008. (in Russ.)].

Printed by Printforce, the Netherlands